THE FUTURE OF
GEOGRAPHY

THE FUTURE OF GEOGRAPHY

Edited by
R.J. Johnston

Methuen
London and New York

First published in 1985 by
Methuen & Co. Ltd
11 New Fetter Lane,
London EC4P 4EE

Published in the USA by
Methuen & Co.
in association with Methuen, Inc.
29 West 35th Street
New York NY 10001

The collection © 1985 R.J. Johnston
Individual chapters
© 1985 the respective authors

Printed in Great Britain at the
University Press, Cambridge

All rights reserved. No part of this
book may be reprinted or reproduced
or utilized in any form or by any
electronic, mechanical or other means,
now known or hereafter invented,
including photocopying and recording,
or in any information storage or
retrieval system, without permission
in writing from the publishers.

British Library Cataloguing in Publication Data

The Future of geography.
1. Geography
I. Johnston, R.J.
910   G116

ISBN 0-416-36690-2
ISBN 0-416-36700-3 Pbk

Library of Congress Cataloging in Publication Data

Main entry under title:
The Future of geography.
Bibliography: p.
Includes index.
1. Geography – Philosophy – Addresses, essays, lectures.
I. Johnston, R.J. (Ronald John)
G70.F87 1985   910'.01   85-7322

ISBN 0-416-36690-2
ISBN 0-416-36700-3 (pbk.)

CONTENTS

List of illustrations		vii
Notes on contributors		viii
Acknowledgements		ix
Preface		x

I THE CONTENT OF GEOGRAPHY   1

Introduction: exploring the future of geography   3
R.J. Johnston

1 Physical geography and the natural environmental sciences   27
Peter Worsley

2 Holistic and reductionistic approaches to geography   43
I.G. Simmons and N.J. Cox

3 Geography has neither existence nor future   59
Michael E. Eliot Hurst

4 The value of a geographical perspective   92
Peter J. Taylor

II PHILOSOPHY AND METHODOLOGY   111

5 Geography as a scientific enterprise   113
John U. Marshall

6 Scientific method in geography   129
Alan Hay

7 Arguments for a humanistic geography   143
Stephen Daniels

	8	Realism and geography Andrew Sayer	159
	9	Individual action and political power: a structuration perspective James S. Duncan	174
	10	Any space for spatial analysis? Anthony C. Gatrell	190

	III	GEOGRAPHY FOR SOCIETY	209
	11	Quantification and relevance R.J. Bennett	211
	12	Geomorphology in the service of society Denys Brunsden	225
	13	Understanding and predicting the physical world Antony Orme	258
	14	Will geographic self-reflection make you blind? Peter Gould	276
	15	Geography and schooling John Huckle	291
	16	Geography, culture and liberal education J.M. Powell	307
	17	To the ends of the earth R.J. Johnston	326

Index 339

LIST OF ILLUSTRATIONS

FIGURES

2.1	A hierarchical scheme of knowledge	44
3.1	A holistic framework for studying transportation	80
4.1	The distribution of geography departments c.1970	94
4.2	The establishment of chairs of geography in German-speaking Europe, 1871–1912	98
10.1	The optimal locations of fire stations in Manchester and Salford	192
10.2	Construction of a hypothetical autocorrelation function	193
10.3	Kriging rainfall in Münster, West Germany	195
10.4	Single-parent families in Crescent Ward, Salford	196
10.5	'Socio-economic status' in Salford	198
10.6	A third-order probability surface: probabilities of noise annoyance in south Manchester	200
10.7	Hypothetical market areas	203
10.8	A cognitive transformation of rural Humberside	204
12.1	West Bay, Dorset	230–3
12.2	Three examples of landslide hazard maps prepared for practical purposes	238–40
12.3	Typical, standard, geomorphological inventory map for an area of soil erosion in Canterbury, New Zealand	241
12.4	Landsat image for part of Somalia and a geomorphological structure-landform survey	242–3
12.5	Effects of flood on Wadi Abha, Saudi Arabia	244
12.6	A sediment yield study for the new Kotmale Dam catchment in the Mahaweli Ganga Development Plan	247

TABLE

12.1	Examples of geomorphological contributions to practical problems	227

NOTES ON CONTRIBUTORS

R.J. Bennett, Professor, Department of Geography, London School of Economics
Denys Brunsden, Professor, Department of Geography, University of London (King's College)
N.J. Cox, Lecturer, Department of Geography, University of Durham
Stephen Daniels, Lecturer, Department of Geography, University of Nottingham
James S. Duncan, Associate Professor, Department of Geography, Syracuse University
Anthony C. Gatrell, Lecturer, Department of Geography, University of Lancaster
Peter Gould, Professor, Department of Geography, Pennsylvania State University
Alan Hay, Reader, Department of Geography, University of Sheffield
John Huckle, Lecturer, Department of Geography, Bedford College of Higher Education
Michael E. Eliot Hurst, Professor, Department of Geography, Simon Fraser University
R.J. Johnston, Professor, Department of Geography, University of Sheffield
John U. Marshall, Professor, Department of Geography, York University
Antony Orme, Professor, Department of Geography, University of California (Los Angeles)
J.M. Powell, Reader, Department of Geography, Monash University
Andrew Sayer, Lecturer, School of Social Sciences, University of Sussex
I.G. Simmons, Professor, Department of Geography, University of Durham
Peter J. Taylor, Senior Lecturer, Department of Geography, University of Newcastle upon Tyne
Peter Worsley, Professor, Department of Geography, University of Nottingham.

ACKNOWLEDGEMENTS

The publisher, editor and author are grateful for permission to reprint figure 10.3 from U. Streit, 'Analysing spatial data by stochastic methods', in G. Bahrenberg and U. Streit, *German Quantitative Geography*, Münstersche Geographische Arbeiten Bd. 11, Ferdinand Schöningh, Paderbon, 1981.

PREFACE

A healthy discipline is one which subjects itself to continual, constructive criticism, involving all its adherents in debates about its goals and procedures. For much of the time that process of appraisal will be low-key and relatively uncontroversial; occasionally, however, it may involve major debates about the subject, or at least – with disciplines as broad as geography – some parts of it. This seems to be the case at the present time for geography in the English-speaking world. Debates are pronounced and profound, and the discipline is being subjected to a great deal of internal scrutiny.

This book has been designed to contribute to those debates, but not in the usual ways: it is not a 'balanced overview' of the protagonists' views, nor is it a series of essays designed for the small number of active participants only and written in terms and contexts that only a few will appreciate. Its goal, rather, is to broaden the debating base by involving students. The essays have been written as a set of forthright personal statements, aimed specifically at an undergraduate audience; an interest in geography and its subject matter is assumed, but there are no prerequisite readings on philosophical or methodological issues. The book is avowedly introductory, to stimulate appreciation of the major issues currently confronting today's geographers by those who will form tomorrow's.

The book makes no claim to be comprehensive. Its structure, the topics covered and the authors selected reflect editorial predilections, whereas the details of each chapter reflect individual views of the issues and interpretations of the brief. The aim is to provide insights into current concerns about the future of geography.

I am grateful to the contributors for agreeing to participate in this venture and for keeping to the timetable, and to Mary Ann Kernan for her enthusiastic support.

R.J.J.
January 1985

THE CONTENT OF GEOGRAPHY

I

INTRODUCTION: EXPLORING THE FUTURE OF GEOGRAPHY

R.J. Johnston

Geographers, not for the first time, are undertaking a critical reappraisal of their discipline – of its contents, its philosophy, its methodology, and its interrelationships with other academic disciplines and with the outside world. To some observers – including participant-observers – such reappraisals are continuous, because they are necessary; a healthy academic discipline, like a healthy society, is one that keeps itself continually under the microscope, occasionally increasing the intensity of the critique as circumstances demand. To others, reappraisals are episodic, resulting from particular crises – often identity crises; during some periods – rarely prolonged ones – reappraisal is unnecessary. Whether the present period of activity is an episode to be followed by quiet consensus or merely a peak in a continuous field of debate is of little relevance here. What is crucial is that debate is intense and self-criticism within the discipline is substantial. That debate and self-criticism, if they are to be fruitful, must engage all geographers in a constructive way. In particular, they must engage those who have chosen to study geography in further and higher education, either for the intellectual enrichment they believe it will bring or for the openings it provides for later career development. It is to this particular audience that the present book is addressed.

There are several ways in which contemporary debates, involving academic practitioners, can be conveyed to a wider audience. The first is by summary over-view volumes (or text-books). These are valuable as signposts but they rarely convey the sense of intellectual cut-and-thrust; the author tends to take a 'neutral' stance, so that the essence of commitment is lost. When, as in the present situation, the debate is many-faceted, a bland summary can obscure the intensity of differences among the viewpoints. A second method is to involve the protagonists in a published debate. This is sensible if the audience comprises the debaters' peers, who appreciate the context of the issues and points, who have access to the materials alluded to in footnotes and references, and who are tuned in to the language of the debaters. But for the wider audience such debates – frequently conducted in the pages of academic journals – lack significance because their origins are unclear. What that readership requires is a set of clear statements introducing it to the main items of debate, laid out by the protagonists in a manner and language which make them readily accessible.

The present book has been designed, therefore, as a series of introductions to the debates, written to inform and to engage those not closely involved in them – at least at present. Given the complexity of the current debates, and the wide range of views held, the book makes no claim to be comprehensive. The aspects of the debates presented here represent a combination of editorial selection and author willingness to participate. Three areas of debate (outlined in detail below) form the structure for the book. Other topics – such as outsiders' views of geography – are excluded, valuable though they may be for a more comprehensive view. Authors were invited to contribute, in the following terms:

> On all of the issues identified here, students will wish to appreciate the major positions and to seek their own resolutions. To aid them in this, they need clear statements of the issues, presented in a forthright way by scholars concerned to see geography develop in a particular direction. Unfortunately, such statements are rare. Most text books seek to present a 'balanced view', whereas journal articles are written for one's peers rather than one's students. The latter, in particular, often provide difficult material for students to assimilate, because they are not necessarily aware of the context in which the articles were written and are almost certainly not aware of much of the specialised literature cited to bolster a case.
>
> The aim of the book, then, is to present a series of short, forthright statements designed to inform and to stimulate.

Within this framework,

> The editorial role is a facilitative one. There will be a general introduction. ... There will be no editing of individual contributions except with regards to possible suggestions for

improving the clarity of the statements and to ensuring that contributions are not too long.

The editorial role has been one of selection, therefore, and to that extent the content of the book – and hence the presentation of the debates – reflects: (1) my interpretation of the major issues; (2) my selection of authors to present particular views; (3) the willingness of those authors to participate; and (4) the authors' interpretations of the brief given. Thus editorial responsibility and input is substantial, and the purpose of the remainder of this chapter is to justify what has been done. This takes the form, first, of a broad discussion of the context of the debates and, second, of an introduction to the remainder of the book. Beyond that, the authors' statements speak for themselves.

THE CONTEXT

Debates, to be appreciated, must be set in context. This is not a rationale for a further essay into the history of geography (see Johnston 1983a). It is, however, necessary to outline briefly the current nature of geography and the forces influencing it.

GEOGRAPHY: WHAT IT HAS BECOME

From the outset, it must be accepted that there is no necessity for a discipline of geography, nor for one with a different name but the same content focus. Academic disciplines are human creations. Once created, they have a defined existence and are invested in by individuals, who wish to protect their capital. For some – such as mathematics, perhaps – it is harder to accept that they may be unnecessary as defined educational and research enterprises than it would be for others. For geography, it may be difficult to accept that a separate discipline is unnecessary. But at one stage of intellectual development it was. Since then, geographers have defined and redefined their discipline, seeking to demonstrate its utility within the corpus of intellectual activity.

Geography, then, is a human creation; the nature of that creation represents the views of its originators and the continuing reinterpretations of those who have succeeded them, inside and outside the discipline itself. As a discipline, it is historically constituted. It is also regionally constituted, for it has several separate (not necessarily independent) origins, reflecting the contextual interpretations of geographers experiencing different situations. Geography as practised in the

USSR at the present time is very different from geography as practised in the USA; geography in the USA is not the same as geography in the UK; and there are several quite distinct interpretations, each with a clear national base, in western Europe (Johnston and Claval 1984). In the UK, for example, geography has a very strong presence in the secondary school curricula, whereas it is very weak in comparable USA institutions. There is no natural reason for this; it merely reflects one of the many nuances which differentiate the separate British and American societies.

Despite the many national and regional differences in the constitution of geography, there is general agreement as to its purpose. Literally defined as 'earth description', geography is widely accepted as a discipline that provides 'knowledge about the earth as the home of humankind'. Such a broadly defined discipline is accepted because that goal is perceived as necessary to society, both in an academic sense because such knowledge is required to advance societal goals and in a more general sense because such knowledge is considered desirable in a well-educated society.

Geographers have been able to convince the powerful groups within many societies of the need for their discipline, therefore, and have won for themselves privileged positions in the 'research and education industries'. But the nature of the commodity that they provide has varied, by place and through time. Their focus is 'knowledge about the earth as the home of humankind'. But what is knowledge? How should it be produced and used? What subject matter should be embraced by 'the earth as the home of humankind'? Such questions have been the focus of past and present debates.

DEBATE RATIONALES

Why debate questions such as these? Why do geographers (and members of other disciplines too) find it necessary to consider their *raison d'être*? Two sets of rationales are suggested here.

The first set comprises *internal reasons*, and provides the base for continuous – though often not strenuous – debate. Primary among these is the organized scepticism many believe should be the hallmark of an academic discipline. In its research work, a discipline is penetrating the unknown, seeking solutions to problems and answers to questions. This involves conjectures, ideas, even guesses, gut-feelings and intuitions, all of which must be submitted to scrutiny. Learning is often a painful experience, with many mistakes made along the way – some of which may be very public. If that learning is to provide valid knowledge, then it must be open and subject to scrutiny; conjectures must be subject to refutations. If something is to be presented as a contribution to knowledge, it should be agreed as such and not imposed. Agreeement is rarely reached without discussion, and so debate is endemic to a healthy academic discipline advancing the frontiers of knowledge. Such debate is usually about the minutiae of research – about methods and meanings. Occasionally, however, a whole body of research may be called into question, and a restructuring of knowledge demanded. When this occurs, the discipline as a whole – its educational goals and materials as well as its research programmes – may be subject to intense scrutiny (see Johnston 1983a, 1984).

These internal debates represent a view of an academic community working in relative harmony; the members share goals and are disinterested participants whose only purpose is the advancement of knowledge. To some extent this is true. But within that community the individual members have their own personal goals, which for many combine the advancement of knowledge with self-advancement. The two can be harmonized, with success in the former bringing the charisma, status and other rewards of the second. But self-advancement can only be at the expense of others, in communities where the rewards are distributed unevenly, and so the promotion of oneself as successful is to some extent only attainable by promoting the (relative) failure of others. Thus, to some extent, academic debate is engendered by interpersonal rivalry (albeit not usually personalized; it is over ideas, not people). It may be beneficial to the discipline, because it 'keeps people on their toes' and ensures that all research is fully scrutinized. But to the extent that decisions are made because of the personal power of certain contestants rather than the quality of their work, then the ruling orthodoxy of a discipline is the result not of disinterested scepticism but of organized scientific politics.

All disciplines keep themselves under scrutiny, therefore, and are also subject to debates generated by those seeking personal status. Such scrutiny and debate is usually about details rather than generalities and can be contained by the 'political organization' of the disciplinary community. Occasionally, it may develop into more open and widespread conflict, involving the entire community in what may have many of the features of internecine warfare.

The second set, of *external reasons*, involves the discipline and its community in scrutiny from the outside, to which it must respond. Such scrutiny usually comes about when the encompassing society is experiencing a substantial crisis. (There are examples of external scrutiny generated in other ways, by a campaign somewhat similar to the internal personal campaigns just described; the McCarthy purges in the USA in the early 1950s illustrate this; see Harvey 1982.) That crisis is probably economic, though others – notably major wars – have stimulated scrutiny of the academic communities.

At the present time, there is a general recession in the world economy, to which political and other groups are responding. Part of their response involves a scrutiny of public investment in research and education, and its role in promoting recovery from the recession. Are academic disciplines such as geography contributing to that recovery? Should they be, and how? Should they be given the resources they ask for? Should they be more accountable to society, and only earn those resources by contributing to national and international recovery? Such questions are being asked in particular by governments, which have diagnosed high levels of public expenditure in non-productive activities as major causes of the economic troubles. They have set out both to reduce this expenditure and to make that which is spent more relevant to the tasks of economic regeneration. Education and research have been subject to general scrutiny and to cuts in resources, therefore, with the cuts being directed especially at those disciplines and areas which have little or no apparent relevance to short-term economic policies – notably those focusing on the promotion of technological developments that should increase productivity and market competitiveness.

A consequence of such policies, which have been initiated in a number of countries in similar forms, is that academic and educational institutions have been subject to scrutiny and cuts. In some cases, they have been 'invited' to participate in the scrutiny and cost-cutting themselves; in others changes have been externally imposed. For disciplines such as geography this has meant considerable potential dislocation, promoting substantial debate. Over and above the continuous sources of contention, therefore, the economic crisis and government reactions to it have stimulated a variety of threats to which the disciplines must respond.

RESPONSES TO THREAT

People, individually and collectively, respond to threats in a variety of ways, reflecting their context and their interpretations of the situation. The exact nature of the response to any particular threat cannot be predicted, therefore, but there are general tendencies typical of the sorts of situation described in the previous paragraph. Two types (or stereotypes) of behaviour can be indentified.

The first type can be categorized as *external posturing* in which one group under threat seeks to promote itself, to the detriment of others. The image is presented of a coherent unit, agreed upon its structures and able to meet the external challenges: the 'opponents', on the other hand, are presented as divided, irrelevant and unprepared. (Often the latter part of the presentation remains unsaid and has to be inferred, perhaps from innuendo. Within an academic institution such as a university, for example, if one discipline/department is to be relatively protected in a period of cuts, others must suffer disproportionately. The politics of such a situation are that disciplines/departments cannot attack others in the contest for resources without losing potential supporters in other 'battles'. They must promote themselves but not overtly denigrate others, hoping that their promotion campaign – perhaps aided by covert activity – will convince the decision-makers.) The discipline has to be presented as in good heart, and relevant.

Such external posturing takes a variety of forms, depending on the local circumstances. Within a country, for example, geographers must promote their discipline as a whole, presenting and furthering its image as a relevant discipline worthy of special treatment because of the contributions it does, and can, make to the achievement of national goals. In this, the discipline is dependent on both skilled publicists, able to use the national media to present a favourable image, and skilled lobbyists. Opportunities must be grasped to ensure that the (unique) talents possessed by geographers are recognized and called upon. For this, national (representative) bodies are required which can 'speak for geographers', and which are increasingly called upon to do so by those of their members who feel threatened. Within particular institutions, such as universities, the same image-processing activities take place: departments must demonstrate to administrators and academic decision-makers that they do good work, are able to attract outside resources, and are worthy of support.

Of necessity, the external posturing promotes the view of a homogeneous, united discipline, skilled in relevant knowledge and with a clear rationale linked to perceived national needs. The discipline is being sold. But internally, the situation is often very different, and the discipline is *rife with uncertainty* as its members debate how to respond to the external pressures. Such debate may be bitter and divisive, for while the external pressure may be real and clear the 'best' response may be far from obvious. Uncertainty reflects a lack of security and guarantees. How can we be sure geography (and so geographers) will survive? What must we do to promote our survival?

Such debate must be kept 'in-house' as far as possible, because to demonstrate division is to invite external criticism: 'If they don't know what they want, why should we give them anything?' Thus geographers – and they are not alone in this – feel they must debate about the nature of their response in order to survive, but at the same time must not show that they are uncertain how to respond.

The biggest internal threat in such situations is not the debate about the detailed response – how far should geography present itself as a technological discipline, for example – but the debate about the need to respond and the nature of the pressure. Should geographers restructure their discipline in order to further the policy objectives of a government and its economic allies, or should they point out that they accept neither the diagnosis nor the cure? Should they continue to promote their discipline as historically constituted, or should they be involved in attempts to create new, more relevant disciplines – or indeed get rid of disciplines altogether? Such queries are almost certain to be raised, because geographers are members of the society-at-large as well as of the disciplinary community, and their interpretation of and response to crises as citizens will influence their interpretations and responses as geographers.

CURRENT DEBATES

Academic disciplines exist to maintain, further and promote knowledge. They are human creations, given a *raison d'être* by individuals and their sponsors. Their work involves debate over their terms of reference – what exactly should they be doing, how, and why? Such debate is continuous, often low-key but sometimes major as disagreement wells up and individuals seek to establish reputations. Such disciplines are part of societies which are continually being changed, as consequences of individual reactions to current concerns. In general, change is incremental and relatively slow. At certain times, however, the nature and severity of current concerns may lead to a more fundamental evaluation by a discipline's sponsors. When this occurs, the members of the discipline may feel under threat, uncertain as to their future and how to safeguard it. They feel obliged to debate their discipline's content, to ensure its future (and theirs), while at the same time promoting the discipline to the outside world.

Part of the rationale for the present book is the belief that we are presently experiencing one of those periods of heightened debate stimulated by a world economic recession and a particular interpretation of this by economists who have access to major governments. Thus geographers feel threatened. What is the future of their discipline in school education? How can they attract sufficient university, polytechnic and college students to ensure their own employment? How can they obtain money to finance their research, and so assist their career prospects? Such questions are translated into: What should we be doing?

What is geography about and what is its value? How do we ensure that its value — and ours — is appreciated?

Academic geographers are always wrestling with such questions, but not always to the extent that they are at present. Much of their wrestling takes place within a closed world, defined by its own particular language and signs, which is difficult for others to penetrate. The goal of this book is to assist that penetration, particularly for the students of geography. Three areas of current debate have been selected, and individuals steeped in them have been invited to contribute — not to the debate itself but to the wider appreciation of its contents. They have been asked to provide a particular view — their own — on the debate. Each has, of course, interpreted the invitation individually and has proceeded accordingly. The result is neither a comprehensive coverage nor a balanced conspectus but a series of forthright, accessible statements which will illuminate the present state of uncertainty for those grappling to come to terms with it. The remainder of this chapter introduces the three areas (which are themselves interlinked) and provides a framework within which a reading of the individual contributions can be set.

1 THE CONTENT OF GEOGRAPHY

What are the boundaries of the academic discipline of geography? What should and should not form the core of a geographical education and training? How does and should geography relate to other disciplines?

Such questions are not new; they have been discussed ever since geography was 'invented', and the answers have always been subject to later questioning. In the UK, for example, a strong case has always been made for geography as a discipline that straddles the arts, the social sciences and the natural sciences, and which therefore synthesizes 'earth knowledge' as opposed to the more specialist, systematic disciplines which fragment knowledge into more easily manipulable parts. Thus a combination of both physical and human geography has been promoted by many as providing both the core and the *raison d'être* of the discipline: there can be no geography without both, and without geography knowledge about the earth is fragmented and unsatisfactory. In education, therefore, geography provides a broad base of 'earth knowledge'. In the USA, however, physical geography was largely eliminated from the discipline in the 1920s and 1930s; university offerings included courses on the geography of landforms, for example, but not on geomorphology — form was presented without discussion of process. More recently, physical geography has been reintroduced in many places (Marcus 1979), not so much — it seems — because of the synthesis arguments outlined above, but rather because it is seen to be relevant to current environmental concerns, and so capable of attracting students and resources. The constitution of the discipline is being changed with a reappraisal of its role, therefore. But the same is not true of the Netherlands, where the two disciplines of physical and human geography have had separate existences since 1907.

The breadth of geography and the wide range of subject matter and skills on which it draws are seen by some as its strength. It is not narrowly vocational, but nor is it necessarily woolly and liberal. It provides a broad base, prevents myopia in the solution of research problems, and ensures a range of skills. Or does it? According to Peter Worsley, writing entirely from a British perspective, physical geography cannot develop within its present disciplinary link to human geography because of an inadequate scientific base, the breadth of its undergraduate courses, and the non-availability of resources.

Worsley's argument is that full development of ability in a particular subject requires specialization within geography; the breadth of the traditional British syllabus precludes the acquisition of the skills and experience necessary to the training of a practitioner. Such a view, according to Ian Simmons and Nick Cox, is reductionist. They do not doubt the need for fully trained specialists, but argue that holistic disciplines are needed as well as systematic specialities because, in a well-hackneyed phrase, 'the whole is greater than the sum of the parts'. As a discipline traditionally concerned with the study of wholes, geography has a valid role, it would seem; but how does one combine the reductionist demands of Worsley with the holist desiderata of Simmons and Cox?

The holistic case can be put in two ways. The first, as expressed by Simmons and Cox, sees the holistic and reductionist strategies as essentially complementary. It is perfectly feasible and proper to study the parts, so a pluralism of disciplines is the consequence. The second case is that the subdivision is counter-productive, because the definition of parts is artificial and precludes understanding. In the context of the social sciences, the apparently complementary viewpoints of the various disciplines – economics, sociology, etc. – are seen by some as distortions, and it is impossible to sustain the case that geography provides the essential spatial and environmental element to social scientific understanding. According to Michael Eliot Hurst, therefore, there is no need for any separate social science disciplines, but only a single social science. Geography, in his terms, must be de-defined.

Why, then, do we have geography if it is hindering rather than assisting the development of knowledge? What is the value of geography? Peter Taylor argues against any concept of a 'natural value' for

INTRODUCTION: EXPLORING THE FUTURE OF GEOGRAPHY 13

geography. The value of the discipline is defined when it is constituted and as it is reproduced.

Within this first area of debate, therefore, four statements are presented which argue very different cases. Is there a case for a holistic discipline of geography? What role would it play? Should it be de-defined and replaced by holistic disciplines which are truly all-embracing and have no particular viewpoint? And is there a need for greater specialization in the elements which currently occupy small niches only within the geographical edifice? There are, as Peter Taylor suggests, no 'right' answers to those questions, only answers which seem right and win support at a given time and place. The existence and nature of geography will depend on the advocacy of the protagonists for the various positions.

2 PHILOSOPHY AND METHODOLOGY

Most definitions of geography refer to it as a discipline concerned with the development of knowledge of the earth as the home of humankind, focusing on the physical environment, on the interactions between this and human society, and on the spatial organization of that society. Whatever the outcome of the debates covered in the first section of the book – unless it is the demise of the discipline – geography is likely to remain linked to such a broad definition and identification of its subject matter. But how to put that definition into operation? What philosophy of science should it adopt; what is its definition of knowledge? And given an agreed philosophy, what methodology is appropriate?

The debates over such questions have raged for at least three decades among Anglo-American geographers (Johnston 1983a) and a range of philosophies has been explored, especially in human geography (Johnston 1983b). Prior to the 1950s, geographers (implicitly) adopted an empiricist philosophy, which places its prime emphasis on depiction of the observed world. It was a discipline that provided descriptions of places. Furthermore, each separate place was presented as singular, as an assemblage of phenomena into a whole (the whole – usually termed a region – being greater than the sum of the parts) which was not repeated elsewhere. The task of geography was to identify those separate assemblages, and describe and account for them.

The debate over this philosophy focused (again usually implicitly only) on the singularity of regions. This representation – termed exceptionalism – was challenged: regions are not singular, but they are unique. The difference is vital. A singular phenomenon is one that cannot be accounted for by any general principles. A unique phenomenon results from a particular interaction of general principles and so can be accounted for by understanding what those general principles are and how they have been combined in that particular instance.

The argument against singularity was in part one of logic: the singular regions are described using terms which have general connotations (rivers, towns, etc.). In other words, regional geography was based on classification, not just of the regions themselves but also of their contents; the use of language ensures the acceptance of general concepts about things. If, then, the contents of regions could be classified according to general principles, then there should be other general principles about their origins: classification involves the statement of substance laws, and assumes the existence of functional and generic laws. Thus, as argued by physical geographers in Britain in the 1950s (Johnston and Gregory 1984), it is possible to develop general laws regarding the elements of the physical environment, which can be combined to provide accounts of the environmental complex making up a region. Thus systematic and regional geography were to be seen as complementary, with the former providing the general understanding that could be synthesized in the latter when it referred to specific, unique places.

In the USA, the absence of a research tradition in physical geography within the university departments of geography precluded the development of a similar argument. Instead, the case against the exceptionalist view drew on two major strands. The first was that it was possible to identify laws of spatial form – of the arrangement of phenomena on the earth's surface. The second, and following from it, was that it is possible to identify general laws of human behaviour, which produce those spatial forms. Again, it was argued that a combination of the relevant laws allowed an understanding of the entire complex – the spatial organization – of a particular place.

The wide acceptance of these two sets of arguments meant that regional geography spawned the development of systematic geographies as the origin of its inputs: regional geographers were those who synthesized the findings of geomorphologists, climatologists, urban geographers, and so on into an explanation of the nature of a place. But the regional component rapidly withered: the systematic specialisms attracted the interests of the new generations of researchers in the 1950s and 1960s. How were they to be studied; how are valid laws of spatial form and spatial behaviour to be developed?

The philosophy adopted in this switch (again, very largely implicitly) was positivism. This builds on empiricism, and assumes the separation of subject (the researcher) and object (the researched): the geographical researcher is a neutral observer, seeing, recording and describing. The goal is more than description, however; it is explanation, which is provided by identifying the empirical links between observations – that one feature is associated with, or never associated with, another, for example. Laws could be developed of the 'if x, then y' variety, which were either laws of coincidence or causal laws. For geographers, the search for laws was the search for spatial coincidences.

How could this search be organized? Associated with the philosophy of positivism, in the mind of many geographers, is a particular methodology – often termed the scientific method. This, it was believed, operates in the following way. One begins with a theory, a general set of beliefs about the subject matter in question, comprising a combination of knowledge (agreed laws) and information (observations taken, often haphazardly, in the context of that knowledge). From this, it is possible to develop a model, a verbal, graphical or algebraic description of the subject in simplified form, organized to aid its understanding and to further the next stage – the development of a hypothesis. In this, one develops a hunch about a law of coincidence, a relationship that should be present because it is consistent with the theory and with the assumed processes and mechanisms underlying it. The hypothesis is a structured speculation, which must be tested empirically. If it proves to be valid, then a positive addition is made to the stock of theory; knowledge has been increased. If it proves invalid, knowledge has also been increased, albeit in a negative sense.

The positivist methodology, then, is a continuous looping sequence of thinking, speculating and testing. Much geographical work in the 1960s focused on the testing, on the empirical validation of hypotheses. The emphasis was on measurement – of the phenomena under consideration and of their empirical coincidences – and on the use of statistical inference to establish the veracity of the hypotheses. Thus the development of positivism was widely known as the 'quantitative revolution'. Geography was to become a law-seeking discipline, using the procedures of the natural sciences and the technology of quantification.

But is it that straightforward? Is the scientific enterprise as simple as it sounds? Deeper philosophical exploration has identified a number of major issues concerning the treatment of geography as a scientific enterprise, some of which are taken up by John Marshall and Alan Hay. John Marshall focuses on the problems of induction, and argues for Popper's fallibilist approach to the development of knowledge. Alan Hay, too, points to the problems of induction and the advantages of a strategy aimed at falsifying rather than verifying hypotheses. But he identifies disadvantages too, because the practice of research and the ideal strategy may not conform. Other problems in applying scientific methods are identified, but Hay's argument is that, notwithstanding such difficulties, geographers must continue to apply scientific methods rigorously if they are to provide useful knowledge and maintain/enhance their academic reputations.

Whereas some geographers have explored the positivist philosophy and associated scientific method, others have extended the search to other philosophies – other definitions of knowledge – and other scientific methods. Within human geography, some attention has focused on what are often termed humanistic philosophies, those which seek to explicate the meanings that lie behind actions. The goal in such work is not explanation, in the positivist sense, but rather interpretative understanding. Explanation seeks to classify events and then account for them in terms of other events, thereby, in the case of work dealing with human subjects, removing individuality and the subjective elements involved in the creation and re-creation of a life-world. Humanistic geography focuses on those subjective elements, applying objective scientific procedures to the identification and transmission of meanings. For geographers, it is meanings about places which attract most attention, and Stephen Daniels argues for the use of narrative to interpret those meanings and bring out the experience of place which is characteristic of daily life. Humanistic geography explicates the 'lived-in world'.

Humanistic geography focuses on the experience of place and environment, and its methods involve means of identifying and reporting these meanings. It is thus opposed to the type of positivist work which stresses regularity and order, and which seeks to explain any event or piece of behaviour in terms of a class of events/behaviours to which it has been allocated by an outside observer. Such work denies the individuality and humanity of people, ignores emotions and meanings, and treats humans in the same way as machines (Harrison and Livingstone 1982).

This argument is accepted by another group of critics of the application of positivist scientific method to human geography, but is at the same time criticized for its own theoretical naivety. Meanings are crucial influences on behaviour, but their generation and development cannot be understood by studying only the individuals involved. To the realist, meanings are produced within structures, modes of organizing society. To understand how the individual agent acts, one must appreciate the mechanisms that agent is operating. This involves setting the study of individual events in their social context, and finding the explanation (or cause) of an empirical outcome in the way in which human agency mediates the operation of the underlying structures. Thus a capitalist society is dominated by profit-making mechanisms, for example, which require individual capitalists to invest money in order to produce profit. But what they decide to invest in, and where, is not determined by those mechanisms; it depends on the interpretations of the situation by the individual capitalists, in their local contexts.

Realism, then, is a philosophy of social science linking the empirical world of appearances to the actions of agents interpreting social structures. That empirical world can be described, and the actions understood: the mechanisms in the underlying structure cannot be apprehended, however, only theorized. Thus, as Andrew Sayer argues, a realist explanation of an empirical event depends not on it being classed as one of the many outcomes of a regular causal sequence but as the result of an intentional action interpreting the mechanisms that govern a society.

Realism is a philosophy of the natural sciences as well as the social sciences, because the empirical observations in the natural world cannot explain themselves (Sayer 1984). The law of gravity is a mechanism that cannot be apprehended, for example. Its existence is assumed from the many experimental and other observations that are consistent with it, and by an appreciation of the conditions in which it operates. But there is a major difference between realist science and realist social science. In the former, the mechanisms are given and unchanging – hence the regularity of the outcomes. In the latter, the mechanisms are not given, they are created by humans and can be changed by human agency. Indeed, they are always being changed, because humans are able to learn, so that every interpretation of a mechanism is influenced by previous interpretations and their outcomes. Regularity in outcomes cannot be anticipated, therefore, because the interpretative context is always varying, as are the mechanisms themselves. Thus a realist social science, Andrew Sayer argues, is a much more ambitious and difficult enterprise than a realist physical science. For physical geographers, the mechanisms are given, and the task is to understand what happens when they are combined in different ways. For human geographers, the mechanisms are changing, and so are the conditions in which they operate.

But what of the individual? The realist argument very clearly gives the individual actor a primary role in the creation and re-creation both of the empirical world – as an interpretation of the structural processes – and of the structures themselves. Nevertheless, it is easy to derogate the role of the individual and to reify the structure: in a materialist (probably Marxist) realism, the individual becomes merely a 'bearer of the structure', or 'economic dupe' (to paraphrase James Duncan), whereas in a cultural realism the individual is a 'cultural dope' whose actions are determined by the culture. To avoid such pitfalls it is necessary, as Duncan argues, to develop a methodology which gives primacy to the individual, acting in the enabling-cum-constraining context of both a structure and the empirical outcome (itself interpreted) of interpretations of that structure. One such methodology, James Duncan contends, is that based on Giddens's concept of structuration, and he illustrates this with a Sri Lankan example, showing how individuals there act out, and seek to modify, a three-stranded ideology.

The positivist-inclined geography of the 1960s focused on spatial form and coincidences: the notion of 'geography as spatial science' was used to present the discipline as that part of the scientific enterprise which emphasizes the *where* component of the general question, 'what, where, how and why?' Physical geographers came to appreciate that where questions were insufficient, and increasingly shifted their attention to how: to understand form one must appreciate the form-producing mechanisms, but their goal was always to integrate the study of mechanism (often termed process: see Johnston and Hay 1983) with that of its empirical consequence. Among human geographers, whereas those committed to the positivist view, as outlined here by Marshall and Hay, continued to focus on spatial form as the outcome of theorized correlations, humanistic geographers studied form only to the extent of recognizing that individuals have their own mental maps which have some type of spatial organization and realists retreated from the analysis of form both because of their argument against empirical regularity as the outcome of causal mechanisms and because of their greater interest in mechanisms than in outcomes. The approach which stresses 'geography as a spatial science' is misconceived within human geography, according to those inclined to the two latter views. It gives space an autonomous status as an independent variable separately influencing human behaviour when, in fact, it should be realized that space is a human creation (according to humanistic geographers), constructed and reconstructed to fulfil their interpretations of the causal mechanisms (according to realists). People create spatial forms, both in their minds and in the empirical world, but the non-positivist views on causation deny the need for spatial analysis – the mathematical and statistical manipulation of data relating to points, lines, areas and surfaces.

For physical geographers, spatial analysis has been downgraded in recent years only because of their needed attention to the form-producing mechanisms: explaining spatial form remains their goal. But for some human geographers, the analysis of form distracts from the real task of understanding. Is there, then, no space for spatial analysis in the latter? Tony Gatrell argues that there is, that there are geometries in the spatial arrangement of human activities, the analysis of which allows interpretation of the origins of such arrangements, because space (or distance) is a major impediment to human organization, and also provides a basis for creating new arrangements, for organizing activity across space according to desired criteria.

INTRODUCTION: EXPLORING THE FUTURE OF GEOGRAPHY   19

3   GEOGRAPHY FOR SOCIETY

Tony Gatrell's discussion of applied spatial analysis raises a series of linked questions which relate back to Peter Taylor's discussion of the value of geography. What role can geographers play within society? This question assumes added importance to many at the present time, because of the problems – social, economic, political and environmental – besetting societies. They wish, as geographer-citizens, to be involved in finding solutions to those problems, thereby advancing human well-being; their roles as academics and as citizens intersect. Further, some see the promotion of geographers as of value in this sense as an important way of ensuring recognition for, and security of, the academic discipline.

This view of the role of a discipline reflects one of three models of science identified by Habermas (1974). This is the *empirical-analytic*, or technical, conception. Its scientific goal is to be able to predict the empirical world successfully (often confusing explanation with prediction), so its focus is on the needed technology. Alongside this must be set the *historical-hermeneutic* sciences, which study not phenomena but meanings and which aim at successful transmission of interpretations of the world, and the *critical* sciences which uncover the real explanations and encourage people to seek a better set of mechanisms governing society. (This realist goal – which involves combining all three sciences – is possible in the social sciences but not in the natural sciences, because the processes in the former are human creations, and so can be changed.)

The *empirical-analytic* goal is one long adhered to by many geographers, some of whom have bemoaned the relatively low standing of their discipline in the 'outside world' where science is applied (see Johnston 1983a). Is this an 'anti-geography' conspiracy, or a reflection on the poverty of geographical work? Bob Bennett focuses on the latter. His own research (e.g. Bennett 1982) has shown how geographers' technical expertise and abilities to focus on just spatial distributions can be used to inform important elements of public policy. But, he argues, we are still technically weak, and need to develop more robust procedures. Much of the work done by geographers under the umbrella term of 'spatial science' was, he argues, philosophically and technically naive. It led also to a fragmentation of the discipline, as geographers discovered links with the practitioners of other subjects. This fragmentation was exacerbated by the twin trends of 'relevant geography' and 'radical geography'. Both, he claims, led to a declining interest in what he sees as the core of geography and he calls for a relevant discipline that combines physical and human geography. For Bennett, geographers must be useful to society, and to do this they must restructure their discipline and its methodology.

In comparison with human geographers, physical geographers have been able to develop a much firmer applied base in recent years (see Briggs 1981). In part, this reflects on their empirical-analytic strengths (though Peter Worsley – chapter 1 – queries whether these are sufficiently developed, and developable), and the greater readiness of outsiders to recognize these. Environmental problem-solving needs environmental scientists, and geographers have been proving (and selling) themselves and their contributions. The sort of work they can do is illustrated by Denys Brunsden. The implications of environmental change should be assessed before it is implemented, which demands the involvement of professional specialists such as geomorphologists, exercising their particular skills in combination with a wide range of others – engineers, scientists and social scientists.

A major problem with a focus on applied work, to some, is that it diverts resources (especially time) from theoretical development: attention to immediate practical problems hinders the growth of fundamental understanding (which in the longer term is necessary for the solution of problems). Clearly a balance must be achieved which protects the essential academic activities while promoting their utility. Denys Brunsden argues that the academic study of geomorphology has benefited substantially from its applied links, and that the forging of such links can only advance the intellectual and political status of its parent discipline – geography.

Following on Brunsden's argument, Tony Orme presents the case for a vital applied physical geography within the larger discipline. Human societies cannot be studied independently of their physical environments, he contends, and the correlative discipline of geography provides this focus on society–environment interrelationships. His is a case for a holistic discipline, which places study of the parts into its wider context, while at the same time alerting societies to the environmental constraints within which they live and the environmental consequences of their actions.

Bob Bennett, Denys Brunsden and Tony Orme all stress the empirical-analytic role of geography, while seeking to maintain its academic status and, for Bennett and Orme especially, its holistic educational value. Indeed, it is in the field of education, broadly defined, that geography has always had its roots in the liberal democracies. As such, its goal has been that of building an informed society rather than the vocational training of specialists. Its focus, now much dimmed, on the study of the regions of the world was set in the historical-hermeneutic mould. What should that focus be today? How can the discipline of geography best serve society via its education systems?

USING GEOGRAPHY

Promotion of geography as an applied empirical-analytic science firmly grounds it, according to some, in the political status quo. The problems identified are at the levels of the empirical and the actual and the solutions advanced do not attack the real, the processes that produce the current problems. Symptoms are treated, not causes. Arguments for a more fundamental applied geography, especially applied human geography, seek to distance the discipline from links with governments, public agencies and commercially oriented enterprises. Thus in 1974, David Harvey asked 'What kind of geography for what kind of public policy?' In answering this, he pointed out that for geographers to contribute to the public policies of the British government involved them working for the corporate state, characterized (Harvey 1974, 20) by 'a relatively tightly-knit, hierarchically ordered structure of interlocking institutions ... which transmits information downwards and "instructs" individuals and groups down the hierarchy as to what behaviours are appropriate for the survival of society as a whole'. Such geographers work, he suggested, 'with an easy conscience' (p. 22), not noting that the corporate state

> operates in the name of the national interest. But if we accept that the only meaning to be attached to an individual's life and existence is that which derives from this national interest then we are close to embracing the ideology of fascism. The corporate state is proto-fascist. (p. 23)

But the moral obligation of the geographer is to resist that fascist tendency, to counter the corporate state with a humanistic concern for the individual.

Harvey (1984) returned directly to this theme a decade later, presenting a powerful manifesto for a 'people's geography'. A discipline linked to 'generals, politicians and corporate chiefs' (p. 7) would be involved in 'making *their* kind of geography, a human landscape riven with social inequality and seething geopolitical tensions', whereas a people's geography

> reflects earthy interests and, [he] claims, ... confronts ideologies and prejudices as they really are, ... faithfully mirrors the complex weave of competition, struggle, and cooperation within the shifting social and physical landscapes of the twentieth century. The world must be depicted, analyzed and understood not as we would like it to be but as it really is, the material

manifestation of human hopes and fears mediated by powerful and conflicting processes of social reproduction.

The goal is to advance 'the transition from capitalism to socialism' (p. 10) by opening people's minds to the reality of their situation, so that they can create their own geographies and histories, not live in those of the corporate state.

Harvey's goal is emancipation, involving an explication of the real structures underlying society; it is clearly a critical programme. (See, however, Eliot Hurst's comments – chapter 3 – on whether a people's *geography*, as against a people's *social science*, can meet the goals that Harvey sets.) For a people's geography to be successful, however, and produce a new form of society without any repressive apparatus, it must lead to mass revolution. Mutual understanding must be advanced, leading to consensus on the necessity for a revolution – for the overthrow of both corporate capitalism and state socialism. Geography has a fundamental historical-hermeneutic role within education, therefore, and the nature of that role is taken up in the four final chapters – not all of whose authors accept Harvey's manifesto, of course.

GEOGRAPHY IN EDUCATION

In some countries, such as the UK, geography is a major subject within the educational system, at secondary and tertiary levels. In others, it is either less important (e.g. the USA) or is suffering from substantial decline in enrolments (e.g. Australia). The latter situation is stimulating much debate over the nature and role of a geographical education.

Traditionally, geography has been viewed as a part – usually not very demanding intellectually – of a liberal secondary education, which provides information about the earth (what is where) and some limited knowledge about processes, especially in the physical environment. Despite general disavowals, environmental determinism remains strongly, if implicitly, embedded in school syllabuses. In the presentation of the material about an area, the physical environment is almost invariably treated first, followed by information on the human use of that environment. The structure implies causation. So, too, does the presentation of material on societies. An examination syllabus for 18-year-olds may claim that the details of an area's human geography reflect its physical environment, its history, and its institutional context, for example, but whereas students will be expected to understand the workings of the physical environment they will not be questioned on the historical and institutional contexts. Further, those contexts are assumed

INTRODUCTION: EXPLORING THE FUTURE OF GEOGRAPHY   23

to be internal to the area being studied, and no reference is made to their linking in a global economic and social system (Johnston 1984).

Alongside its 'factual' base, geography provides other elements of a liberal education associated with the collection, processing and presentation of knowledge. As well as literacy and numeracy – the latter being given increasing prominence – this includes skills relating to field data collection and to the visual presentation of material (sometimes termed graphicacy).

In some countries, such as the UK, the teaching of geography in the tertiary institutions (then almost entirely either universities or teacher-training colleges) was for most of the students preparation for teaching the subject in secondary schools. (The universities provided the graduates to teach the more able school pupils.) This produced interdependence; universities 'controlled' the school syllabuses, for example, especially those for the external examinations taken at 16 and 18 which were examined by university staff, and the later years of the school syllabus were seen as preparation for study of geography at university. This nexus has been broken somewhat in recent years – school teaching is no longer a major destination of university graduates; university control of school syllabuses has been slackened; and teacher-influenced syllabuses have focused more on general educational needs than on the requirements of a small proportion who intend to read for degrees in geography. Nevertheless, education at the school level is still strongly influenced by developments in further and higher education: the debates and changes in the latter stimulate discussion of the purpose and content of geography in general education.

Such discussion takes place in its societal context and is influenced by government and other statements on the role of education within society. Thus if governments promote technical and vocational rather than liberal education, subject specialists feel that they must respond accordingly – otherwise their discipline will decline. But should educationalists accept the views of society presented by governments, and structure their syllabuses accordingly, or should they debate with governments what those syllabuses should contain? If the latter, what do they see as desirable course content?

At the tertiary level, autonomy is somewhat greater and academics have more freedom to define curricula and syllabuses. Nevertheless, they are under pressure from governments, other sponsors, and their consumers (i.e. students) to create degree and other courses that are 'relevant'. But what is relevant? Should the various groups involved – teachers, students, paymasters – negotiate over course structures? On what terms, and to what ends?

Debates on the content of a geographical education stem naturally from the debates introduced in the earlier sections of the book. Not surprisingly, given the vitality of those other debates and the stress on many educational institutions, issues regarding the nature and role of geographical education are very alive at the present time. Many of the earlier contributions to the book highlight some of those issues. Others are developed in this final section.

The initial task, tackled by Peter Gould, is identifying the nature of education. He portrays it as self-liberation, through thinking. Educational institutions create the environment for that thinking, through the provision of resources – including instruction in techniques, the mechanics of describing which aid thinking. But that environment is structured by society as a whole, to further its goals and to stimulate patterns of thinking which support its mode of organization. Thus, as Peter Taylor argues in his essay (chapter 4), geography as a discipline has been created and re-created to perform certain educational functions. In an analysis of geography at the school level in the UK, John Huckle develops this theme and shows how those responsible for the growth and status of geography in the British curriculum reflected the ideology of the British social formation, whereas ideas that ran counter to such ideology failed to take hold. Geography, according to Huckle, should be emancipatory, enabling students to appreciate the real mechanisms operating in the world and stimulating them to act constructively to alter those mechanisms and the many injustices that they promote. Education, in this context, is a medium whereby agents uncover the realities underlying the empirical world and encourage attempts to alter the mechanisms. Huckle's analysis is taken up in another context by Joe Powell, in the light of recent Australian experience where traditional school disciplines have been substantially replaced in a teacher-led revision of curricula. Powell, too, sees the role of geography as increasing self-awareness and argues for the development of tertiary-level syllabuses that incorporate perspectives from the humanities, social sciences and natural sciences, build on the secondary school experience, and encourage education in the widest sense, as proclaimed by Gould.

Geography's origins lie in the need to present material about the world to its citizens – in a packaged format acceptable to the powerful vested interests in society. Initially, this involved emphasizing the differences between places and the singularity of regions. More recently, the positivist orientation within geography has stressed commonalities among places, whatever their creations, environments, histories and cultures. Geographers have disengaged themselves from studying and promoting the uniqueness of place, and consequently have contributed to a general ignorance of the world as a complex mosaic. This disengagement must be corrected, I argue (in chapter 17), and geographers must once again take the lead in portraying the complex variability of peoples and environments, avoiding both the generalization trap of treating the empirical outcomes (as against the real mechanisms) as the consequences of general laws of behaviour and the singularity trap of considering each place as a separate entity. Such a task, of description-in-context, is necessary to human survival. Human societies now possess the awesome power to destroy themselves and their environments in seconds. To avoid unleashing this power, peoples must accommodate to each other – a process which requires a mutual understanding to which geographers must contribute.

IN SUMMARY

This book is not a comprehensive coverage of current debates over the future of geography and geography in the future, neither in the issues raised nor in the points of view expressed. No apology is offered for that. The book has been designed to inform and to stimulate, to make students and others aware of some of the major issues and points of view that concern academic geographers. It offers no recipe, no eclectic mix of the many options for the future shape of the discipline. It does, however, lay out some of the ingredients.

As has been stressed here, there is no 'natural' discipline of geography. Further, it is not a fixed element within the educational structure. It is continually being re-created, by the practices of those who call themselves geographers. Many of those practices disturb the status quo very little, if at all. Some are major shocks to the system, however, as individuals interpret the discipline presented to them (i.e. its past and present), identify problems with it, and seek to introduce change. Their interpretations are coloured by their socialization as geographers, and by their general social context. Thus, to paraphrase, 'Geographers make their own discipline of geography, but they do not make it just as they please; they do not make it in circumstances chosen by themselves, but under circumstances directly found, given and transmitted from the past.' In this context, the future of geography and geography in the future depends on its future practitioners, not only in academic life and in education. It is to them that this book is addressed.

REFERENCES

Bennett, R.J. (1982) *Central Grants to Local Governments*, Cambridge, Cambridge University Press.

Briggs, D.J. (1981) 'The principles and practice of applied geography', *Applied Geography*, 1, 1–8.
Habermas, J. (1974) *Theory and Practice*, London, Heinemann.
Harrison, R.T. and Livingstone, D.N. (1980) 'Philosophy and problems in human geography: a presuppositional approach', *Area*, 12, 25–31.
—— and —— (1982) 'Understanding in geography: structuring the subjective', in D.T. Herbert and R.J. Johnston (eds), *Geography and the Urban Environment*, vol. 5, Chichester, John Wiley, 1–40.
Harvey, D. (1974) 'What kind of geography for what kind of public policy?', *Transactions, Institute of British Geographers*, 63, 18–24.
—— (1982) 'Owen Lattimore: a memoir', *Antipode*, 15 (3), 3–11.
—— (1984) 'On the history and present condition of geography: an historical materialist manifesto', *Professional Geographer*, 36, 1–10.
Johnston, R.J. (1983a) *Geography and Geographers*, London, Edward Arnold.
—— (1983b) *Philosophy and Human Geography*, London, Edward Arnold.
—— (1984) 'The world is our oyster', *Transactions, Institute of British Geographers*, NS 7, 443–59.
—— and Claval, P. (eds) (1984) *Geography since the Second World War: An International Survey*, London, Croom Helm.
—— and Gregory, S. (1984) 'The United Kingdom', in R.J. Johnston and P. Claval (eds), *Geography since the Second World War: An International Survey*, London, Croom Helm.
—— and Hay, A.M. (1983) 'The study of process in quantitative human geography', *L'Espace géographique*, 12, 69–76.
Marcus, M.G. (1979) 'Coming full circle: physical geography in the twentieth century', *Annals, Association of American Geographers*, 69, 521–32.
Sayer, A. (1984) *Method in Social Science: A Realist Approach*, London, Hutchinson.

1
PHYSICAL GEOGRAPHY AND THE NATURAL ENVIRONMENTAL SCIENCES
Peter Worsley

Attempts at prediction of future trends within a discipline must be hedged with reservations and in this respect physical geography is no exception. In order to gain some indication of possible magnitudes of change which might occur within a discipline, an often useful exercise is to examine the previous course of events. The writer is venerable enough to have had experience of physical geography as an undergraduate some twenty-five years ago whilst reading a joint geography-geology degree. In the intervening period first-hand familiarity of the changes affecting these disciplines has been gained through teaching and research in both. Notwithstanding the impact of plate tectonic theory on geology and the ensuing revolution in thought, the most profound changes have undoubtedly affected geography. Whereas modern geology has some semblance to the subject twenty-five years ago the same cannot be said of geography, at least at the higher-education level.

Within geography the most radical changes have probably occurred outside physical geography although the changes affecting the latter are nevertheless deep-seated. Twenty-five years ago British physical geography was at least an equal partner to human geography within the discipline and it is not unreasonable to suggest that it was the more powerful of the two components. At that time physical geography was dominated by geomorphologists, to a much greater degree than is the case today. Geomorphology was a rather strange animal as the largest group of practitioners was trained by just two people, namely D.L. Linton and S.W. Wooldridge. This is not the place to elaborate on the effects of this for it has been chronicled previously (e.g. Dury 1983), other than to note that the well-known adverse effects of inbreeding were manifest. Today we find that the position of physical geography within the broader geographical discipline has slipped from its former balanced status. Although the character of geomorphology has changed significantly its relative strength within the sub-discipline has declined. In part this is because geomorphologists have generally widened their horizons. It is not without significance that no more than a quarter of the contributors to this book can be identified primarily with the physical side of the discipline. Such a proportion underestimates the relative numbers of professional physical geographers in higher education if one examines staff interests within Departments of Geography throughout the UK, but it is hard to escape the conclusion that currently it is no longer an equal partner in the coalition which constitutes the discipline of geography.

In this contribution we shall consider the background to current concern over the destiny of physical geography. Unlike the remainder of geography it faces substantial difficulties in:

a   its relationship to bordering disciplines;
b   its status with respect to funding levels from government agencies; and
c   its role in the school curriculum.

The situation is becoming so serious that it is not an exaggeration to claim that physical geography is at the cross-roads if not exactly facing a crisis over its future. Central to the problem is the academic context within which physical geography should proceed; should it remain firmly integrated as of now within geography or should it loosen its traditional ties in order to form stronger links with the other natural environmental sciences, or simply go it alone? It is essential that future graduates in physical geography receive recognition as professionals fully capable of applying their craft alongside other environmentally orientated scientists.

In order to remove any possible initial doubts, it should perhaps be declared from the outset that the writer's perspective of physical geography as such is that of one who views the subject *primarily* as a component of the earth sciences. Indeed there is much merit in the notion that physical geography would be better served if it were renamed earth surface science. This standpoint is not one which will necessarily meet with the approval of all those who are currently employed as physical geographers in higher education. On the other hand it has probably helped qualify the writer as a contributor to this book! Although one's main activities and interests have been in the field of geomorphology as traditionally defined, concern with Quaternary-orientated studies has led to an appreciation of the complex natural environmental system which is the focus of physical geography. This experience has led to the conviction that physical geography has its most important (but not only) affinities with the earth sciences.

PROBLEMS OF PHYSICAL GEOGRAPHY TEACHING

Earlier an attempt was made to highlight the problems confronting geomorphological teaching and research (Worsley 1979) along with the proposal that the existing geographical context might not be the most appropriate for its future vitality. This brought forth generally defensive responses from both Doornkamp (1979) and Sugden (1979). In order to understand the reasons why there is resistance towards proposals which might result in fundamental restructuring of geography, it is useful to be aware of the manner in which the subject became established as a university discipline. Until recently it was fashionable for certain leading geographers to give addresses which were devoted to the pursuit of the essence of the discipline, together with attempts to define its uniqueness. One can only speculate, but it does seem that these introspection exercises were an expression of a deep-rooted need to regularly reaffirm the faith which still lurks in parts of the geographical soul and thereby boost confidence in the coherence of the traditional subject. By doing so revisionists would be kept in check.

A hindrance to totally objective assessments of the geographical heritage is that normally the examiners are the products of the system and accordingly their viewpoints are not unbiased. Thus it is refreshing to encounter a recent analysis by someone outside geography. This was undertaken by Goodson (1981) who, taking a socio-historical approach to the study of contemporary accounts of school subjects, uses the evolution of geography as a case example. This insight into geography could, with advantage, be prescribed reading for all students of the subject who wish to understand the nature of the subject to which they have been attracted.

Initially geography was first formalized in the schools as a cover for what appears to have been little more than a collection of geographical facts and figures, taught by non-specialists. Soon, however, the more inspired teachers developed this into a 'homes in many lands' theme. In 1893 the Geographical Association was founded with the objective of actively promoting the growth and dissemination of geography within the schools and at the same time started to lobby for its establishment in the universities. Such an upwards pressure from the schools on the higher education sector is rather unusual. A champion of the cause emerged, namely H.J. Mackinder, and he campaigned with considerable vigour for the acceptance of geography in the educational system. For instance, at the 1903 British Association for the Advancement of Science meeting he put forward a four-point strategy:

1 Encouragement for the establishment of university geography 'where geographers can be made'.
2 Get the schools to accept the need for specialized teachers of geography.
3 Develop the machinery for examinations.
4 Make sure that the examiners are practical geography teachers.

As Goodson observes, this scheme is tantamount to a closed shop manifesto and we can see here the seeds of the protective attitude which still prevails in some quarters today. Thus university geography arose primarily in response to a need for advanced teacher training and even now the main vocational outlet for new graduates in the subject remains in education. Through the first part of the century a balanced physical, human and regional geography formed the core of the subject both in the schools and universities.

The nature of school geography has changed substantially since the 'new' geography revolution was stimulated by the Madingley Lectures in 1963 (see Chorley and Haggett 1965) although the transition has not been smooth. A consequence of this has been revision of the GCE syllabuses in order to adjust to the new approaches, but a victim has been physical geography which has become (in many instances) a subsidiary theme rather than a component of the central core as was previously the case. Some traditionalists have responded by bitterly regretting the changing emphasis and abandonment of the former human, physical and regional subdivisions. As Goodson observes, the overwhelming worry was that the myth of the core discipline would be exposed. Yet the encouragement of systematic studies in geography at the expense of the traditional approach led to specialization by university practitioners and inevitably they started to discover in many instances that a more receptive environment for their own particular interests no longer lay within the discipline boundaries. Thus in physical geography the specialists encountered rewarding interfaces with the component disciplines of the natural environmental sciences, botany, geology, hydrology, meteorology, soil science and zoology. This in turn has resulted in several consequences which include:

a Research activities and publications being syphoned off from the established geographical literature.

b New external links through teaching and research have generated dissatisfaction over the levels of funding in comparison with those over the discipline divide.
c The inadequacies of the science background of the average geography student have been highlighted.
d That physical geography has much to offer in the multi-disciplinary natural environment field yet its reception by outsiders is often hampered by ignorance of the nature of modern developments in the subject.

These factors have inevitably contributed to a questioning of the appropriateness of the present disciplinary divide and a consideration of alternatives. After all, subject boundaries are created for administrative convenience and as such are artefacts.

A reasonable case can be made which argues that the present destiny of university-based geography is determined in the schools. This partially arises through the demand by most Departments of Geography that their entrants have a GCE A-level in the subject, normally stating that it should also be accompanied by a high if not the highest grade. Such a stance immediately restricts the field from which potential entrants can be drawn, for clearly those who have taken exclusively basic sciences – mathematics, physics, chemistry and biology – are going to conclude that they have inappropriate qualifications and accordingly will be discouraged from applying. Significantly many Departments of Geology as a matter of policy actively discriminate against those who have taken A-level geology since they wish to encourage a broad science background prior to university training. This attitude to admissions is absent in geography because of its historical development, as noted previously. It is, alas, a fact of life that most schools match geography with non-scientific subjects when compiling sixth-form timetables and the result is that many of those who seek university entrance do so with social science and arts A-level combinations. For the physical geographer matters are exacerbated by recent trends within A-level geography syllabuses towards a greater emphasis on matters other than the physical side of the subject.

This basic problem has naturally been appreciated by those who have to teach physical geography although it has only rarely been discussed in print. A masterly analysis was undertaken by Professor Stan Gregory, a climatologist – and incorporated into his presidential address to the Geographical Association, the body which, as we have seen, is the main lobby for the maintenance of the subject within the schools. Whilst discussing the role of physical geography in the curriculum, Gregory (1978) defined the purpose of physical geography at university as a training so that students should be competent to investigate earth surface problems and so extend our understanding of the processes and the end products that result. Later he stated, 'The physical properties of the earth's surface thus form the particular field of study, that can only be comprehended and explained through the medium of physics, chemistry, biology and mathematics.' Perhaps here the thinking of a climatologist overlooks the role of geology but nevertheless there is much here with which one can wholeheartedly agree. Following this reasoning further, Gregory examined the nature of A-level subjects taken in combination with geography and discovered that only one in seven candidates took two science subjects. In the light of this and other data he was forced to conclude:

> Thus we are in the position of attempting to provide a degree level training in physical geography on the basis of, at best, O-level science. . . . It is no wonder that other environmental scientists are at times sceptical about our potential role in the field.

Here we have a clear prognosis of the problem, but he followed this up with a proposal which, in the writer's view, fails realistically to tackle it. The notion of a 'two tier' teaching of both human and physical geography was advanced. Under Gregory's concept the teaching of physical geography would be handled at two levels of intensity, one soundly based on full-blooded science (i.e. taken by those students with science A-level profiles) and the other with rather different objectives; these being of course a simplified version of physical geography avoiding any advanced material. It is difficult to avoid the conclusion that the latter would have very little academic purpose. A converse scheme would operate for human geography. These proposals were followed by the suggestion that in large schools there ought to be encouragement for teachers with specialism in either human or physical geography with respectively social science and science backgrounds. Overall one derives the feeling that Gregory's penetrating analysis clearly identified the fundamental problem arising from a growing incompatibility between the effective teaching of human and physical geography yet the proposed solution appears to fall short of the logical outcome, perhaps in response to a perceived need to draw back from 'the oft-threatened rupture between physical and human geography'. If we are solely concerned with the training of geography teachers then this may be acceptable. However physical geography as an earth science cannot be sustained on this basis.

In many universities student entry to read for a degree in geography is possible through two channels, arts/social sciences and science. Whilst recognizing that the chicken and egg situation may in part prevail, the preponderance of candidates enter with an arts/social science profile. One consequence of this has been for the entry standards, in terms of A-level grades demanded by the universities, to be higher in the arts/social sciences than in the sciences. However, it must be noted that there does not appear to be any good correlation between final degree classification and A-level grades. This is an important finding since some would argue that the 'higher' entry standards through arts/social sciences is a good reason for increasing this kind of intake at the expense of those coming via the sciences.

Having gained entry into university geography via either route the normal situation is for the students to be treated alike as far as previous backgrounds are concerned, and often no qualifying limitations are imposed on the array of courses available. Usually in the first year the syllabus consists of two components, human and physical, and both are taken by the entire student intake. The immediate effect of this is that the introductory courses have to be pitched at a level appropriate for the diverse backgrounds of the students and this frequently means that they have to be adjusted to the lowest common denominator. The general problem is expressed by the publication of books which are designed specifically to serve geographers who do not have the appropriate background (e.g. science for geographers – see Davidson 1978). Restricting discussion to the field of physical geography we find several problems:

a   The physical enthusiasts become bored since the course is not challenging enough and not significantly different from work which they have undertaken at A-level.
b   This also applies to those intending scientists who are taking geography as an ancillary/subsidiary. Not infrequently they find the human component totally unattractive. More seriously their impressions are transmitted to staff in parent departments who conclude that the geography course is non-scientific and therefore unsuitable. Later students are discouraged from selecting geography as a supporting subject and progressively teaching links with the science departments become more and more tenuous.
c   Understandably the 'arts' students often struggle because of their poor appreciation of even simplified science-based concepts and soon become disillusioned.

Distressingly, we have a 'no-win' situation since no particular group is satisfied. It is difficult to escape the conclusion that the provision of a first-year physical geography course suitable for all comers, arts, social sciences and science, is an impossible task yet this is precisely what is attempted by most geography departments. The most serious loser is physical geography itself. Is there any wonder therefore that undergraduate physical geography is increasingly becoming a minor component of its current parent discipline? A radical change in the entry rules is going to be necessary if an advanced modern physical geography is to emerge.

Apart from the purely academic problems arising from the scholastic organisation of physical geography, matters are further exacerbated by government perception of the subject. The principal resource allocation to the universities generally is through the University Grants Committee (UGC). In the United Kingdom the total recurrent grant to be distributed amounts to £1123 million in 1984–5. The committee acts as a buffer between the Department of Education and Science and the universities, which like to view themselves as independent establishments. As is well known, the current trend is for the 'arms length' distancing of the universities from government to be progressively reduced and replaced by de facto direct intervention in university decision-making. In order to calculate the yearly allocation of university finance to each university, a primary consideration is the number of what are termed 'full time equivalent' student numbers. According to the type of student (e.g. arts-based) the level of support per student varies, since the teaching costs are going to change in light of the kind of subject being studied.

Unfortunately for these purposes, physical geography is classified as a social science and hence the unit of resource accorded to it is almost the same as that of say economists or sociologists. The qualifying 'almost' is used since, along with psychology, geography is now recognized as a slightly different social science as it has a requirement for laboratories and consequently is given a small additional resource allowance. Nevertheless the specific needs of the kind of geographer being trained are not fully acknowledged, with the result that physical geography as a teaching and research activity has to manage on a norm which is clearly below that accorded to their fellow biological and earth scientists. This unsatisfactory arrangement arises simply because of the outdated perception of modern physical geography whose financial needs are so obviously different from those of human geographers. The root cause is a unitary geography.

PROBLEMS OF PHYSICAL GEOGRAPHY RESEARCH

Having examined the difficulties confronting physical geography at the undergraduate level, it is now necessary to follow through to the research field where extra resources can be won in open competition with other disciplines. On face value the lay person might consider that physical geography is very fortunate in that the UK government supports an organization called the Natural Environment Research Council (NERC). This council was established by Royal Charter in 1965 with a responsibility to 'encourage, plan and execute research in those sciences physical and biological that relate to man's natural environment and its resources'. Bearing in mind that the 1982–3 budget was some £87 million an outsider might reasonably expect an excellently funded physical geography, especially when the rubric continues:

> such investigations seek to provide a better understanding not only of the nature and processes of the environment in which we live and on whose resources we depend, but also of their influence on man's activities and welfare and, of growing importance today, of man's influence on them.

In reality we find that our initial optimism is unfounded for physical geography as such has no separate identity within the structure of NERC. The fields of research recognized by NERC have been defined as:

The Solid Earth (geology, geophysics and geochemistry)
The Seas (physical oceanography and marine ecology)
The Terrestrial Environment (terrestrial ecology and soil sciences)
Inland Waters (hydrology and freshwater ecology)
Atmosphere
Interdisciplinary Studies

The persistent optimist might rejoice, thinking that in this schema the last might largely correspond with physical geography. Alas this is far from the case since the interdisciplinary studies specifically relate to the Antarctic environment and physical geography as such is not incorporated. Yet it is apparent that the scope of physical geography embraces elements of all the fields as defined above and hence ought to have a vital role in the functioning of NERC.

The Council pursues its aims of promoting research and training through the operation of its own institutes and grant-aided institutes and by awards to universities and other higher education establishments. There are thirteen of these institutes and only two, oceanographic science and hydrology, have sections which employ physical geographers. In terms of the total staff in institutes the physical geographical element is below 1 per cent. NERC finance is derived from two main sources, grant-in-aid through the Department of Education and Science – this is known as the Science Budget – and commissioned research from government departments. The latter is the response to the so-called Rothschild principle of customer–contractor relationships, supposedly to encourage 'relevant' research. This has been, and continues to be, a major policy issue attracting strong criticism from leading scientists. Currently the Science Budget is approximately twice that derived from commissioned work. Of the total 1982–3 budget of £87 million mentioned previously, almost £78 million goes to fund NERC institutes and the largest component – £23 million – is given to the Institute of Geological Sciences (this has been renamed British Geological Survey since 1984). Hence, the remaining budget for non-institute support is relatively small in comparison with the whole. Nevertheless, despite this, NERC funding of postgraduate activities and faculty research is the most important contributor to advanced studies of the natural environment in higher education. Thankfully NERC does recognize that physical geography research is its prerogative. Without it well over half of the work in the field would cease and unfortunately, if present trends continue, with progressively reduced directly funded (UGC) university budgets, the relative importance of NERC support is bound to increase. Although official literature refers to NERC contributions to universities and polytechnics, the latter organizations received only a few per cent of the total resource allocation; thus under this heading we are mainly considering the role of NERC in the university sector.

In 1982–3 a total of just under £10 million was disbursed by NERC in support of universities. Operationally, with the university support area, the natural environmental sciences are subdivided into four groups:

1 Aquatic and atmospheric physical sciences
2 Aquatic life sciences
3 Geological sciences
4 Terrestrial life sciences

The allocation of support is largely determined by the recommendations emanating from committees (each of fifteen or so members) which serve each of these groupings. If a conservative approach is adopted and committee members' affiliation taken as representative of discipline, then we find that physical geography is relatively poorly represented. Normally, only two physical geographers serve on each of the aquatic-atmospheric and geological committees (there is a complication with the latter which will be explained later). If we consider the traditional components of physical geography we find that the former group embraces the fluvial, hydrological and climatological elements and the latter the remainder of geomorphology and what might loosely be termed Quaternary studies. It is reasonable to ask who represents the geographical interest in pedology and biogeography? The straight answer is that there are no geographically based representatives in these fields and consequently those seeking support are often obliged to bias their applications for support so that they might be considered by the two committees with geographical members, otherwise the feeling is that a sympathetic consideration cannot be taken for granted. Here is the making of what is popularly known as an 'endangered species' and obviously does not help to establish a more balanced physical geography.

As already noted, the committees have responsibility for making recommendations to Council on the distribution of available funding. There are two quite separate types of resource allocation, research grants and research studentships. A minor part of the latter are fellowships for post-doctoral studies but since they are of comparatively minor importance they will not be considered further. Apart from the geological sciences committee the decisions on grants and studentships are dealt with by the same groups in the three other subject areas. The exception in geological sciences arises since the scope of its area of responsibility is roughly twice that of the others and the sheer volume of business dictates the existence of two separate committees to independently handle the two types of support. One might reasonably wonder that if this is necessary, would it not be more logical to split geological sciences into new 'hard' and 'soft' rock committees? Just as geography is often seen as consisting of two components, human and physical, in geology there is a natural tendency to separate into two groups those interested in igneous and metamorphic rocks and those orientated towards sediments and fossils. If a hard–soft realignment were to occur it is possible that definite advantages might accrue to those whose interests lie in the realms of surface processes and obviously physical geography would gain from this.

Research grants are awarded in order to provide financial help for research workers so that they may initiate and develop their own research projects and ideas. An important proviso is that the grants are not intended to meet the entire costs of such projects, for an infrastructure provision is assumed, provided by the UGC recurrent grant to universities. Here it should be recalled physical geography is classified as a social science. This is the basis of what is termed the 'dual funding' of university-based research. Another problem for university researchers is that NERC policy is against long-term support and this is a serious restraint on environmental monitoring programmes. Applications are considered twice yearly and each proposal is subject to 'peer review' by several workers who are active in the general field but their identity is not revealed to the proposer. In the light of this external advice and committee judgement all the proposals received within each six-month period are ranked and the best of these are funded as far as the budget allocation will allow. Some one-fifth of the total grant allocation is distributed by another committee – the University Affairs Committee – which is composed mainly of the chairmen from the sub-disciplinary committees and Council members. However the awards are made to applications which are initially made via the conventional route. An examination of the latest available figures relating to grant expenditure (1982–3) shows that a mere 5 per cent of the total was attributable to grants tenable in geography. Of the 5 per cent, most was allocated by the aquatic-atmospheric and geological science committees and a smaller component by the terrestial life sciences. The latter was entirely accounted for by two remote sensing projects whose subject matter fell outside that of the other two committees. The small proportion of funds attracted by geography can be explained by several factors including:

a   The cost of physical geography proposals which tends to be relatively modest.
b   A lesser tradition of seeking this type of support and a corresponding lesser experience in writing good proposals.
c   Low representation on the committees which inevitably results in a certain degree of bias against the subject.

A rather different procedure is followed with research studentships. First, these are only considered annually, linked to the yearly flood cycle of new graduates, and second the assessment is done entirely by members of the committee without an external input. There are two kinds of postgraduate studentship, those for research and those for advance courses, and each will be briefly considered. Proposals for research studentships (of three years duration) are formulated by university staff who wish to promote a research topic and offer supervision and they are submitted through the departments within which the work would be conducted. Again, all the proposals are ranked and available resources determine the cut-off point. In contrast the advance course studentships are allocated to taught M.Sc. degree courses which normally are of one-year duration only. A major objective is to encourage training in advanced applied techniques in the hope that the masters' level graduate will be attractive to industry. Physical geography currently has one lone representative amongst the fifty plus approved courses and somewhat surprisingly this is not the geomorphological field but in climatology and applied meteorology. A new course in remote sensing is partly accommodated in a geographical milieu. However, at best only a lamentable 1–2 per cent of the course studentships are in the field of physical geography. The relative costs of research and course studentships are in a ratio of 5 : 1 and even the twenty-five or so research awards made each year leaves a total allocation out of the research training budget to physical geography of below 10 per cent. It is therefore difficult to deny sympathy to those who feel that physical geographers receive the scraps which fall off the rich man's table.

FUTURE PHYSICAL GEOGRAPHY

Having outlined the background to the current issues confronting physical geography at the undergraduate teaching level and in the field of research, it is necessary to consider what, if anything, can be done in order to improve matters. Experience has shown that although recognition of the issues involved will be given, when it appears that a solution might encroach upon established territories, the forces of inertia promote conservative attitudes. Admittedly the atmosphere generated by government-imposed cut-backs in university funding does not aid flexibility but it is nevertheless disappointing to encounter hostility when the implementation of rational proposals is discussed. After all, an ideal view of universities is that they are places where liberal attitudes and progressive thinking should flourish. Hence the key constraints upon any adaptations to cater for the problems of physical geography are going to be pragmatic.

It seems that the only institution which has so far developed an unrestricted physical geography is the University of East Anglia at Norwich where, within a School of Environmental Sciences, geographers and geologists of all persuasions work under a single umbrella. In this instance, it has to be recalled, a new university was founded and, as a matter of policy, the usual disciplinary boundaries were not adopted throughout its structure. The evidence available suggests that this is a successful combination. Yet a rather similar attempt at the University of Lancaster eventually resulted in the creation of separate environmental science and geography departments. On face value this experience does not encourage excessive optimism for the future but very recent changes affecting the University of London appear to be the precursor to a new physical geography.

Somewhat ironically these changes have their source outside geography. An enquiry into the possible rationalization of geology within the University of London collegiate federation recommended a revolutionary programme of transfers and closures. Historically, many of the colleges had developed their own separate Departments of Geology but, with the exception of Imperial College, these had never grown into large units. Consequently they were relatively expensive to operate and had difficulty in coping with an expanding area of knowledge. Hence in the interests of greater efficiency, availability of a wider range of specialized expertise and critical mass, a drastic restructuring plan was proposed. Arising from this, adjustments to the established structure of geography have occurred in a manner which was probably not even contemplated.

First, the planned merger of Queen Mary College and University College geology departments with concentration on the latter's site has only partially been achieved. Some of the 'applied' Queen Mary geologists have managed to evade transfer to engineering by seeking refuge in their adjacent geography department, resulting in the creation of a new Department of Geography and Earth Science. An immediate consequence has been the strengthening of physical geography and the emergence of a separate degree in the sub-discipline. In addition a newly launched degree in engineering geomorphology has become more viable. Second, associated with the move of Bedford College from its long-established Regent's Park location to the campus of the Royal Holloway College at Egham south-west of the capital, is the creation of a new earth science complex which will house the combined geology departments from Bedford, Chelsea and Kings College *plus* Bedford College geography which unlike many other geography departments has a concentration in physical geography. Clearly the potential for future innovation in this earth science milieu will be considerable. Third a plan for closer co-operative linkages between Birkbeck College (which is unique in that it specializes in part-time evening teaching) and the physically close University College has encouraged the formulation of proposals for a new physical geography degree to be supported by the two departments. Thus in three different parts of the University of London a new breed of physical geography is emerging and it is difficult to envisage these growth points not having a national impact. This is especially so since the opportunity arises to alleviate many of the basic problems with physical geography which have previously been discussed.

If we accept the premise that a physical geography capable of full acceptance as an equal constituent of the earth sciences is desirable then obviously it must modify its current position within the university geographical system. It would seem that this will only be possible where sufficiently large departments exist and where the right kind of political climate also prevails. There are a number of ways in which the operation of a science-based physical geography can be achieved and generally a policy of flexibility of course structure is going to be essential. Perhaps the least traumatic accommodation can arise if a degree of autonomy can be granted to physical geography within its existing departmental affiliation. Such an arrangement would enable the continuation of much of the present pattern of geography alongside the new with shared staff and resources. Another possibility is to explore the feasibility of mergers between geography and geology in order to create stronger and more efficient units within which modular degree course structures would permit a range of possible degree combinations. Another arrangement might seek total independence for physical geography although in the context of current rationalization trends this would seem ill-advised.

Syllabus adaptation and modification of departmental boundaries within universities are going to be easier tasks than those arising from the issues to be confronted externally. Both government departments and the research councils are not readily going to revise their wary perception of even a radically reformed physical geography. For instance the British Geological Survey has recently undergone a major revision of its internal structure, yet in the new arrangements no recognition is given to Quaternary geology. As a result there is no pool of specialized expertise available within the national institute to advise on the most recent part of the geological record and it will not be lost on physical geographers that this is precisely where the interface with geomorphology occurs. As a rare geologically based voice in the wilderness recently observed, 'Quaternary geology is relevant geology'. The reasons for this anomalous situation are not known but the suspicion is that a preponderance of classically trained British geologists see the end of respectable time at the close of the Cretaceous some 65 million years ago! Here the traditional system of British geological education could serve the national interest better yet those officially appointed to advise NERC are largely the products of this unbalanced approach to earth study. Another illustration of the establishment's 'blinkered' attitudes occurred at a recent multi-disciplinary conference. After hearing the delivery of an excellent paper an eminent physical scientist expressed total surprise after it had been revealed that the contributor was a physical geographer, for it was clear that real competence as an earth scientist was simply not expected of geographers. Alas this type of attitude is common and only the naive will expect it to quickly disappear.

It is evident therefore, that the achievement of a sound basis upon which a vital physical geography can thrive is going to be no easy matter. If the changes which have occurred in the last two to three decades are a guide to the potential scale of future developments, then there is hope. The sustained application of our convictions is going to be necessary but surely a more rational situation will ultimately emerge.

REFERENCES

Chorley, R.J. and Haggett, P. (eds) (1965) *Frontiers in Geographical Teaching*, London, Methuen.

Davidson, D.A. (1978) *Science for Physical Geographers*, London, Edward Arnold.
Doornkamp, J.C. (1979) 'Comment on whither geomorphology', *Area*, 11, 307–9.
Dury, G.H. (1983) 'Geography and geomorphology: the last fifty years', *Transactions, Institute of British Geographers*, NS, 8, 90–9.
Goodson, I. (1981) 'Becoming an academic subject: patterns of explanation and evolution', *British Journal of Sociology of Education*, 2, 163–79.
Gregory, S. (1978) 'The role of physical geography in the curriculum', *Geography*, 63, 251–64.
Sugden, D.E. (1979) 'Comment on whither geomorphology', *Area*, 11, 309–11.
Worsley, P. (1979) 'Whither geomorphology', *Area*, 11, 97–101.

2
HOLISTIC AND REDUCTIONISTIC APPROACHES TO GEOGRAPHY
I.G. Simmons and N.J. Cox

A SCHEME OF KNOWLEDGE

The picture provided by science of the natural world is hierarchical: the units or systems identified at any level or scale of investigation are made up of smaller units or subsystems, which in turn can be regarded as systems at a different level. Somewhere at the smallest end are the fundamental particles not yet fully elucidated by the physicists; but together they form atoms, atoms in consort constitute molecules, molecules of different kinds make up cells, and so on. Each level includes all the phenomena from the layers below (organisms incorporate organs, which in turn are composed of cells, themselves aggregates of molecules, and so on) but may have unique properties which cannot be predicted from current knowledge of the level below: for example the freezing properties of water cannot be predicted from the normal behaviour of hydrogen and oxygen. At some stage we can add to the purely material our own species with its property of self-reflexive consciousness, and after that the sky's the limit or, rather, the next level. Such a hierarchy of integrative levels of matter also corresponds (Figure 2.1) to our way of dividing academic disciplines one from another, and so it is relevant here to discuss geography in such a context. At once we notice that geography is quite high on the scales of both complexity of unit and absolute size: it deals neither with the simplicities of the few fundamental particles and their binding forces (these may be very difficult and expensive to investigate but that is a different matter) nor with the unimaginable distances of the galaxies that appear to compose the universe.

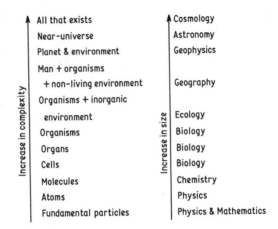

Figure 2.1 A hierarchical scheme of knowledge

In the world of western ideas the comprehension of the phenomena we perceive around us has, since the seventeenth century, most frequently been organized into two types of knowledge. The first of these is an analytical approach, most highly developed within science. Science wishes to explain, to generalize and to determine causes, so hypotheses are formulated and tested by experiment and comparison, and general theories are developed which include statements of the laws of nature. The great characteristic of a theory in science is that it is temporally predictive with a high degree of probability; for example physical laws such as those of gravity or thermodynamics appear to apply with a probability of 100 per cent tomorrow just as much as today. In another instance, astronomic physics has predicted the existence of phenomena such as black holes, which have then been shown to exist.

Now, in a majority of instances, the scientific method achieved its success by tackling the problem at the next lower level in the hierarchy of complexity: the properties of atoms are best explained in terms of their fundamental particles, the life of cells by study of their chemistry, and the classification of organisms such as plants by reference to their flowering parts – these are all examples of this approach. In geography at one time it was considered the ideal to catalogue separately every facet of the physical environment and all the known facts of human distribution and activity about a particular place; thus the geography is determined by a disaggregation into a multitude of component parts. This general approach is termed *reductionism*, defined more formally as concepts or statements redefined in terms which are more elementary or basic. In many cases the communication of knowledge thus gained cannot be expressed in ordinary verbal language, and two other methods must be added. The first is that of the use of words with specialized meanings (I use technical terms, they use obscure jargon), and the second is the language of pure symbols associated with mathematics and symbolic logic. In practical terms there is no doubt whatsoever of the successes of the method; to give but two obvious examples: the technology opened up by knowledge of how to manipulate recombinant DNA and the advances made by drug therapy in humans and other animals come from reductionist investigation and experiment.

The other basic category of knowledge has been sought outside science as well as within. It attempts to comprehend a given level of phenomena in its own terms rather than the level below. (It may even relate to the level above in terms of eventual evolution.) This means that it seeks to understand the whole rather than the parts and suggests that the whole is more than the sum of its parts, that is that it has emergent properties not predictable from knowledge of the constituent parts. This type of knowledge is termed *holism*, defined as the view that wholes have properties that cannot be explained in terms of the properties of the individual constituents; the apparent uniqueness of every human being is a good example. The comprehension of wholes in the upper levels of complexity, once life becomes part of the schema, is very difficult: the distinguished biologist Ernst Mayr (1982) has even suggested that he was 'not aware of any biological theory which has ever been reduced to a physico-chemical theory', and that 'a philosophy of science restricted to that which can be observed in inanimate objects is deplorably incomplete'.

Attempts to mount a scientific investigation of complex wholes have in general centred on the description and explanation of the relationships between constituent parts of the whole, believing that in these relations lies the key to understanding the emergent properties. Systems analysis is an umbrella label for several fields which have sprung up within science to tackle complex wholes: their relevance to geography is discussed below. But the analysis and description of wholes has throughout history often relied upon intuition ('immediate apprehension by the intellect alone' – *Shorter Oxford Dictionary*) in which the integratory powers of the human mind are the key instrument and the key mode of communication has been verbal language, including that of poetry. In geography such intuitive synthesis is explicit in the work of those who have tried to set down 'regional character' or 'regional personality': the French school of the 1920s and 1930s is one outstanding example, some humanist geographers of the present another. Further, when we get to the emergent properties of the whole cosmos, then language itself becomes strained in the effort to convey the possibilities, and Rabindranath Tagore vividly expressed a conviction of the limitations of analytical science when he wrote,

> For the world is not atoms or molecules or radioactivity or other forces, the diamond is not carbon, and light is not vibrations of ether. You can never come to the reality of creation by contemplating it from the point of view of destruction.

It has to be said that where western intellectual frameworks are used, then those who employ the methods and languages of reductionism in general look down upon any branch of knowledge which relies upon intuition, and apply the epithet 'soft'. Many non-western philosophies and systems of knowledge are of course less worried by the desire to be so apparently rational and to verbalize or mathematize everything.

REDUCTIONISM IN GEOGRAPHY

Studies in purely physical geography have, in general, no option but to be carried out using the methods of contemporary science, whether these be reductionistic or attempts at holism like systems analysis. We seek an understanding of landslides or glacier motion in terms of mechanical principles, and apply chemistry in the study of weathering processes and soil formation. The behaviour of larger, more complex systems is routinely analysed with elaborate computer models, as in drainage basin hydrology or weather forecasting. But at this point we ought to ask whether at the next level of complexity, that of human societies together with their environment both living and non-living, reductionist approaches are appropriate or applicable, and with what degree of success. We are perhaps asking, basically, should geography try to be like physics? Should it be possible to express everything in laws (themselves communicated by mathematical rather than verbal symbols) which are 100 per cent applicable in the sense of temporal prediction? (Actually the laws of physics do not exhibit 100 per cent probability but they are not far short for practical purposes: see p. 55.) Some of the attempts made since the 1960s do seem indeed to want to partake of the aura of physics and its mathematical language: the very terms used, like gravity models, spatial autocorrelation and probability surfaces, point that way, as does the umbrella term 'Quantitative Revolution' which has been used to describe them, a term which is often used rather more widely also to cover attempts at that time to develop theory in geography. The techniques of the mathematician and statistician have been paramount in, for example, the elaboration of models of spatial interaction, the use of ideas from catastrophe theory and Q-analysis, and the widespread employment of bivariate and multivariate linear models for the analysis of data.

Where has this movement carried geography? In general, although the methods may derive from mathematics or statistics, the applications which geographers have taken up or fostered have been from the social sciences, notably economics, but also sociology, history and political science, which have on the whole not become very much like physics since they have been forced to acknowledge that they are permeated with values and that true objectivity is scarcely possible: the choice of subject for study, according to some writers, reflects a basic ideological attitude. Even so, the editors of *Models in Geography* (Haggett and Chorley 1967) offered the thought that 'those subjects which have modelled their forms on mathematics or physics ... have climbed considerably more rapidly than those which have attempted to build internal or idiographic structures'. In terms of results, however, it is possible to be equivocal. There have been remarkable advances in the development and adaptation of modelling and analytical techniques but whether empirical studies which have used these have provided any more generalizations and accurate predictions than in non-reductionist approaches is arguable. What is most striking is the great emphasis on space as the central element in geography; to the end that if space only is considered then geography can indeed become much more like physics, even to the extent that Wilson (1981) used the idea of entropy-maximization, standard within statistical mechanics, in modelling flows in urban settings.

Naturally enough, not everybody is content to equate geography with spatial science and leave it at that. The question of subjective knowledge has to be considered, no matter how quantitative and rigorous the treatment. 'What is the point,' ask Bennett and Wrigley (1981), 'in setting up a null hypothesis to test whether a relationship does exist when the relationship is part of prior theory and existing empirical knowledge?' This viewpoint takes us in the direction of a quantitative approach in geography that will include examination of goals and norms, which will inevitably take it in the direction of social policy. The criteria for success will be less those of statistical or mathematical elegance and more those of usefulness in social and political questions, and in answering the 'fundamental research questions of the sub-disciplines involved'.

Writers unsympathetic to quantification, with its frequent association with 'hard' science, have often criticized the view for its apparent lack of humanity: its cold objectivity (if it should exist) does not appeal to all. Others have seen it as a tool of vested interests, that is of the powers that be, in maintaining the status quo. But yet, we might not accept reductionism of a different variety which views all human patterns in terms of a single-factor explanation such as the class struggle at the heart of classical Marxist theory, which seems to be too simple to explain the great variety of society–environment relationships which have been observed on the face of the earth.

It is perhaps too early to deliver a verdict on the success of reductionist methods in geography. In any case, how do we measure success? Is it to be by the criteria of physics, meaning the symbolization of laws bearing a precise probability with a known degree of error? Or, given the world as it is, and scholars as they are, do we see them in terms of social engineering, meaning the chance to affect the public polity and to help remove injustice? Any interim assessment may perhaps be forgiven for coming to the conclusion that in terms of laws, geography has not had a great success. Some regularities have been pointed out (rank-size rules of cities; the spatial patterning of towns as service centres) but the bulky literature concerning them contains evidence of numerous exceptions and assertions of the culture-bound nature of the findings. Equally, geography seems to have brought about a more just world no more and no less than any other division of learning with which it shares a reluctance to be committed to single-element solutions, especially those of an ideological character. But we may write off neither approach, for they are both relatively new and deserve more time in which to make their contribution. Wilson (1981), however, sums up many of these developments by suggesting that 'it would be good to think that geography could continue to be a synthesizing discipline and that a particular theoretical concern would be to put together effective theories for whole-systems of interest to geographers'. For one scholar, therefore, the aim is holism even if the current methods are reductionist.

HOLISM IN GEOGRAPHY

The term holism was invented by the South African statesman-scholar Jan Smuts and for him, as for some later writers, it was suffused with ideas which belong equally to mysticism and vitalism, that is an element of a metaphysical reality had to be involved. This has in some eyes discredited the term, but happily it can now be used as defined above (p. 45), purely in terms of matter and its internal relationships, stressing that at any hierarchical level, the dissection into component parts always leaves unexplained an unresolved residue. The idea of geographical wholes, as in the integral nature of place and people, is by no means new.

HOLISTIC AND REDUCTIONISTIC APPROACHES TO GEOGRAPHY 49

As Glacken (1967) has described, the ancient Greeks elaborated a number of concepts to explain what they saw as consistent and explicative relationships between climate and society, for example, or soil and personality. In later times, the German word *Gemeinschaft* has come to express the implication of an organic unity both within a community and with its external environment. The quasi-organic affinity which some French geographers felt for the regions which they lived in, studied and described has already been mentioned. Some scholars have always felt that such a procedure was too purely personal and did not come anywhere near the scientific canon of replicability (François's account of rural Brie is bound to be different from that of Jean-Marie) and so work on the nature of place has been subject to studies influenced by behaviourism, that is the study of unambiguously observable and measurable behaviour, where the observer as in physics is said to be completely neutral. Such studies are, not surprisingly, prone to be reductionist rather than holistic and in contemporary humanistic geography the observer's role is explicitly acknowledged by stating that only an individual mind can apprehend the whole and come anywhere near communicating the synthesis which it perceives to other people. The controversy over the validity of this procedure in any wider framework is lively and as yet unresolved.

In parallel with the intellectual progression just described, there have been attempts to develop a more scientific holism. Greek philosophy had its early versions of an all-embracing explicatory holism, particularly in the notion of teleology, the theory that the evaluation of anything can only be validated by consideration of the purpose to which it is in the end directed. In the Greek case, the holistic *telos* was that of a designed earth, made by divine will to be a fitting home for the human species, where all things fitted neatly into the ultimate plan. Such an all-embracing concept has, understandably, suffered some vicissitudes but its latter-day successor is perhaps the science of ecology, though this lacks the metaphysic of the original. One of the most fertile contributions of ecology to geography (and other fields as well) has been the concept of the ecosystem: an interactive, dynamic system linking together plants, animals and inorganic matter over a given area, at virtually any scale. (Its virtues and relevance were demonstrated very clearly in a paper by Stoddart (1965)).

Ecosystems are among the best examples for geographers of systems in general. The idea of a system is basic throughout science: in the simplest cases, a system is just a bounded unit used for drawing up a mass balance or an energy balance, as might be done for a soil profile or a drainage basin or a city region. But what has particularly engaged many geographers has been a cluster of new fields which have sprung up since about 1950 to tackle the problem of understanding systems, including cybernetics, control engineering, information theory, general systems theory and computer simulation modelling, often referred to collectively (and rather loosely) as systems analysis or the systems approach. The introduction into geography of the general idea of systems analysis was signalled by Blaut and by Chorley in 1962, with an influential paper by Ackerman following in 1963. Ackerman in particular was enthusiastic that here was the possibility of viewing a set of problems at different levels in a way never before possible, that it made available new problem-solving methods, and placed geography in proximity to other disciplines with similar concerns. The actual application of some of these approaches can be seen in the new approach to physical geography taken by Chorley and Kennedy (1971), and in the massive compilation by Bennett and Chorley, *Environmental Systems: Philosophy, Analysis and Control*, published in 1978. In this book, they distinguish hard systems (those capable of specification, analysis and manipulation in a more or less rigorous and quantitative manner) and soft systems which are not amenable to mathematical methods. These systems are interfaced in discussions of 'physico-ecological' and 'socio-economic' systems and material on environmental intervention and symbiosis. In general, though, the formal symbolization of the hard systems is replaced by verbal discussion of a more conventional kind when it comes to examples of 'interfacing'. We might take away the conclusion that emergent holism was at that time elusive of capture by the language of the 'hard' systems. Similarly, Coffey's book (1981), though using mathematics where possible, resorts to words where whole systems are concerned; we must go to Wilson's (1981) book for an unyielding formality of approach which maintains its rigour throughout, except that in his example of moorland management, the mathematics extend only to the biological components of the system: the main land use interactions and opportunities for change are described verbally in two large tables. Urban models and water resources are however more reducible to equation sets. As Wilson concludes, such techniques are the outcome of a relatively young field and although the handling of complexity has been improved by these approaches, it seems certain that improvements will be forthcoming. So the ability of systems analysis to handle holistic ideas in geography is not yet proven, nor is it yet disproven: as with other quantitative approaches it is premature to deliver final verdicts.

One way which the ecosystem idea has led is that of trying to characterize the functional relationships of the whole by certain measurable features such as energy flow and matter cycling, following the Lindeman (1942) model of a trophic-dynamic approach to ecosystem function. Given that energy flow and matter cycling are just as much features of the man-made world as they are of the natural environment, then the chance exists to study the whole interaction. It is possible, for example, to characterize agriculture according to the balance of the energy fixed by green plants and the energy added to the food production system in the form of machinery, fertilizers, pesticides and packaging, all of which represent fossil fuels in some transformed state. It is equally possible to investigate such flows for an entire region-state as has been done for Hong Kong, where its increasing dependence upon outside sources of energy not only for industrial growth but to maintain more intensive food production has been charted, and policy initiatives made (for example to reduce the need for imported energy as fertilizer by using night-soil once again instead of allowing it to contaminate inshore water). At a wider scale still, E.P. Odum (1975) has suggested a four-fold classification of the globe's surface according to the type of energy system which occupies it, with the open oceans and upland rain forests at one end (with no added energy subsidy for human societies), and Manhattan Island at the other, with virtually nothing but fossil fuel-based energy.

We need not be surprised, therefore, that this type of holistic approach has led to a view of geography as a form of human ecology. This latter term has been appropriated by a number of disciplines of which geography is only one and as a type of geography has been strongly influenced by the French school of the 1920s and by the anthropological leanings of the Berkeley school founded by C.O. Sauer. In both these ways it has been intuitive rather than quantitative and so is not perhaps a springboard for a scientific holism. Yet in the hands of a gifted scholar such as K.W. Butzer (1982), a verbal synthesis of the totality of society–environment relations in ancient times becomes a piece of scholarship that nobody need be ashamed of.

Has holism been successful? As discussed above in relation to its similarity to physics, not spectacularly so, though the regional studies of energy and matter (unfortunately not conducted by geographers) seem to promise predictability of a certain kind, though never at the 100 per cent probability level. In terms of social engineering, we can point to the great successes of the environmental movement of the 1960s and 1970s in bringing to the fore all kinds of issues ranging from the impacts of population growth, to environmental contamination, the environmental impact of high technology, and the need for high environmental quality. Above all, the movement succeeded in planting the seed of the idea that the planet was one whole (aided by some powerful iconography from the space programmes) and that actions in one place might produce unexpected outcomes elsewhere, and that the whole had not an unlimited resilience to manipulation by human society.

HOLISM BEYOND GEOGRAPHY

Other contributors to this book are to deal with the potential for further applications of reductionist methods in geography, so in this section we will look further at some of the work by scholars outside our field who have been concerned to try and find a way of study of the society-and-environment level of hierarchy which will lead to understanding how it functions and what might be wrong with current interactions. One such idea is the Gaia concept of Lovelock and Margulis (1974; see also Lovelock 1979). They have hypothesized that important physical features of the earth such as the gaseous composition of the atmosphere are not a condition to which life has adapted, but a consequence of the co-evolution of life and inorganic matter. Their hypothesis has brought to light, for example, the crucial role of organically produced methane in regulating the oxygen content of the air, and the dangers to the atmosphere of overproduction of methyl iodide if kelp were to be farmed. In policy terms, they point out that the living systems of the earth survived the Pleistocene glaciations, when much of the temperate zone of the northern hemisphere was lifeless. Thus the loss of living matter due to human activity in northern industrial zones might not be so important: the regulators of a global organic-inorganic stability might more likely be the organisms of the continental shelves and of the biologically highly productive lowland equatorial forests. The protection of those two biomes might thus assume an even greater importance than is now accorded to such an endeavour.

One of the central ideas in the Gaia hypothesis is that of co-evolution of both inorganic and living matter, with breaks in symmetry as new levels of complexity are achieved. Emergent properties can be observed and some holistic criteria discerned. E. Jantsch (1980) notes that each level of complexity possesses its own self-organizing properties (not predictable from those at lower levels of complexity) and that certain holistic measures can be elaborated. One of these is the ability to cope with the unexpected: in humans the quantity of genetic information stored is about 10-15 times that needed and the human immune system actually improves during the whole life-span of the organism. Is there, then, a holistic measure of a society's ability to cope with the unexpected which might be added to other spatial characterizations and which would by its very nature include facets of man as an animal with both an ecology and a psychology? The theme of evolution occurs throughout Jantsch's work, where in discussing the types of economy possible, he recognized not only the historic categories such as (a) dominance of the environment by energy and matter, (b) adaptation of the social structure to the environment, (c) symbiosis, using closed systems to mimic biological processes, and (d) colonization of parts of the solar system; but also (e) an evolutionary extension of consciousness, with information technology not only serving the steering of energy and production processes but helping with the diversification of our own personal experience. 'Learning,' writes Jantsch, 'would become a creative game played with reality . . . creative processes would be permitted to unfold and form new structures.' There is even some extravagance allowed since evolution is never purely functional.

Another measure which can be used in the study of society together with its environment is that of entropy. A concept initially derived from thermodynamics, in its most general form it allows the quantification of the extent of disorder in a system. In the case of energy exchanges, the second law of thermodynamics tells us that entropy can never decrease in an isolated system, since energy is dissipated in any process, and the temperature differences which permit work to be done tend to diminish, so that the inevitable end result is a system with homogeneous temperature and thus complete disorder, or maximum entropy. One of the triumphs of life as a process seems to be the building up of complexity in the face of such a progression towards complete disorder, but it can only be achieved at the expense of increasing consumption and dissipation of energy. The economist Georgescu-Roegen (1976) has drawn our attention to the ways in which human societies have increased the rate of production of entropy by wasteful practices which do not build up complexity, such as weapons, planned obsolescence of consumer goods and hedonistic consumption of energy. True and self-evident as much of this must be, it is nevertheless also the case that all evolutionary processes consume energy: the residual difficulty is sorting out sheep from goats at the ovicaprid stage of domestication. However, some overall measure of the entropy-creation rate of a society would indeed be an integrative assessment of its environmental relations.

On a converging track, but coming from the direction of the humanities rather than science, is the work of the social critic Theodore Roszak, starting with *Where the Wasteland Ends* (1973) and including, more recently, *Person/Planet* (1979). The central theme of the latter book is that the systems of the planet can feed back to life in forms such as man messages about their behaviour; and that the current uncertainties of the type raised by the environmentalist movement are a sign that the planetary health is threatened by resource depletion and environmental contamination. That a collection of biophysical systems can 'communicate' with *Homo sapiens* is an odd idea to most of us, but if we believe in co-evolution of the Gaian type and add some of the ideas outlined in the last section of this essay, then the level of madness is probably no greater than that of, for example, the weapon holdings of the super-powers.

An attempt to combine scientific discourse with a humanism based upon a planetary awareness is seen in F. Capra's *The Turning Point* (1982). In essence, this book suggests that there are two paradigms for viewing the world. The first is the dominant reductionist view which is a result of post-Cartesian thought over the last four centuries. This in effect gives us a mechanistic view of the world. By contrast there are the organic or holistic world-views which in his case are based on a 'systems view', or 'the essential interrelatedness and interdependence of all phenomena – physical, biological, psychological, social and cultural'. In this view, reductionism and holism, analysis and synthesis are complementary approaches but neither is complete without the other and it is to be hoped (and if possible planned for) that cultural evolution will keep the balance between the two:

Science of nature has one goal
To find both manyness and whole

writes Goethe, and it is perhaps not a bad aspiration for geography. Capra's essential conclusion is that we are currently at a 'turning point' when the reductionism prevalent in post-Cartesian times is about to be caught up into a new whole, of a different order from anything which previously existed. It is an optimistic view and we may all hope, at any rate, that he is right.

CONVERGENCE

The kind of knowledge of the world gained in this century has in one way effected a perceptual transformation, for it is no longer possible to believe in the sort of mechanistic kind of global system which is analogous to a snooker table: hit the cue ball properly and you will pot the black and then another red will be lying in just the right position. Rather, the world (and even perhaps the cosmos) is more like a hanging mobile where a slight touch imparted to one component causes the rest to readjust their positions: some a great deal, others relatively little. And the comprehension of the totality of a 'mobile' world is a task of far greater complexity than one we had supposed was more like a game of snooker. One conclusion that can be argued is that true holism is too difficult for our brain power; J.B.S. Haldane appositely remarked that the world is not only queerer than we suppose, it may be queerer than we *can* suppose. If we as a species are unhappy at accepting our own limitations then where is help to come from? The obvious source in terms of current thinking is the electronic computer and here it is the so-called 'Fifth Generation' machines, able not only to store vast quantities of data but to manipulate them in a self-organizing and adaptive fashion, which may be of most help, though exactly how is not yet known nor likely will be in the lifetime of this book. An alternative is to accept that the apprehension and comprehension of wholes is the proper business of intuitive thinking and that the knowledge industry should re-value the power and value of the consciousness of humans, both as individuals and as groups. In certain types of eastern thought, notably that of Taoism, such a way of proceeding is the normal way: reductionist analysis is not necessary or desirable. It is unlikely that the west can forget its heritage of rational, post-Cartesian, science, but some might sneakingly agree with Kenneth Boulding (1981) who averred that the 'hard' sciences were the social sciences because in fact they knew much more about the set of facts and relationships they set out to investigate, whereas the 'soft' sciences like physics and chemistry know very little of their chosen universe. The great difference was simply the ability of the one to make accurate temporal predictions.

Are we, then, on the horns of a dilemma? If so, the best solution is to throw sand in the bull's eyes. Is it conceivable that we have been asking the wrong questions or using the wrong criteria for evaluation? Is physics for example not all that we have taken it to be? If we turn to some of the writers on the epistemology of contemporary physics then we see that any naive picture of a value-free mode of investigation, in which the observer merely records what is there and produces laws which have a total certainty of operational success, needs considerable modification. In his earlier book called *The Tao of Physics* (1975), F. Capra not only drew attention to the similarities between particle physics and eastern philosophy, but argued that since the enunciation of Heisenberg's Uncertainty Principle, the observer and his consciousness had to be counted as part of the system: 'Nature does not show us any isolated "basic building blocks" but rather appears as a complicated web of relations between the various parts of the whole. These relations always include the observer in an essential way.' Thus in looking at the wave motion of particles, experiments which measured position very precisely could not measure momentum very precisely, and vice versa. What was known about a particle therefore depended upon the choice of experiment, and further, there was always an error term so that the laws of physics are, strictly speaking, probability laws not deterministic laws. The apparently radical notion that in physics the observer's consciousness is an integral part of the system has been taken further by R. Jones who suggests that the fundamental concepts of physics (space, time, matter and number) are intimately related to human consciousness and cannot be guaranteed to have any objective, external status beyond our minds. So physics becomes a metaphor, an idea or symbol with meaning; if this transcendence of the duality of person and nature can hold here, then what implications there must be for holistic approaches in geography.

BLUE SKIES

In recent years, geographers have brought to the core of their subject many methodologies and philosophies from the social sciences and humanities. By way of ending this chapter let us look at two pioneer pieces of scientific thought from biology and physics which may in future become more obviously relevant but which at present are admittedly 'blue sky' concepts near the frontiers of knowledge.

Since biology has featured from time to time in this account, it may be appropriate to mention here the views of Rupert Sheldrake (1981) which have caused some stir in the life sciences. Sheldrake suggests that regularities in nature are not caused by laws which are invariant through time but that the occurrences are more like habits which depend on what has happened before and how it happened. Experimental work showed that if laboratory rats learned a new pattern of behaviour, subsequent rats all over the world tended to learn the same pattern more easily. This process he called 'formative causation', with the transmitting pattern being labelled 'morphic resonance'. Thus living organisms can be said to tune in to the experience of their predecessors. Like other bold and sweeping ideas, Sheldrake's proposals have been subject to keen scrutiny and scathing criticism, but they have the advantage that they can be tested experimentally, and this is now in progress. If acceptable support is found, and it is admitted that the behaviour of living organisms is subject to influences across space and time, then the consequences for geography's ways of thinking are indeed profound and perhaps would take us past the point where a dichotomy between holism and reductionism has any meaning at all.

The second set of ideas derives from the physicist David Bohm and is an extension of the idea that consciousness and physicality cannot be objectively separated. Bohm proposes a new notion (1980) which he calls the 'implicate order' where everything in the cosmos is enfolded into everything. This contrasts with the idea of explicate order where each thing exists only in its own region of space and time and outside the regions belonging to other things. An analogy can be made with the hologram: if only a portion of a hologram is illuminated we still get a full image of the object since each part of the hologram contains information about the whole. So what is explicate is a projection of a higher-dimensional reality which is a common ground to both mind and matter but which is of a kind that is beyond both. 'It follows, then, that the explicate and manifest order of consciousness is not ultimately distinct from that of matter in general,' says Bohm. If this is indeed so there are many implications (not least of a metaphysical nature) but for geography it would seem to imply an ultimate holism since there is in reality no dualism to be observed. The complexities of this set of ideas might make us think that, for the present, holism can only be properly experienced by means of intuition, whose highest form of verbal expression is poetry. In this immediate context we might indeed be in the position inspiringly experienced by William Blake, able

> To see a World in a Grain of Sand
> And a Heaven in a Wild Flower
> Hold Infinity in the palm of your hand
> And Eternity in an Hour.

REFERENCES

Ackerman, E. (1963) 'Where is a research frontier?', *Annals, Association of American Geographers*, 53, 429–40.
Bennett, R.J. and Chorley, R.J. (1978) *Environmental Systems: Philosophy, Analysis and Control*, Methuen, London.
—— and Wrigley, N. (1981) Introduction, in Wrigley, N. and Bennett, R.J. (eds), *Quantitative Geography: A British View*, London, Routledge & Kegan Paul, 3–11.
Blaut, J.M. (1962) 'Object and relationship', *Professional Geographer*, 14 (6), 1–7.
Bohm, D. (1980) *Wholeness and the Implicate Order*, London, Routledge & Kegan Paul.
Boulding, K.E. (1981) *Evolutionary Economics*, Beverly Hills, California, Sage Books.
Butzer, K.W. (1982) *Archaeology as Human Ecology: Method and Theory for a Contextual Approach*, Cambridge, Cambridge University Press.
Capra, F. (1975) *The Tao of Physics: An Exploration of the Parallels Between Modern Physics and Eastern Mysticism*, London, Wildwood House.
—— (1982) *The Turning Point: Science, Society and the Rising Culture*, London, Fontana.
Chorley, R.J. (1962) *Geomorphology and General Systems Theory*, United States Geological Survey Professional Paper, 500B, US Government Printing Office.
—— and Kennedy, B.A. (1971) *Physical Geography: A Systems Approach*, London, Prentice-Hall.
Coffey, W.J. (1981) *Geography: Towards a General Spatial Systems Approach*, London, Methuen.

Georgescu-Roegen, N. (1976) *Energy and Economic Myths: Institutional and Analytical Economic Essays*, New York, Pergamon Press.

Glacken, C.J. (1967) *Traces on the Rhodian Shore: Nature and Culture in Western Thought from Ancient Times to the End of the Eighteenth Century*, Berkeley and Los Angeles, University of California Press.

Haggett, P. and Chorley, R.J. (1967) 'Models, paradigms and the new geography', in Chorley, R.J. and Haggett, P. (eds), *Models in Geography*, London, Methuen, 19–41.

Jantsch, E. (1980) *The Self-Organizing Universe*, Oxford, Pergamon Press.

Jones, R. (1983) *Physics as Metaphor*, London, Wildwood House.

Lindeman, R.L. (1942) 'The trophic-dynamic aspect of ecology', *Ecology*, 23, 399–418.

Lovelock, J.E. (1979) *Gaia: A New Look at Life on Earth*, Oxford, Oxford University Press.

—— and Margulis, L. (1974) 'Atmospheric homeostasis by and for the biosphere: the Gaia hypothesis', *Tellus*, 26, 2–10.

Mayr, E. (1982) *The Growth of Biological Thought: Diversity, Evolution and Inheritance*, Cambridge, Mass., Harvard University Press.

Odum, E.P. (1975) *Ecology: the Link Between the Natural and the Social Sciences*, 2nd edn, New York, Holt, Rinehart & Winston.

Roszak, T. (1973) *Where the Wasteland Ends: Politics and Transcendence in Post-industrial Society*, London, Faber & Faber.

—— (1979) *Person/Planet*, London, Gollancz.

Sheldrake, R. (1981) *A New Science of Life: The Hypothesis of Formative Causation*, London, Blond & Briggs.

Stoddart, D.R. (1965) 'Geography and the ecological approach: the ecosystem as a geographic principle and method', *Geography*, 50, 242–51.

Wilson, A.G. (1981) *Geography and the Environment: Systems Analytical Methods*, Chichester, John Wiley.

3
GEOGRAPHY HAS NEITHER EXISTENCE NOR FUTURE
Michael E. Eliot Hurst

The history of our discipline cannot be understood independently of the history of the society in which the practices of geography are embedded. (Harvey 1984, 1)

The fragments of social science as we now know them, history, economics, anthropology, geography, and so on, emerged as concomitants to the development of a new socio-economic system, capitalism. They were the products of a particular historical moment. For example, 'sociology' was a term coined by Comte in 1838 after the bourgeois revolutions to represent a new counter-revolutionary conservatism (Therborn 1976, chs 2 and 3). Foucault has argued that these nineteenth-century discourses represent 'monuments' to past divisions of 'epistemological space' (1970, 1972; also Claval 1980). Hudson (1977) has noted that in this period geography 'was vigorously promoted ... largely if not mainly to serve the interests of imperialism in its various aspects, including territorial acquisition, economic exploitation, militarism, and the practice of class and race domination'. By 1870, Hudson argues, geography as we now know it had crystalized out as a response to a particular set of demands brought about by the socio-economic relationships determined at a particular conjuncture – gazetteer for the ruling class, explorer under the guise of so-called royal geographical societies, apologist for the inhumanities of the industrial revolution, colonialism and imperialism. One only has to glance back at the handbooks, the *Commercial Geographies* of the period, to affirm the veracity of such arguments. In the twentieth century such a relationship still exists, although the language is often more subtle than merely 'white man's burden', 'development aid', 'comparative advantage', 'balance of trade', and so on. What 'development' really means, and to whose advantage particular trade flows accrue, is rarely enunciated. Now, having weathered the stagnation after 1920 in which its services in a post-exploration period were less in demand, geography functions to rationalize the defense of the status quo, to help plan the effective location of global units of a corporate capitalism, and, in Blaut's words, to 'lie to little children' (1974, 44). The technical practice of the 1980s may be infinitely more elaborate and sophisticated than that of the 1880s, but the same ideological functioning of this particular fragment of 'knowledge' remains.

The main thrust of this chapter is not a critique of geography *per se*, but a concern with a far more fundamental question concerning the ontological status of the discipline and of the social sciences as presently constituted. It is simply not sufficient to criticize or interrogate geography with the hope of finding some new definition or reformulation (e.g. Harvey 1984) since such a task is self-defeating. Rather the necessity is to dissassemble, de-define, the social sciences as currently constituted – whether as geography, sociology, political science, or whatever – in order to understand society within the knowledge framework of the only holistic scientific theory of human praxis: historical materialism. It is in this sense that there is no *need* for geography and the other presently constituted fragments of social science, since they *must* be rejected. The questions and answers that historical/dialectical materialism tackle do not lie within the disciplinary contexts as we know them, however much they are changed in response to limited critiques which *begin by accepting their pre-existence*, a trap which Harvey, Lee, and other critics have unfortunately not recognized or admitted.

Lee (1977) commented that geographers now needed

> a genuinely interdisciplinary approach to the analysis of the social production of material life in which Geography, Economics, Sociology, Political Science, and Law cannot be distinguished, but are restructured and reconstituted into a coherent attempt to uncover, from beneath superficial appearances, the true relationships between economy and society in the totality of social existence. (Lee 1977, 75)

Such notions were paralleled about the same time by Godelier (1977) concerning anthropology, Bernstein (1976) about political science, and Therborn (1976) with regards to sociology and economics. They all argued against a fragmentation of knowledge that dates back to Comte and Spencer. These commentators posited, directly or indirectly, a transcendence of their discipline's particular way of fragmenting that whole, which Foucault has called 'epistemological space'. They saw the necessity of overcoming the limitations of separate disciplines and practices, much as Marx had seen the necessity of working toward a transcendence of Adam Smith's or David Ricardo's political economy in the nineteenth century. But such calls today are fraught with contradiction, as for example when Lee urges a genuine interdisciplinary approach *within* geography or when Harvey's call for a meta-disciplinary approach is still rooted in a disciplinary paradigm (1982, 1984). Such appeals are paradoxical since the artificial fragments of 'epistemological space' we call disciplines are *virtually antagonistic*. To argue for merger, cooperation, articulation, fusion, within and between such practices is still to be trapped within bourgeois social science.

The thrust of this chapter is therefore to argue *not* for a *re*defining of geography according to the latest philosophical or technological bandwagon (see for example the supermarket of philosophies presented by Harvey and Holly (1981)) but for its total rejection and de-definition.

Similarly Marxist practice cannot be simply laid alongside non-Marxist practices as alternative approaches to the same thing, because they organize and understand their subject matter in a manner which challenges not only the theory and methods of more traditional approaches, but the very classification of their subject matter, however defined, which is built into the curriculum and the bourgeois intellectual division of labour. Thus Harvey correctly identifies disciplinary boundaries as counter-revolutionary and that reality has to be approached directly and conceptualized in 'non- or meta-disciplinary terms' (1972, 40) based in Marxism (1973a, 1973b, 1974a, 1974b, 1982). The city, for example, he argued was to be studied without the interference of disciplinary frameworks. But by 1984 the Marxist approach becomes only one of a range of 'radical' images, citing also 'anarchism, advocacy, geographical expeditions, and humanism' (1984, 5) in the creation of a 'people's geography' (p. 7) which inserts 'spatial concepts into social theory' (p. 8). Such legerdemain is insufficient, because outside bourgeois practice lies a completely different route in which all geographies are nullified (Slater 1978, 4) and all the disciplines are transcended in one science of society in evolution (as all science must be). The analytical fronts are many, as are the political tactics, and a rich variety of points of departure. This is no monolith that replaces the bourgeois discipline, but a plural terrain which is not subscribed by one system, is inconclusive and indeterminate (philosophically = undecidable). A field of multiple struggles on many related fronts, always at work on many planes, always exceeding the axioms of its theories, never completed 'neither *at one go* nor *once and for all*. The work involved is constant and repetitive like, as Gayatri Spival puts it, keeping a house clean' (Ryan 1982, 117).

IDEOLOGY

Ideology ... is characterized by posing problems where solutions are preordained, produced outside the cognitive process. Ideological questions constitute a mirror in which the ideological subject can recognize its own ideological solutions. (Therborn, 1976, 60)

Ideology is a general concept which involves both material practices and the ideas intertwined with them; it is a system of representation of the world, but one which does not refer to the world as an object external to that system. It is a lived relationship with the world in terms of practical and institutional ideologies; the former might be labeled 'common sense', an unquestioned common pool of 'wisdom' available to a class or classes in a particular society, that requires no logic or thought to help one steer a way through daily life. This common pool, transmitted through an array of *institutional state apparatuses* (Althusser 1971) such as the family, school, religion, law, *r*epresent more or less systematic frameworks and models about the natural and social world in terms of *theoretical ideologies* (O'Donnell *et al.* 1978). Although both ideological practices can be defined in terms of 'the imaginary way in which people experience the real world' it should not be taken to mean that ideology is merely false consciousness, error, or illusion; ideology is authentic, but *closed*. Ideologies investigate and relate to a certain part of social reality and cannot produce knowledge of anything outside it. The limitations are not innocent of course, but function to support and *r*eproduce that limited reality. Thus ideology is *produced* by class interests to serve the material interests of those classes and to help *r*eproduce their position in the class structure of a given society. The ideology of the dominant class(es) penetrates all classes in the interests of those who dominate. Those material interests, and hence ideologies, change somewhat in structure according to the dynamic of capital accumulation and in the course of class struggle. The means by which this is achieved and the historical course of ideology in any one social formation are very complex and can be analyzed both from various viewpoints (for example, Gramsci's notion of hegemony by non-coercive means; see Boggs (1976, ch. 2)) and from various points of departure (for example, *Jump Cut*'s attempt to understand Hollywood-type films, Hess *et al.* (1978)).

Ideology comes to represent the interests and views of only one class and its supporters, although it is presented to us as universally true and valid, like the divisions of epistemological space we call departments or disciplines. But those representations — ideas, disciplines, the legal systems, films, etc. — are mediated. Between the dominant ideology and its expression occur individual and group thought, experience, needs, creativity, which may open up what Ryan calls 'fissures' and 'tensions', or even call attention to contradictions between aspects of this ideology (the inclusions and exclusions, lapses and silences, of a bourgeois discourse for instance). Thus ideology is never a total dominating force. Clearly the more intense class struggle becomes, the more the questions are raised about the views of the dominating class or classes. If people were totally dominated by metaphysics and ideology there would be no trade unions, strikes, or revolutions, nor deconstructionist or critical papers from academics! Class struggle, though not always recognized as such, takes place in all areas of life — in the struggles for better education, health care, housing, and jobs. Marx referred in fact to the ideological forces in which people become conscious of this conflict and fight it out. The struggle against bourgeois geography in this case includes the tools and theories to analyze more clearly the world around us so that it can be changed, including the practice of that geography.

THE PROBLEMATIC: GEOGRAPHY

> To most [people], even educated ones, geography is simply a matter of knowing which city is the capital of which country and where the largest rivers run. (Harvey and Smith 1984, 1)

'Geography' exists at one level as an objective discourse within the ideological practices we now commonly identify as 'social science'. Geography has 'texts', 'theories', 'teachers'; a range of technical practices (areal display and manipulation of data, remote sensing, etc.); institutions (departments, graduate programs, funded research, 'refereed' publications, etc.); and an ideology, 'for understanding . . . [the] twin fields of concern: the organization of space, and the relation of humanity to the environment' (Harvey and Smith 1984, 5–6): *ipso facto* it exists. Many of these structures and the mechanism of daily *re*production are spelt out in more detail in earlier publications (Eliot Hurst 1973a, 1980, 1981a) and are equally applicable to *all* social science disciplines.

The problematic, geography, designates the specific framework within which the discourse's basic concepts and approaches are perceived to have significance and function. In determining what is *included* geographers have also determined what is *excluded*! Excluded have been 'imperialism', 'class', 'capital', and all the Marxist conceptions that surround mode of production, ideology, and social formation. Included are 'space', 'environment', 'free-market system', 'culture', and the cartographic display of such data. To understand geography's trajectory over the past one hundred years one must account for concepts that are excluded, and for problems that are posed inadequately or not at all; these *lacunae, lapses,* and *silences,* to use Althusser's terms, are as much a part of the geographic problematic as those practices which are explicitly performed. Geography is thus practiced as a day-to-day discourse at a direct ideological level; but it also exists at an 'archeological level', conjointly with the explicit, as a silent discourse.

The ideas, knowledge, and technical practice in which geographers trade are *produced* (that is, an active work, ideas neither fall from the sky nor are they innate). Working within a theoretical ideology this geographical production is *closed,* that is it can only investigate a certain limited domain of social reality and is unable to produce knowledge of anything outside this domain. Thus geographers can produce knowledge which is descriptive of 'spatial patterns', but they cannot give us knowledge of how these patterns are generated, and are thus severely constrained in advancing our understanding of the world.

METHODS, TECHNIQUES, AND 'THEORIES' OF GEOGRAPHY

> Most geographers theorize as little as possible ... some ... make no secret of their low regard for 'abstract considerations'; ... in fact, they make a virtue of it and proclaim their preference for the 'concrete'. (Lacoste 1973, 243)

If there exists any framework as to the scope and practice of geography it is largely a *negative* one, based on its technical *not* theoretical practice. Geography is merely descriptive of self-selected phenomena and is scientifically bankrupt. Some geographers, like Tricart, Buchanan, Santos, Quaini, and Harvey, have attempted to look at *process,* but for a variety of ideological and economic reasons have chosen to remain *within* geography's parochial boundaries. Even with these exceptions and the increasing sophistication of debate in the past twenty years the history of geography is littered with inaccuracy, misrepresentations, red herrings, ideological distortions (conscious and unconscious), polemics over definitions, spatial fetishisms, and deliberate internal obfuscation (the creation, for example, of adjectival geographies).

In the late 1950s and throughout the 1960s ideography (description as such), which had been the mainstay of geography since its earliest incarnation by Erosothenes to describe the contents, peoples, and resources of the Alexandrian oecumene, came under critical scrutiny. The 'space age' was paralleled by a search in the various social sciences for exactness, mathematical modeling, and eventually the utilization of computers. It was to the natural sciences, particularly physics and chemistry, that social scientists looked enviously, borrowing piecemeal concepts like Newtonian gravity, systems analysis, entropy, and certain ecological notions in a latter-day Spencerian homology. 'Efficiency' criteria, what things 'ought' to be like (the normative), were to be used – though efficient by what or whose standards were questions rarely raised; 'objectivity' became the goal,

> hypotheses were tested, paradigms traded, models proposed, theories suggested, explanations offered, systems simulated and laws sorely sought after. This search for *order* was dictated more by available techniques than by a set of coherent objectives: reality was ransacked in search of ... methods and techniques. (Smith 1979, 356, emphasis added)

This was the period Burton (1963) was to call the 'quantitative revolution'; beginning in the other social sciences first, by the mid-1960s it had taken a firm hold in transport geography, and later in other areas of geography itself. The ideological rationale, and the pinnacle of this quantitative approach, was to be the publication in 1969 of David Harvey's *Explanation in Geography*.

> Commonly and correctly seen as a milestone in [post 1940s] geographic thought ... [it] was more important for what it did not do than for what it did. Harvey explicitly refused to define geography or to acknowledge that such a definition was necessary. ... He asked not whether geography was a science but how ought geographers to practice their science. (Smith 1979, 356–7)

Ironically now repudiated by its author it is still widely quoted as a 'philosophical' foundation for logical positivists operating in geography; it was in fact a watershed for the empiricist approach (a 'milestone' and a 'gravestone' in Smith's words) and there has been little intellectual development on this front since.

The consensus that geographers have achieved since Harvey's book was published amount to the creation of a 'science', or to what Albrow has epitomized as the *categorical paradigm*, or more commonly logical positivism. It involves the idea of a science of society which has:

a particular attitude to what constitutes 'data', an emphasis on the
use of formal logic and a theory of explanation ('hypothetico-
deductive'), which, in combination, result in a concentration on
strictly delimited units of analysis, on concepts of widest possible
application in time and place and on the production of systems
of universal propositions, ideally laws of society. (Albrow 1974,
184)

At a very minimum, geographers, by testing hypotheses, using particular data, and pursuing 'explanation' in the most restricted sense, believed they were being 'scientific', as also when they could 'no longer ignore the scientific criteria according to which the discipline's findings are internally judged' (Smith 1979, 357).

This concentration on method, technique, rules, and the hypothetico-deductive approach, with some sorties into functionalism, and the *belief* in the 'scientificity' of what they were practicing, was simply a *re*presentation on a day-to-day basis of an ideology which justified the mechanics of global corporatism, the *re*production of 'capitalist landscapes', and the perpetuation of ruling-class hegemony. Premised on keeping apart 'theory' and 'practice', 'philosophy' and 'methods', 'values' and 'objectivity', geographers would claim, none the less, that 'since about 1954, scientific work in geography has been almost exclusively logical positivism' (Amedeo and Golledge 1975, iii–iv). Paradoxically as that judgement was being made there were cries from other geographers questioning the 'faith' (e.g. Zelinsky 1975), and the subsequent decade has seen a literal ransacking of philosophical movements, as the promise of logical positivism never matured: hence Harvey and Holly's (1981) compendium of functionalism, phenomenology, existentialism, idealism, realism, and Marxism, or Ley and Samuels' (1978) pursuit of metaphysical concerns, which even includes a prolegomenon to a Christian economic geography! But to no avail, as this unseemly search simply repeated on many more fronts those who had earlier tried to achieve 'scientificity' in the 1950s and 1960s. Outside of Anglo-Saxon geography there were some attempts to discard the discipline (for example in the pages of *L'Espace géographique*), but within that tradition only Slater (1975, 1977, 1978) made a rejection of the moribund discipline explicit. Thus despite ransacking a supermarket of philosophical packages, in the 1980s geography remains straitjacketed and blinkered by its own algophobia.

Many of the critics of 'scientific' geography questioned some of the tacit assumptions of logical positivism, queried the uncritical way in which the latest assembly-line techniques were applied, and even to some degree questioned positivism's undue abstraction from its socio-political context. But, repudiation aside, they had few well worked-out alternatives to offer.

Within geography one problem has been the transformation of a ragbag of borrowed methods and techniques as a substitute for 'theory'. These essentially technical practices, borrowed or not, are by themselves producers only of various technological know-how. In technical practice, subservient always to economic, political, ideological, and theoretical practices (though rarely acknowledged), knowledge is a borrowing from outside, from existing science, or is generated within the technique-process itself.

> In every case the relation between technique and knowledge is an *external*, unreflected relation, radically different from the internal, reflected relation between science and its knowledges. . . . Left to itself a spontaneous (technical) practice produces only the 'theory' it needs to produce the ends assigned to it: this 'theory' is never more than the reflection of this end, uncriticized, unknown . . . a byproduct of the reflection of the technical practice's ends on its means. A 'theory' which does not question the end whose *byproduct* it is remains a prisoner of this end and of the 'realities' which have imposed it as an end. Examples of this are many branches of psychology, sociology, economics, politics, [geography]. . . . This point is crucial if we are to identify the most dangerous ideological menace: the creation and success of so-called theories which have nothing to do with real theory but are mere *byproducts* of technical activity. A belief in the 'spontaneous' theoretical virtue of technique lies at the root of ideology, the ideology constituting the essence of Technocratic Thought. (Althusser 1969, 171n)

In the problematic, geography, we can distinguish *at least* three aspects of this technical practice: first, technical analysis for the interests of capital (e.g. the spatial analysis of complex data sets for corporate interests and state agencies); second, ideological 'theories' about the spatial tendencies of capitalist development (e.g. spatial diffusion theory, growth pole models, central place theory, or the 'metropolis/hinterland' concept); third, the transmission of notions and ideas from the dominant segments of social science cast into a spatial perspective for use in populist media ('geography texts' aimed at grade schools, the *National Geographic* including its television programs, the *Canadian Geographical Magazine*, etc.).

Samir Amin (1976, 10) has criticized this technical set of practices in geography, accusing the discipline of simply spatially juxtaposing socioeconomic phenomena, without the ability to understand and explain such patterns. Despite all its current sophistication geography is still formulated in terms of a description of spatial differences. Yet as Slater (1977, 21) has asked: on what basis do geographers decide what to describe? How do they account for what they have chosen to describe? And for whose interest do they describe? 'Techniques allied to description constitute an excellent vehicle for the transmission of ideology, since as Althusser . . . explains, the overt absence of scientific theory leads to the covert transmission of ideological theory, which, although being latent, is nevertheless determinate.'

In addition, of course, descriptive empirical data are not just *collected*, they are *produced*, just as research results are not *findings*, but *creations*. These data are not meaningless artifacts, but their production is a social process which is carried out for specific reasons and in specific ways.

> The particular concepts and techniques used in each piece of research are chosen and developed within a structure of interests encompassing those of both the funding agency and the institution in which the work is carried out – to say nothing of the individual commitments of the researcher. The complex . . . end product is therefore no arbitrary creation: it is rather more in need of being explained than either being taken for granted or dismissed . . . .
> The techniques used to produce and to process data, are like the data themselves, social products: statistical techniques are developed, and are continually transformed, under particular historical circumstances. . . . There is no such thing as a universally applicable method of going about doing . . . statistics.
> . . . Furthermore, as a social practice it is not necessarily the case that quantitative analysis makes research any more 'scientific', let alone 'objective'. The use of statistical techniques poses its own problems, as in the same way do other methods of presenting and assessing knowledge of society; it is inherently no more theory-free or value-neutral than other approaches. (Irvine *et al.* 1979, 5–6)

To compound all of these problems, by legerdemain, these questionable technical practices themselves become transmitted through alleged 'theories'. Thus Christaller's limited case empirical study of certain types of service center in southern Germany in the 1920s became a 'theory of central places', which in translation into English and North American usage becomes a means of analyzing shopping centers within cities! An essentially heuristic device becomes a latent ideological 'theory'. As Althusser has commented about history, but of equal application to geography:

> History lives with an illusion that it can do without *theory* in the strong sense, without a theory of its object and therefore without a definition of its theoretical object. What acts as its theory, what it sees as taking the place of this theory is its *methodology*, i.e., the rules that govern its effective practices. (Althusser and Balibar 1970, 109)

In summary, geography is nothing more than a theoretical ideology based on technical practice alone.

GEOGRAPHY IN THE 1980s

> Geography is too important to be left to geographers.
> (Harvey 1984, 7)

The discourse here called 'geography' is in the 1980s like a somnambulist, sleepwalking its way through a plethora of socio-political realities of late capitalism. Despite the appearance of many debates on several fronts it is also probably at its most conservative, which is hardly surprising given the nature of the discipline, the impact of monetarist policies by Bennett,[1] Reagan, or Thatcher and their feedback into the academic system, and the current crisis phase of world capitalism. Such factors are, as we know from historical evidence, represented through the social sciences. Thus the 1980s are marked within academic geography by:

1 *retrenchment*, methodologically and philosophically;
2 *incorporation/recuperation* of those Marxists of the 1970s via their work, methods, and teaching; and
3 *regrouping* of those still committed to a Marxist analysis.

The first factor is evident through a 'people-the-barricades' attitude, not just as a counter to financial assaults, but from methodological and philosophical viewpoints too. If, at times in the 1970s, those espousing some radical criticisms had the upper hand in some of the philosophical debate, now is the moment for counter-attack. Although saner than some of the comments of the earlier decade (Berry's description of those who criticized his technocratic approach as a group 'of malcontents and kooks and freaks and dropouts' (Halvorson and Stave 1978, 223) and Carter's attack on Marxist and gay geographers, with his advice that geographers stay out of moral and political controversies and confine themselves to professionalism, 'to geography by geographers and for geographers'

(Carter 1977, 102)), behind the academic facade lurks the same anti-intellectualism and reactionary stance. Szymanski and Agnew's (1981) attempt to redefine geography within a realist/order-skepticism 'dialectic' is humorously naive, though probably well intentioned; Eyles (1981) argues why it is impossible for geography and geographers to be Marxist in the course of which his own idealism is revealed; and Duncan and Ley (1982) seriously caricature structuralism and then, as if in parody, present it as a 'strawman' to be demolished by metaphysics. These antiscientific programs of retrenchment are a certain indication of a discourse in deep crisis.

The incorporation/recuperation tactic is no less unsubtle. The move is to incorporate notions which are seemingly associated with the radical criticism of yesterday, and in that incorporation in effect to emasculate them. Thus Szymanski and Agnew's use of the dialectic gives the appearance of a critical approach, whilst in fact being Hegelian in tactics. A whole host of Marxian concepts and language can be parachuted into a text, which is itself non-Marxist, thus rendering them meaningless and unscientific. Now more subtle than the eclectic welfare approach favored by Smith (1977) it none the less plays the same game: logical positivism garbed in apparent Marxist clothes! The illusion of assimilation is present, but only in the most superficial sense, and the crucial *praxis* of Marxism is absent in this new form of 'repressive tolerance'. Thus Peet (1981, 107) writes of the 'inherent tendency', particularly under capitalism, 'for various kinds of uneven development' in all social processes which passes as some 'universal law'. As Smith notes (1981, 117) this is pure 'abstract metaphysics' which complements the spatial fetishism inherent in Peet's notion of 'spatial dialectics' (see p. 106).

Parallel to this incorporation is the strategy of recuperation; the former 'radicals' of the 1970s have had their 'youthful indulgences', but in their 'mature' years their writings can be published, analyzed, and reified (e.g. Russell King's appendix to Quaini (1982)) or worst of all canonization, as in Paterson's (1984) recounting of David Harvey's career. Much of Paterson's treatise is no more than a hagiography, a painful and detailed recounting of virtually every note, review, and toilet jotting of one of 'the discipline's leading scholars'. Despite Harvey's own rejoinder, 'the important question for me is where I am going, not where I have been' (Paterson 1984, 1), the publisher and editors elevate a pedestrian Master's thesis to function as ideological incorporation. The book relates in tedious detail an individualistic geographer (note, not an account of a Marxist working within a discipline and what that means) whose roots lie in a 'genuine philosophical pluralism in geography'. Here Paterson's own bias becomes apparent, but only in the closing pages of the book.

Pluralism and 'the integrity of each ... ought to be recognized and allowed to develop within the discipline ... a neorealativistic ... perspective which the researcher finds most in accord with his or her view of reality' (p. 175). The mind boggles!

As frequently recuperation is entered into more willingly by allowing a more critical paper to appear in an otherwise orthodox compilation, to allow an Uncle Tom in every department (none the less Harvey and Smith (1984, 35) estimate there are only six Marxists amongst all those with tenure in American geography departments), or the publication of some work in translation (e.g. Quaini 1982). These parallel processes are important to the ideological functioning of the discipline; what was 'radical' in one decade can be brought back into the mainstream *and* at the same time gives the appearance of a liberal and 'up-to-date' discourse (Ryan 1982, 121–2).

There are of course those who are Marxists and who resist incorporation. The treatise by a liberal on Harvey's geographical and Marxist practices is symptomatic of the geographical problematic, not Harvey's own praxis, as the awesome scholarship of *Limits to Capital* (1982) so clearly attests. However, the *sui generis* anti-left atmosphere of American universities and habitus have shifted much of the current focus of work in Marxism and geography to Europe. There is a longer historical tradition of European intellectual thought which has recognized both the necessity of philosophy in general and of Marxism in particular. The period since 1950 has been marked by a whole series of debates centered on Althusser, Anderson, Balibar, Barthes, Castells, Hindess, Hirst, Lacan, Lefebvre, and Poulantzas, to name a few. The heady excitement of these debates spilt over into geography and the work of Claval (1977), Frémont (1976), Van Beunigen (1977), Dematteis (1980), Lacoste (1973, 1976), and the journals *Hérodote* and *L'Espace géographique* (for a review of some of this work see Quaini 1982, ch. 6). Whilst it is true that in the USA, *Antipode* began publishing in the same time period, its early years were marked by a left-liberal pluralism (Peet 1977), and the Union of Socialist Geographers founded in Canada a little later has been racked by internecine disputes and seems to have lost much of its initial impetus. In Britain, on the other hand, the work of Anderson (1973, 1975, 1980), Boddy (1976, 1980), Burgess (1976, 1981), Cosgrove (1978, 1983), Gregory (1978, 1982), Lee (1976, 1979), Massey (1978), Massey and Meegan (1982), Sayer (1979, 1981), and others comes to mind; although not all Marxism *sensu stricto*, it carries through the European tradition of critical deconstruction *within* geographical practice. To this point in time no clear consensus has emerged on either side of the Atlantic however, and, it is the essential argument of this chapter, could never *really* emerge within the geographical discipline, although some essential debates have occurred, particularly concerning the role of 'space' (see following section).

Interestingly perhaps, most recently there has been a revival of 'populist geography' (Harvey 1984; Harvey and Smith 1984). Historically there have been a number of attempts to relate geography to a 'grassroots', literal or otherwise. Kropotkin and Reclus in the anarchist tradition (Breitbart 1981; Dunbar 1981), Vidal de la Blache's mosaic of 'genre de vie' (1911; also Buttimer 1971), Wooldridge and East's geography learnt through 'the soles of your boots' (1958), and Bunge's geographical expeditions into the inner neighborhoods of Detroit and Toronto (1978, 1979; and Horvath 1971), are some attempts which come to mind. Following perhaps E.P. Thompson's idealistic calls for 'cross-class' cooperation in halting nuclear war, Harvey's call to a populace-at-large to recuperate geography within 'Geography' is equally perplexing. The continuance of geography, with or without a capital 'G', may be laudable from the point of view of certain vested interests (that is to ensure the viability of an academic discipline with a guaranteed position in university curricula in the latter part of the twentieth century) and a particular class position (new petty bourgeois) but what relevance it has to other classes is not made clear. Geography/geography to most people is nothing more than a recollection of neocolonial gazetteering, capes and bays, world capitals, and the socio-ecological determinism of *National Geographic* television specials. Harvey's criticism of those who are historical materialists who have also undervalued the spatial dimension, a covariable amongst many, is apt. But by ignoring the ideological and class dimensions of geographical practice, real or potential, Harvey, like Lee earlier, submits to an idealism. In part this stems from a trajectory which sees the reconstitution of 'geographical thought [as] a first step down a very long and difficult road' which would ultimately see a 'fusion of geographical studies with history, sociology, politics, geology, biology, and other subjects under the umbrella of historical materialism' (1981, 209). Such appeals as 'build a popular geography' or 'create an applied peoples' geography, unbeholden to narrow or peaceful interests, but broadly democratic in its conception' (1984, 9) are fine as rhetoric but dangerous in both short- and long-term interests. If your goal is to transcend geography why delude people (including yourself) as to a current viability when geography's class orientation and usefulness to a particular class are so evident and intranscendable without its total abolition. Geography in the 1980s when it is simply more-of-the-same is as resilient as the Great Auk.

IS 'SPACE' A SUFFICIENT JUSTIFICATION FOR A SEPARATE DISCOURSE?

Geography, including so-called Marxist geography, 'treats space in *practice* as a relatively autonomous thing or field, a separate realm of existence'. (Smith 1981, 112)

Definitions in both the 'common sense' and 'academic' realms frequently invoke geography's distinctiveness around the concept 'space', endowing 'space' with an ontologically separate existence. Whether recycled as spatial interaction, or spatial differentiation, spatial dialectics, regional geography, etc., the importance and separateness of 'space' is raised to that of a unique variable. A whole series of definitions through the history of the discourse attest to this, the following being merely a brief selection:

1. 'the science that describes the earth's surface with particular reference to the differentiation and relationship of areas' (Kinvig 1953, 158).
2. 'Geography is concerned to provide accurate, orderly, and rational descriptions and interpretations of the variable character of the earth's surface' (Hartshorne 1959, 21).
3. 'the study of spatial distributions and space relations on the earth's surface' (Ackerman *et al.* 1965, 8).
4. 'to understand the earth as the world of man, with particular reference to the differentiation and integration of place' (Broek 1965, 79).
5. 'Marxist geography is that part of a whole science dealing with the interrelationship between social processes on the one hand and spatial processes on the other hand' (Peet 1977, 254; see also Peet and Lyons 1981, *passim*).
6. 'Geography ... [offers] ... a broad synoptic view of spatial relationships in human affairs' (Smith 1977, 2).
7. 'The study of the earth's surface as the space within which the human population lives' (Haggett 1981, 133).
8. 'The focus of all geographical enquiry is place. This implies location on the earth's surface, the relationship between it and other locations, and the processes affecting changes in those relationships' (Jones 1984, 5).

There are very serious problems with such definitions. Anderson (1973) labeled the fallacy involved *spatial fetishism*, since 'space' becomes reified and endowed with an ontological autonomy it does not possess, and social relations between people are treated as spatial relations between places. By transforming an abstraction into a monocausal relationship, whether the word 'relative' is added to it or not, creates an idealist, unscientific, and wholly ideological conception. Far from being a 'separate' distinct object, 'space' is simply a covariable, a conflationary relationship, in which unevenly developed patterns on the earth's surface can direct us to that abstraction, space (which then can be a pedagogic tool, an heuristic device, but no more), and not vice versa. The notion that 'space' produces spatial patterns by auto-dominant autogenesis, whether referred to as 'spatial differentiation' or 'uneven development', is simply a confusion of *causa causans* with *causa causata*. This is not to dismiss 'space' as a covariable out of hand – obviously its usefulness depends on the theoretical and technical practices within which it is contained – but it does make nonsense of the attempt to build a 'discipline' around such a purely descriptive category. Marx and Lenin both used the category of territorial division of labor but always in a descriptive sense, and although Harvey and others have argued that they may have underutilized it, they never imbued it with ontological status. Their usage of the term did allow them to clarify certain spatial manifestations of capitalism's structure, but it was never part of historical materialism's terrain to invoke a separate area of study.

The recent debates around a new adjectival geography, 'Marxist Geography', illustrate the ideological hold which 'space' has over various commentators and the other contradictions of fragmenting a social whole. Peet and Soja have been the principal advocates of a Marxian spatialism. Peet has argued that we need to develop a 'spatial dialectic' as a means to understanding the uneven development of capitalist landscapes, and that this spatial dialectic forms in fact also the foundation for a distinctive adjectival geography. Continuing from definition 5 above, Peet compounds the contradictory nature of his stance:

> Marxist Geography thus looks at one area of the set of interactions surrounding social processes. It is so immersed in process that it merges with the other [Marxist] sciences also dealing with social process and is distinguishable from them only by its degree of specialization in their environmental and spatial aspects. (Peet 1977, 254)

In a series of sometimes bitter and personal debates, Peet's position has been revealed as 'degenerate' (Smith 1981; see also Eliot Hurst 1979b).

> Some fundamental issues are at stake. ... They concern not only the concrete relationship between Marxism and geography, but also

the very conception of space and the way in which we develop in theory an integration of space and social process. ... The problem with 'spatial dialectics' is ... that in practice it is unable to achieve what it sets out to do [i.e. to make 'social process' and 'social form' two components of a dialectical unity]. ... In fact, there is nothing dialectical about 'spatial dialectics'. It must be seen for what it is – a piece of mechanically constructed jargon with no theoretical impact. It begins from a dichotomy that it never does transcend: space on the one side, social process on the other. The theory conceptualizes space as separate ... 'social relations' are still said to occur 'across space'; social relations are still said to 'inject material content into space', and so on. Far from achieving a dialectic, the theory treats space in *practice* as a relatively autonomous thing or field, a separate realm of existence. ... Peet is able to *assert* that space and society comprise a dialectic; he cannot show exactly how they do so. ... In practice [his conception is] little more than a banal reaffirmation of the idea that space and society *interact*. Crude *spatial interaction* masquerades as spatial dialectics. ... It refines but does not remove the orthodox fetishism of space. (Smith 1981, 111–13)

Soja (1980) has sought to steer his way through these arguments in the usual 'left-liberal tradition' and has argued for a 'socio-spatial dialectic':

The basic structures and contradictions in the capitalist (and any other) mode of production are expressed simultaneously and dialectically in both social and spatial relations ... not only does the organization of space express social relations but ... social relations (and hence class structure) are themselves, to an important degree, expressions of the spatial relations of production. (Soja, quoted in Peet and Lyons 1981, 200)

Soja's solution is to invoke 'relative autonomy' somewhat more forcefully than Peet, but the whole notion of the 'last instance' remains but one small step from determinism, as critics of Althusser have been quick to point out (Hirst 1976a, b; Thompson 1978; Lovell 1980; Clarke *et al.* 1980; Smith 1980).

Conjointly with this neospatial fetishism, these same arguments are used to justify the continuance of a separate discipline: traditional geographic objectives, an ontological status for space, process as a mere descriptive tool, reliance on positivist philosophical categories, and the preservation of unique regions of epistemological territory. If this were not sufficient, these concerns are then cast in polemical terms to make social revolution the basis of the solution to 'endemic spatial problems' (see also Peet and Lyons 1981, 202). This uncritical acceptance of so many 'degenerate formulations' condemns these attempts to remain within the confines of bourgeois social science. These confusions obviously are not just attributable to Peet or Soja, but are repeated many times over in the pages of *Antipode* and elsewhere (by Corraggio 1977; Buch-Hanson and Nielson 1977; Santos 1980; Walker and Storper 1981; etc.). By not reconstituting outside geography, by using idealist notions, and by not using spatial differences as merely a point of departure, such interventions leave the core of the discipline unscathed. 'Space and time, mind and matter, economics and culture, history and geography' are 'philosophically defined into separate boxes and reality ordered to fit them'. The notion of 'spatial dialectics' simply adheres to this 'pigeonholing of reality' which remains idealistic in scope. 'The point of the dialectic is not to accept the different boxes and squeeze a separate dialectic into each one, but to break down the artificial walls in favour of a more synthetic, integrated understanding of reality' (Smith 1981, 113).

What we see as the built landscape is merely some facets and results, in capitalist-dominated social foundations, of decisions concerning, amongst other things, capital. What we observe, and attempt to understand, are not autonomous, no matter what form, spatial or otherwise, they assume. Space, transportation networks, production and consumption facilities, etc., treated as *isolates*, are inert and insignificant, since they are merely facets of one relationship, that generated by a mode or modes of production in a given social formation. That relationship will involve power-centralizing institutions like private ownership, the concentrative structure of capital itself, various forms of state planning and legislation (a response to and supportive of the first two), the particular nature of the class struggle, relationships, consciousness, etc. Landscape patterns are however only one expression (albeit a 'spatial' one) of the dominant mode of production. At issue is not that we should study these relations, but what our point of entry is, how we proceed, practice, and reach conclusions from the historical materialist perspective.

> If Peet truly expects that adhering 'materially' to inherited boundaries (i.e., remaining obedient to the discipline) will guarantee a 'modicum of [academic] freedom' for Marxism along with the rest of academia (Peet, 1979, 1981), I suggest that his political memory is short and his analysis of the current situation hopelessly utopian. McCarthyism knows no such subtleties. (Smith 1981, 113)

This lack of scientificity and the inability to transcend the barrier of ontological space, technical practices, and ideologies rooted in geography since the mid-nineteenth century, condemns its practitioners, no matter how sincere and well-intentioned they may be, to a schizophrenic and thus unbalanced practice. As Slater (1978, 3) commented, geography as a branch of bourgeois social science cannot be incorporated into Marxism, it can only be *transcended* and *superseded* by it. For geographers to become Marxists, they have as a 'professional or academic group to commit suicide'.

MARXISM AND HOLISTIC SOCIAL SCIENCE

Marx's dissection of the capitalist mode of production continues to provide the skeleton on which can, in Thrift's words, 'be hung the fact of other, more contingent determinations'. . . . One must learn to think in terms of various dialectics – structure and agency, continuity and change, value and use-value, function and contradiction – and not succumb to simplistic dualisms and accusations based on them. The problems of historical analysis and good social science, of how one thinks, what theories one should hold to, and how one can verify his/her notions are not easy ones. (Walker and Greenberg 1982, 42)

Marxism is a mode of analysis and explanation (*historical materialism*) linked to a particular theory of the production of knowledge, science, and theoretical practice (*dialectical materialism*). Like all scientific theory and philosophy it is always in evolution; there are areas not yet fully worked out, others are still to be discovered and examined: all the seemingly invisible networks of relationships linking together forms, functions, modes of articulation, and the hierarchy of appearance and disappearance of particular social structures; the different tempo of development of the interrelated areas of economic, political, cultural, and ideological work are only very incompletely understood; and although the economic practices within a mode of production have been analyzed fairly systematically for the past century, the other practices, particularly the ideological, have only come under serious scrutiny in the past twenty-five years. When such work is further developed it may be possible to uncover the types and mechanisms of networks, relationships, articulations, and so on, in a more systematic way. However, already, in the process of developing this science of society, the point has been reached where the nineteenth-century distinctions between anthropology, geography, history, sociology, and so on, are an encumbrance, and such fragmentation must be transcended. Current attempts at cross-disciplinary fertilization are doomed to failure, just as geography's long history of borrowing freely from every other discipline created an eclectic pluralism of methods, techniques, and concepts without theory in the strong sense. Borrowing in the reverse sense has also created problems:

> the insertion of space, place, locale and milieu into any social theory has a numbing effect upon that theory's central propositions. Microeconomists working with perfect competition find only spatial monopoly, macroeconomists find as many economies as there are central banks and a peculiar flux of exchange relations between them. ... The insertion of spatial concepts into social theory has not yet been successfully accomplished. (Harvey 1984, 8).

And neither will it ever be. A particularly convincing example of the impossibility of such an approach is Giddens's structuration theory which attempts to parachute in 'primitive' notions of space, and time/space from Christaller, Hagerstrand, Berry, Thrift, and others. Validly arguing that space and time cannot be ignored as covariables, Giddens's demonstration of incorporation whilst retaining certain sociological imperatives is unconvincing. The autonomous fetishized domains of the past have long outlived any usefulness, even to academia under capitalism itself.

More convincing was Godelier's attempt to transcend anthropology and economics (1977). He begins by reminding us that a historical materialist approach must not be 'confused with the observation of its visible objects nor with the interpretation of spontaneous representations peculiar to the economic agents of this system and which, by their activities, reproduce it'. In geography, a comparable example would be the empiricist approach which looks only to immediately identifiable 'landscape objects', or when transportation geographers proceed from there to examine network location decisions in isolation or with reference only to some 'spatial strategy'. Godelier continues:

> it is a commonplace that capitalists appropriate to themselves the use of a worker's labour against the payment of wages and they also spend money in order to appropriate the other means of production – machines, raw materials, etc. Everything happens *as if* the wages paid for the work and *as if*, in the value of the goods produced and the end of the process of production, many other elements apart from human labour are involved. On the surface,

therefore, capitalist profit has nothing to do with a mechanism exploiting the workers' manpower, since they receive a wage which seems equivalent to the cost consideration represented by the work. (Godelier 1977, 77)

Within the closed problematic, geography, a researcher observing the 'facts' and the individual behavior of industrialists, entrepreneurs, and employees, 'cannot analyze the deep, invisible logic' of the capitalist mode of production and instead 'merely reproduces' the obvious aspects of those relations, whether less abstract, less complex as in the case of 'commercial geography' or more abstract and complex in the case of 'regional science'. At the same time such a reproduction of surface features alone plays the *ideological role of* hiding the real capitalist/worker relationship and the *source of surplus value*.

The final pattern of economic relations as seen on the surface, in their real existence and consequently in the conceptions by which the bearers and agents of those relations seek to understand them, is very much different from, and indeed quite the reverse of, their inner but concealed essential pattern and the conception corresponding to it. (Marx 1971, 20)

As Marx noted elsewhere 'all science would be superfluous if the outer appearance and the inner essence of phenomena directly coincided'. It was Marx's particular strength to look beneath such surface appearances to identify the hidden connections between phenomena. Marx was thus able to demonstrate that industrial and commercial profits, financial interest and income from land, although seemingly all from different sources and the results of differing behaviors, were in fact simply distinct but different forms of surplus value which accrued to one particular class, the bourgeoisie, rather than to the workers themselves. Going further, he was able to show that the labor of that working class had no price, only their *labor power* as such had a price

equivalent to the cost of all that is socially necessary for its reproduction. From this he could criticize the empirical categories of political economy and show that if wages are not equivalent to the value created by the use of manpower, but equivalent to the cost of reproducing this manpower, then the surplus-value is nothing more mysterious than the difference between the total value created by the use of manpower and the fraction of this value which is handed to the producer in the form of wages. Far from corresponding to reality, the 'facts', or at least their appearance and representation, the ideas corresponding to these 'facts', conceal this deep invisible reality, thereby revealing precisely the *contrary*. (Godelier 1977, 23)

Historical materialism endeavors to *discover* (uncover) in its totality all the structural relations which exist between apparent surface appearances and the deeper currents. If one does not do this, then whatever aspect of reality one is attempting to comprehend becomes a fetishized ideological domain, in the sense that one sees the analysis of socio-economic relations to be confined to the analysis of what are or appear to be, one bounded set of relationships. This is to underline the existence of only one science of society which entails both the theoretical/scientific practice needed to unravel socio-economic relationships and their explanation in a historically specific concrete situation. This science of society submerges the extant problematics of history, anthropology, political economy, sociology, geography, psychology, etc., and transcends them in the creation of an objectively superior way. As an example, figure 3.1 illustrates a tentative heuristic device for examining transportation as one point of entry. Of course this diagram *re*presents a four-dimensional world on a two-dimensional plane, but that aside it

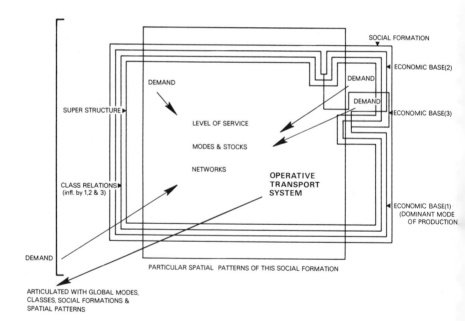

Figure 3.1 A holistic framework for studying transportation

offers a more comprehensible and scientific approach than that within geography as illustrated by Rimmer (1978).

Transportation has received much attention from geographers (see also Hay 1977, 1981, etc.) and yet, despite being in the vanguard of applying logical positivist techniques to transport phenomena, has remained in a cul-de-sac since the mid-1970s moving neither backwards nor forwards but forever, it seems, reproducing more of the same. What they cite as 'new' in the 1980s on closer examination is simply recycled 'sixties geography' with the empirical detail and technical practice at various scales far outweighing even 'theory' in the weakest sense. In an earlier critique this kind of continuing emphasis on technical practice was forecast with obvious ease, although an attempt was made to suggest redirection *within* the geographical discourse (Eliot Hurst 1973b). Despite this attempt to include factors of wider range than those usually analyzed by transport geographers, it remained firmly entrenched in the discipline, and therefore by its very nature unable to transcend it and thus never able to understand transport's role in various modes of production. Repudiation, no matter how correct or devastating without the presentation of a soundly worked-out alternative, only leads to pessimism, confusion, and ultimately more of the same. What was needed was not more ideography, not refined arithmomorphism, nor exposés of boardroom decisions, even though grounded in a 'conflict paradigm' and correctly criticized by both Hay (1977) and McCall (1977), but a realistic and scientific approach which transcended the limitations of one hundred years of *re*search in transport geography. Figure 3.1 was a first attempt made in 1979 to present an alternative configuration.

The concept represented by this tentative framework is that transportation is not autonomous, it does not have a life of its own, but is seen to develop in consistency with the demands of a particular economic structure at a particular point in time and space (a social formation). At a particular conjuncture, such as Britain in the mid-1980s, several modes of production may be represented of which one, in this case capitalism, dominates. The tensions and priorities which pulsate through the spheres of production and exchange are transformed and transmitted to and from the political and ideological practices, thereby leaving their mark on the way (in this instance where the point of entry is transport) state policies are formulated and created, the private sector articulates with these, and on the way we think about 'economic relations', 'efficiency', 'transport' *per se*, and so on. Ideological and political patterns feed back and forward to more basic decisions concerning the circulation and accumulation of capital, micro- and macro-level policy decisions, as well as to the ways we envisage transport to operate in a given society. These demands and conceptions generated within a social formation are mediated by state and private sector policies which give rise to a supply of transportation – what geographers have observed and measured as mere stocks, modes, networks, volumes of freight, etc. Of course no social formation exists in isolation, but is part of a chain of socio-economics (of greater or lesser importance) which can influence and be influenced in a complex interlocking whole.

The dominant mode of production overdetermines transport, the built environment, 'space' as geographers have tended to examine it, and influences directly and indirectly various ideological and political practices, as well as institutions like the state itself. But reference to these structures and patterns, including the spatial layout of transport, is much more complex, and represents merely one facet of reality created in the evolution of a social formation. Although figure 3.1 can only be a crude representation of such an approach, it is an alternative to Ullman's (1956) typology of the mid-1950s which is still the basis of most transport geography. Ullman's whole analysis took for granted production-for-exchange, ascribed universal validity to his three-factor typology (complementarity, transferability, and intervening opportunity; the 'theory' equivalent to the social physics of that era), and assumed spatial distributions to be the 'natural order of things'. But the spatial distributions of 'supply' and 'demand' used by Ullman as causes (*causa causans*) are in fact secondary (*causa causata*). By confusing cause and effect he overlooked *vera causa*: the striving to create new opportunities for capital accumulation; a struggle to realize surplus value; the crystalization of patterns of overdevelopment and underdevelopment; the need to create global markets, and so on. The present, or past, world mosaic of economic landscapes and circulatory flows are *not* the results of 'natural' areal differences, but rather reflect the historical movements of social formations as a whole, for example changes in modes and relationships of production. Cocoa beans, for example, are not just grown in Ghana because it is a specialized tropical region of supply for a western European region of demand. Why cocoa beans? Why Ghana? Why western Europe and not North America? How many Ghanaians eat or drink chocolate? Would the Ghanaians be better off with 'self-reliance' and less international trade in a non-essential commodity like cocoa? There is nothing inevitable, natural, or innate about trade in cocoa, yet geographers in dealing with commodity flows have so assumed. 'Supply' and 'demand' treated in the orthodox manner are trapped in ahistoricity and ideography, whereas what geographers term 'spatial' or 'areal differentiations' is an active *process*, the pattern of which is specific to particular conjunctures.

This point cannot be overstressed since geographers have been trapped in this *empirical fallacy*, to use Husserl's admonition, for many decades. Marx's comments on India also underline this argument. It might appear at first sight that the circulation of cotton from India to Britain in the nineteenth century was based more or less on the areal differences of 'supply' and 'demand', the basis of Ullman's conception of spatial interaction. India specialized in the production of raw cotton and Britain the production of textiles. However, for most of the eighteenth century *Indian* textiles had supplied the markets of Asia and even Europe. It was the British imperial conquest of India, a reflection of Britain's burgeoning capitalist economy, that turned India into an exporter of raw cotton and now an importer of manufactured textile commodities! Undoubtedly India has special advantages (notably climatic) for growing cotton, but there is no inherent reason (no *causa causans*) why it should specialize in the export of raw cotton. What geographers ignored was that the British social formation dominated by a capitalist mode of production in its violent expansionary phase destroyed existing economies in many parts of the world, of which India is only one example, and imposed upon them a different set of relationships which contained that apparent 'supply/demand' equation. As Marx noted:

> by ruining handicraft production in other countries, mechanization forcibly converts them into fields for the supply of its raw materials. In this way East India was compelled to produce cotton, wool, hemp, jute, and indigo for Great Britain. ... A new and international division of labour, a division suited to the requirements of the chief centres of modern industry springs up and converts one part of the globe into a chiefly agricultural field of production for supplying the other part, which remains a chiefly industrial field. (Marx 1972, vol. 1, 424–5).

World landscapes are dramatically transformed as the development of capitalism makes the production of commodities *per se* widespread and as it simultaneously destroys the self-sufficiency and/or preexisting external links of contemporary non-capitalist modes of production. Obviously, therefore, the spatial differences, the patterns of 'supply' and 'demand', and the connective circulatory flows that arose in the nineteenth or twentieth century cannot be accounted for in terms of natural/climatic factors, Newtonian analogies, or Ullman's typology. Production for a market, accumulation at a global scale, and the power and force of a particular dominant mode of production crystalized through certain social formations, explain the dramatic increase in commodity flows, means of transport/communication, and the new circulatory patterns that characterize this historical period (Curtin 1976; de la Haye 1980).

Even this simple holistic framework suggests a range of questions for further work utilizing transport as a point of departure, some of which were raised in 1973, but at that time left untheorized. For whom is transport provided? Who benefits in practice, in both the short and long term, from particular transport development strategies? What is the particular range of objectives to be achieved by state or private sector decisions in a social formation dominated by capitalism? Or since these objectives are to foster dependence and concentrate power and capital, by what order of things does Rimmer's (1978) plea to the behavior of the individual in transport decision-making have meaning? What role do the decisions (and the struggles around them) of particular classes, that is those with the power and capital, have in the production and reproduction of the built landscape? What does this mean for strategies of capital investment? What role *does* transport play in the production process? But to answer these questions, basic though they may be to holistic social science, presupposes many more studies of the transport process from within the Marxist approach than are currently under way or complete. Curtin's (1976) pioneering work is one of only a few exceptions, and this author (Eliot Hurst 1979a, 1981b) after a few tentative forays abandoned the whole area of transport in order to interrogate the theoretical terrain and practice surrounding 'ideology' as *re*presented through film, landscape symbolism, and foodways.[2]

In order to grow and expand, renew and redevelop the theory, method, and practice of this holistic approach, as any science must to match a changing reality/production of knowledge, it is always necessary to radically criticize, and deconstruct if need be, any current practice. The central problem in creating this science of society,

> is to explain the circumstances behind the appearance of different social structures, articulated in a determined and specific manner with the circumstances of reproduction, for change, and for the disappearance of these structures and their articulation. At the same time it is the problem of analyzing the specific causality of overlying structures of their particular role and different meanings in the processes of the appearance, reproduction, and disappearance of the various social relations which are the content . . . (Godelier 1977, 20) of both society, social existence, and being human. . . . 'A science is obtained on condition that the domain in which ideology believed that it is dealing with the real is abandoned, that is by abandoning its ideological problematic . . . and going on to establish the activity of the raw theory "in another element", in the field of a new, scientific problematic.' (Althusser 1969, 192–3)

CONCLUSION: THE IRRELEVANCE OF GEOGRAPHY TO CONTEMPORARY SOCIETY

> Marx's analyses comprise a coherent (if incomplete) whole. . . . One cannot treat them as one would a supermarket by picking and choosing at random from the well stocked shelves until one's basket is full. The latter process is haphazard. . . . If our aim is the development and extension of Marxist theory to provide an understanding of . . . capitalism . . . then it is incumbent upon us to begin from the coherent analysis of capitalism *as a totality*, and not from a degenerate spatial interactionism and radical eclecticism. (Smith 1981, 117)

Geography is irrelevant to contemporary society on two basic grounds. *First*, it has been argued that historically it arose as a discourse within 'epistemological space' to meet certain specific needs of a then burgeoning socio-economy. Colonialism and the Industrial Revolution, and later the growth of global corporations, called forth certain kinds of ideography, method/technique-strong, but lacking theory in the strict sense. Those needs, where they exist today to protect capitalism's imperialism from contradiction and collapse, are better served by better-organized, more innovative and responsive fragments of the so-called social sciences. *Second*, it has been argued that from a philosophical and epistemological viewpoint, geography is an untheorized point of entry to knowledge. As such, categories of concern to the claimed discourse, like 'space', 'spatial differentiation', 'spatial interaction', or even that preferred by so-called Marxist geographers, 'uneven development', are mere fethisims and unscientific. A completely new and radical (*sensu stricto*) focus has to emerge that is grounded in the praxis of Marxism but which transcends all of the current 'divisions' of epistemological space. In total this scheme of radical reconstruction cannot be carried out without a revolution. It is not just geography as currently or potentially practiced that must be buried, but also the bourgeois academy and the oppressive socio-economic system of which it is but one part.

NOTES

1 W. Bennett is the Premier of the Province of British Columbia who leads a far-right-wing populist party, Social Credit. Milton Friedman has recently lauded Bennett as the best current exemplar of his economic policies.
2 For example by teaching as an Associate Faculty Member in the Centre for the Arts on the history and aesthetics of dominant and alternative cinemas;

co-editing a new journal, *Opsis*, which focuses on avant-garde and political film; and by expanding courses taught 'within' geography to include the culture and political economy of food, diet and nutrition (Eliot Hurst 1984a), the semiotics of landscape symbolism (Eliot Hurst 1983, 1984b), as well as continually interrogating the needs and uses of lower level introductory courses in 'economic' and 'cultural' geography.

REFERENCES

Ackerman, E.A., et al. (1965) *The Science of Geography*, Washington, DC, National Academy of Science/National Research Council (report of the Ad Hoc Committee on Geography, Earth Sciences Division).

Albrow, M. (1974) 'Dialectical and categorical paradigms of a science of society', *Sociological Review*, 22, 183–210.

Althusser, L. (1969) *For Marx*, Harmondsworth, Penguin.

—— (1971) 'Ideology and ideological state apparatuses', in *Lenin and Philosophy and Other Essays*, New York, Monthly Review Press, 127–86.

—— and Balibar, E. (1970) *Reading Capital*, London, New Left Books.

Amedeo, D. and Golledge, R. (1975) *An Introduction to Scientific Reasoning in Geography*, New York, John Wiley.

Amin, S. (1976) *Unequal Development: An Essay on the Social Formations of Peripheral Capitalism*, New York, Monthly Review Press.

Anderson, J. (1973) 'Ideology in geography', *Antipode*, 5 (3), 1–6.

—— (1975) 'The political economy of urbanism: an introduction and bibliography', London, Architectural Association, mimeo.

—— (1980) 'Towards a materialist conception of geography', *Geoforum*, 11, 171–8.

Bernstein, R. (1976) *The Restructuring of Social and Political Theory*, Oxford, Basil Blackwell.

Blaut, J. (1974) Commentary on *Values in Geography*, Annette Buttimer, Commission on College Geography, Resource Paper, 24, Washington, DC, 44–5.

Boddy, M. (1976) 'The structure of mortgage finance: building societies and the British social formation', *Transactions, Institute of British Geographers*, NS 1 (1), 58–71.

—— (1980) *The Building Societies*, London, Macmillan.

Boggs, C. (1976) *Gramsci's Marxism*, London, Pluto Press.

Breitbart, M.M. (1981) 'Peter Kropotkin, the anarchist geographer', in D.R. Stoddart (ed.) *Geography, Ideology, and Social Concern*, Oxford, Basil Blackwell, 134–53.

Broek, J.O.M. (1965) *Compass of Geography*, Columbus, Ohio, Charles E. Merrill Books.

Buch-Hanson, M. and Nielson, B. (1977) 'Marxist geography and the concept of territorial structure', *Antipode*, 9 (1), 1–11.

Bunge, W. (1978) 'The first years of the Detroit geographical expedition: a

personal report', in R. Peet (ed.) *Radical Geography*, London, Methuen, 31–9.
—— (1979) 'Comment on "theoretical geography"', *Annals, Association of American Geographers*, 69 (1), 169–74.
Burgess, R. (1976) *Marxism and Geography*, University College, London, Department of Geography, Occasional Paper 30.
—— (1981) 'Ideology and urban residential theory in Latin America', in D.T. Herbert and R.J. Johnston (eds), *Geography and the Urban Environment*, vol. 4, Chichester, John Wiley, 57–114.
Burton, I. (1963) 'The quantitative revolution and theoretical geography', *Canadian Geographer*, 7, 115–62.
Buttimer, A. (1971) *Society and Milieu in the French Geographic Tradition*, Chicago, Rand McNally.
Carter, G.F. (1977) 'A geographical society should be a geographical society', *Professional Geographer*, 29, 101–2.
Clarke, S., et al. (1980) *One Dimensional Marxism: Althusser and the Politics of Culture*, London, Allison & Busby.
Claval, P. (1977) 'Le marxism et l'espace', *L'Espace géographique*, 6 (3), 145–64.
—— (1980) 'Epistemology and the history of geographical thought', *Progress in Human Geography*, 4, 371–84.
Corraggio, J.L. (1977) 'Social forms of space organization and their trends in Latin America', *Antipode*, 9 (1), 14–28.
Cosgrove, D. (1978) 'Place, landscape, and the dialectics of cultural geography', *Canadian Geographer*, 22, 66–72.
—— (1983) 'Towards a radical cultural geography: problems of theory', *Antipode*, 15 (1), 1–11.
Curtin, B. (1976) 'Aspects of Karl Marx's theory of circulation and some of its geographical implications', MA Thesis, Simon Fraser University.
de la Haye, Yves (1980) *Marx and Engels on the Means of Communication*, New York, International General.
Dematteis, G. (1980) 'La nascita dell'indirizzo marxista nella ricerca geografica italiana', in G. Corna Pellegrini and C. Brusa (eds), *La ricerca geografica in Italia 1960–1980*, Varese, Ask Edizioni, 781–92.
Dunbar, G.S. (1981) 'Elisée Reclus, an anarchist in geography', in D.R. Stoddart (ed.), *Geography, Ideology, and Social Concern*, Oxford, Basil Blackwell, 154–64.
Duncan, J.S. and Ley, D. (1982) 'Structural Marxism and human geography: a critical reassessment', *Annals, Association of American Geographers*, 72, 30–59.
Eliot Hurst, M.E. (1973a) 'Establishment geography: or how to be irrelevant in three easy lessons', *Antipode*, 5, 40–59.
—— (1973b) 'Transportation and the societal framework', *Economic Geography*, 49, 163–80.
—— (1976) 'The railway epoch and the North American landscape', in J.W. Watson and T. O'Riordan (eds), *The American Environment: Perceptions and Policies*, London, Wiley, 183–206.
—— (1979a) 'The political economy of transportation, some redefinitions for geographers', mimeo, circulated but not offered for publication.
—— (1979b) 'On criticism, counter-criticism, and counter-productive criticism',

Union of Socialist Geographers, *Newsletter*, 4 (4), 57–66.
—— (1980) 'Geography, social science and society: towards a de-definition', *Australian Geographical Studies*, 18 (1), 3–21.
—— (1981a) *Human and Inhuman Geography: An Autocritique*, with comments by Mary Hall, John Holmes, Frank Williamson, Ron Horvath and R. Peet, and a rejoinder: 'The inauthenticity of geographical practice', University of New England, Armidale, NSW.
—— (1981b) 'Cars and capitalism: the political economy of urban transportation', *Geoforum*, 12, 71–84.
—— (1981c) 'Vestiges of colonial and post-colonial food in regional English-Canadian cooking', in Alan Davidson (ed.), *National and Regional Styles of Cookery*, London, Prospect Books.
—— (1983) 'Fast food merchandizing – its origins and significance', in A. Davidson and C. Smith (eds), *Food in Motion: The Migration of Foodstuffs and Cookery Techniques*, London, Prospect Books.
—— (1984a) 'Foodways in multicultural societies, with particular reference to Israel and Canada', mimeo, Simon Fraser University.
—— (1984b) 'Films of our own: representations of landscape in Canadian film', in F.M. Helleiner (ed.), *Cultural Dimensions of Canada's Geography*, Occasional Papers 10, Department of Geography, Trent University, 109–36.
Eyles, J. (1981) 'Why geography cannot be Marxist: towards an understanding of lived experience', *Environment and Planning*, A, 13, 1371–88.
Foucault, M. (1970) *The Order of Things*, Pantheon, New York.
—— (1972) *The Archaeology of Knowledge*, Pantheon, New York.
Frémont, A. (1976) *La Région, espace vécu*, Paris, Presses Universitaires de France.
Giddens, A. (1981) *A Contemporary Critique of Historical Materialism: Power, Property and the State*, London, Macmillan.
Godelier, M. (1977) *Perspectives in Marxist Anthropology*, Cambridge, Cambridge University Press.
Gregory, D. (1978) *Ideology, Science and Human Geography*, London, Hutchinson.
—— (1982) *Regional Transformation and Industrial Revolution*, London, Macmillan.
Haggett, P. (1981) 'Geography', in R.J. Johnston *et al.* (eds), *The Dictionary of Human Geography*, Oxford, Basil Blackwell, 133–6.
Halvorson, P. and Stave, B.M. (1978) 'A conversation with B.J.L. Berry', *Journal of Urban History*, 4, 209–38.
Hartshorne, R. (1959) *Perspective on the Nature of Geography*, Chicago, Rand McNally.
Harvey, D. (1969) *Explanation in Geography*, London, Edward Arnold.
—— (1972) 'A commentary on the comments', *Antipode*, 4 (2), 36–41.
—— (1973a) *Social Justice and the City*, London, Edward Arnold.
—— (1973b) *A Question of Method for a Matter of Survival*, University of Reading, Department of Geography, Geographical Paper 23.
—— (1974a) 'Ideology and population theory', *International Journal of Health Science*, 4, 515–37.
—— (1974b) 'Population, resources, and the ideology of science', *Economic*

Current Check-Outs summary for Caldwell,
Fri Dec 05 15:33:11 GMT 2014

BARCODE: 30114003921940
TITLE: Humanistic geography.
DUE DATE: 05 Jan 2015

BARCODE: 30114008229943
TITLE: The future of Geography / edited
DUE DATE: 12 Dec 2014

Geography, 50, 256–77.
—— (1981) 'Marxist geography', in R.J. Johnston et al., *The Dictionary of Human Geography*, Oxford, Basil Blackwell, 209–12.
—— (1982) *The Limits to Capital*, Basil Blackwell, Oxford.
—— (1984) 'On the history and present condition of geography: an historical materialist manifesto', *Professional Geographer*, 36 (1), 1–11.
—— and Smith, N. (1984) 'Geography: from capitals to capital', *The Left Academy*, vol. 2, New York, Praeger.
Harvey, M.E. and Holly, B.P. (eds) (1981) *Themes in Geographic Thought*, London, Croom Helm.
Hay, A. (1977) 'Transport geography: progress report', *Progress in Human Geography*, 1, 313–18.
—— (1981) 'Transport geography', *Progress in Human Geography*, 5 (2), 263–7.
Hess, J., Linton, J., Rosenthal, M., Buscombe, E. and Powers, T. (1978) 'Film and ideology', Special Section, *Jump Cut*, 17, 14–28.
Hirst, P. (1976a) *Problems and Advances in the Theory of Ideology*, Cambridge, Communist Party Pamphlet.
—— (1976b) 'Althusser's theory of ideology', *Economy and Society*, 5.
Horvath, R. (1971) 'The Detroit geographical expedition and institute experience', *Antipode*, 3 (1), 73–85.
Hudson, B. (1977) 'The new geography and the new imperialism: 1870–1918', *Antipode*, 9 (1), 12–19.
Irvine, J., Miles, I., and Evans, J. (1979) *Demystifying Social Statistics*, London, Pluto Press.
Jones, E. (1984) 'On the specific nature of space', *Geoforum*, 15 (1), 5–9.
Kinvig, R.K. (1953) 'The geographer as humanist', *The Advancement of Science*, 38, 157–68.
Lacoste, Y. (1973) 'La Géographie', in F. Châtelet (ed.), *La Philosophie des sciences sociales*, Paris, Presse Universitaire de France, 242–302.
—— (1976) *La Géographie, ça sert, d'abord, à faire la guerre*, Paris, Maspero.
Lee, R. (1976) 'Integration, spatial structure and the capitalist mode of production in the EEC', in R. Lee and P.E. Ogden (eds), *Economy and Society in the EEC: Spatial Perspectives*, Westmead, Saxon House, 11–37.
—— (1977) 'Anti-space: geography as the study of social space writ-large', in R. Lee (ed.), *Change and Tradition: Geography's New Frontiers*, London, Queen Mary College, 69–75.
—— (1979) 'The economic basis of social problems in the city', in D. Herbert and D.M. Smith (eds), *Social Problems in the City*, Oxford, Oxford University Press, 47–62.
Ley, D. and Samuels, M. (eds) (1978) *Humanistic Geography: Prospects and Problems*, Chicago, Maaroufta Press.
Lovell, T. (1980) *Pictures of Reality: Aesthetics, Politics, and Pleasure*, London, British Film Institute.
McCall, M.K. (1977) 'Political economy and rural transport: an appraisal of western misconceptions', *Antipode*, 9 (3), 98–110.
Marx, K. (1971) *A Contribution to a Criticism of Political Economy*, London, Lawrence & Wishart.

—— (1972) *Capital*, 3 vols, Moscow, Progress Publishers.
—— (1973) *Grundrisse: Foundations of the Critique of Political Economy*, trans. by M. Nicolous, London, Penguin/New Left Review.
Massey, D. (1978) 'Regionalism: some current issues', *Capital and Class*, 6, 102–25.
—— and Catalano, A. (1978) *Capital and Land: Landownership and Capital in Great Britain*, London, Edward Arnold.
—— and Meegan, R. (1982) *The Anatomy of Job Loss*, London, Methuen.
O'Donnell, R., Stevens, P. and Lennie, I. (1978) *Paper Tigers: An Introduction to the Critique of Social Theory*, Department of General Philosophy, University of Sydney, Sydney, NSW, Pilot Edition.
Paterson, J.L. (1984) *David Harvey's Geography*, London, Croom Helm.
Peet, R. (1977) 'The development of radical geography in the United States', *Progress in Human Geography*, 1, 240–63.
—— (1979) 'On comradely criticism and Marxist geography', Union of Socialist Geographers, *Newsletter*, 4 (3), 20–6.
—— (1981) 'Spatial dialectics and Marxist geography', *Progress in Human Geography*, 5 (1), 105–10.
—— and Lyons, J.V. (1981) 'Marxism: dialectical materialism, social formation and the geographic relations', in Harvey and Holly (1981), 187–205.
Quaini, M. (1982) *Geography and Marxism*, Oxford, Basil Blackwell.
Rimmer, P. (1978) 'Redirections in transport geography', *Progress in Human Geography*, 2, 76–100.
Ryan, M. (1982) *Marxism and Deconstruction, a Critical Articulation*, Baltimore, Johns Hopkins Press.
Santos, M. (1980) *The Shared Space*, London, Methuen.
Sayer, R.A. (1979) 'Epistemology and conceptions of people and nature in geography', *Geoforum*, 10, 19–43.
—— (1981) 'Defensible values in geography: can values be science-free?', in D.T. Herbert and R.J. Johnston (eds), *Geography and the Urban Environment*, vol. 4, Chichester, Wiley, 29–56.
Slater, D. (1975) 'The poverty of modern geographical enquiry', *Pacific Viewpoint*, 16, 159–88.
—— (1977) 'Geography and underdevelopment', part 2, *Antipode*, 9, 1–31.
—— (1978) 'Beyond geography's ideological domain, provisional notes for a paper in progress', mimeo, paper presented at a USG regional workshop, Montreal.
Smith, D. (1977) *Human Geography: a Welfare Approach*, London, Edward Arnold.
Smith, N. (1979) 'Geography, science, and post-positivist modes of explanation', *Progress in Human Geography*, 3 (3), 356–83.
—— (1980) 'Symptomatic silence in Althusser: the concept of nature and the unity of science', *Science and Society*, 44, 58–81.
—— (1981) 'Degeneracy in theory and practice: spatial interactionism and radical eclecticism', *Progress in Human Geography*, 5 (1), 111–18.
Soja, E. (1980) 'The socio-spatial dialectic', *Annals, Association of American Geographers*, 70.

Szymanski, R. and Agnew, J.A. (1981) *Order and Skepticism: Human Geography and the Dialectic of Science*, Washington, DC, AAG Resource Publication.
Therborn, G. (1976) *Science, Class, and Society*, London, New Left Books.
Thompson, E.P. (1978) *The Poverty of Theory*, London, Merlin.
Ullman, E.L. (1956) 'The role of transportation and the bases for interaction', in W.L. Thomas (ed.), *Man's Role in Changing the Face of the Earth*, Chicago, Chicago University Press, 862–80.
Van Beunigen, C. (1979) 'Le Marxisme et l'espace chez Paul Claval: quelques réflexions critiques pour une géographie marxiste', *L'Espace géographique*, 8 (4), 263–71.
Vidal de la Blache, P. (1911) 'Les Genres de la vie dans la géographie humaine', *Annales de Géographie*, 20, 193–212.
Walker, R. and Greenberg, D. (1982) 'A guide for the Ley reader of Marxist criticism', *Antipode*, 14 (1), 38–43.
—— and Storper, M. (1981) 'Capitalism and industrial location', *Progress in Human Geography*, 5 (4), 473–510.
Wooldridge, S.W. and East, W.G. (1958) *The Spirit and Purpose of Geography*, revised edn, London, Hutchinson.
Zelinsky, W. (1975) 'The demigod's dilemma', *Annals, Association of American Geographers*, 65, 123–43.

4

THE VALUE OF A GEOGRAPHICAL PERSPECTIVE
Peter J. Taylor

The title of this chapter was provided for me by the editor. I want to make this quite clear from the outset because it is just the sort of topic that is designed to send even the most enthusiastic geographer to sleep. I envisage that this is the section of the book where the harassed undergraduate student will be tempted to write 'boring' (in pencil please) in the margins. For many older readers it will be a topic that conjures up memories of pompous presidential addresses in which geography is unnecessarily justified before an audience of the converted. And yet I agreed to write this chapter, not because of the persuasive powers of the editor, but because it is a question which cannot be avoided. Past treatment of the issue must not lead us to neglect such a topic. We certainly do not need any further 'proofs' of the innate worthiness of geography but we do need self-conscious analyses and evaluations of our collective role in the society that supports and distorts us.

GEOGRAPHY'S 'HIDDEN SELF-CONTEMPT'

I take it as self-evident that a geographical perspective does have some value. Millions of dollars, pounds, francs, yen, marks and other currencies are paid out every year as salaries to geographers. The maximum spread of geography as a university discipline so far (that is at the beginning of the current world recession) is shown on figure 4.1. Either we have perpetrated one of the greatest hoaxes in modern history or else there is a value to the geographical perspective. Assuming the latter to be the more reasonable of these alternatives I concentrate in this chapter on the nature of that value. Of course, the dollars, pounds, francs, etc. being spent on geography have been declining in recent years. This has led to some very pessimistic prognostications: 'geography may not survive the next several decades as a subject that is taught in most American colleges and universities' (Wilbanks and Libbee 1979, 1). My view is that in a recession we can expect public expenditure cutbacks; these will include education cuts and geography will not be able to avoid some reduction in state support. This is a crisis of public expenditure not of geography. I have great faith that the current generation of geographers will produce their fair share of influential spokesmen in the various corridors of power to ensure continued support of geography at relatively modest levels as in the past (Jumper 1984). This chapter will not be a contribution to that propaganda exercise.

I do believe, however, that there is a crisis in geography today but it is only tangentially related to our getting a fair share of the academic cake. It is far deeper than the question of current government support. Geography is going through another identity crisis in which such questions as 'Is our venerable discipline coming apart?' (Mikesell 1979, 358) are being asked. Hence, despite the continued willingness of governments throughout the world to support geography, its practitioners have exhibited a perpetual lack of self-confidence (David 1957). This is the real message of those awful presidential addresses, and to me it represents one of the most interesting features of the discipline. The current crisis is merely the latest manifestation of what Bartels (1982) has identified as geography's 'hidden self-contempt'. Geographers more than any other discipline keep worrying about the value of what they do. This paranoia can be traced back to the unusual history of the discipline which has often left geography out of step with general intellectual trends. This is the story I will tell in this chapter – why the question of the value of a geographical perspective keeps rearing its ugly head.

A NEW INTELLECTUAL DIVISION OF LABOUR

Geography is a social institution. Like all such institutions its value to society varies over time and place. The creation of any social institution is the result of a group of people who identify a particular need and are able to find the resources to meet that need. As needs change the institution has to adapt to survive. All this is true of modern geography as a university discipline. In our discussion, therefore, we must consider

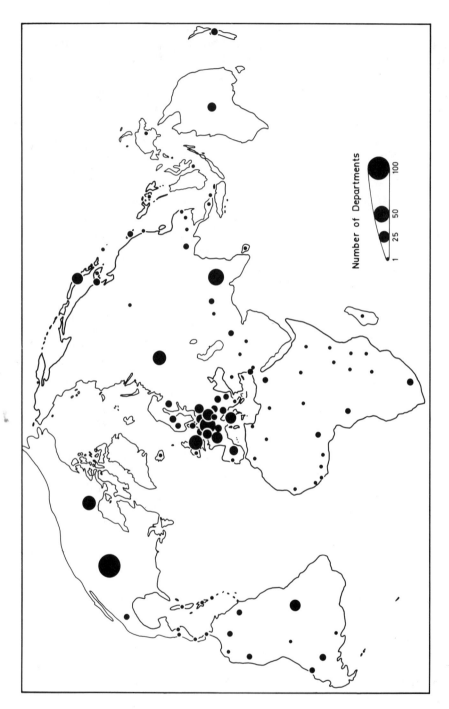

Figure 4.1 The distribution of geography departments c.1970 (derived from data in International Geographical Union 1975)

both the creators and the societal need they imagined geography could fulfil. We shall start with the needs before identifying the creators, and later adapters.

If we accept that 'the history of our discipline cannot be understood independently of the history of the society in which the practices of geography are embedded' (Harvey 1984, 1) then we need to consider the characteristics of the societal change of which geography is a part. Two types of change can be identified, secular trends and cyclical movements (Research Working Group 1979). The latter will be considered in some detail in the second part of this essay; at this stage we concentrate on the cumulative processes which are approximately linear over time. In this context Shaw (1975) identifies two key trends emanating from the fundamental nature of the capitalist mode of production. First there is the unprecedented expansion of intellectuals as a group in modern society. The social division of labour has always incorporated a small intellectual sector in class societies but this has qualitatively changed in the last century or so as intellectual workers have become a major component of the division of labour. Old intellectual categories have not expanded, however, but new specializations have been spawned and continue to be created. That is the second key trend to emerge. This growth of an army of specialized intellectuals is part of the dynamic of capital accumulation where the division of labour is being constantly revised in the search for increased productivity. In the private sector this is termed research and development; in the public sector we refer to higher and other education.

Habermas (1972) has characterized the rise of the specialist in slightly different terms. His *constitutive-knowledge interests* are deep-seated rules whereby societies are able to maintain production and generate reproduction. This is achieved in two distinctive types of cognitive interest – a technical one for instrumental purposes and a practical one for communicative purposes. In modern society these are represented by the empirical-analytic sciences and the historical-hermeneutic sciences respectively. These operate through different methodological frameworks: the former produces predictive knowledge for technical control, the latter interprets meanings to understand actors and promote consensus. The expansion of intellectuals has not occurred equally in these two groups of sciences. The specialization discussed by Shaw (1975) clearly favours the development of empirical-analytic sciences and Habermas recognizes the rise of the technical to subsume the practical. The modern discipline of geography should be seen therefore as part of a secular trend of increasing numbers of intellectuals biased towards specialization and technical interests. Geography has never been strongly represented in the private sector and so we will concentrate on the education part of the 'knowledge industry'.

The modern university system was created in Germany in the nineteenth century. Ben-David and Zloczower (1962) have described how the traditional faculty of philosophy evolved from its weak service function for theology and law to become the major growth area centring on the empirical natural sciences. Ben-David and Zloczower employ a competitive career model to produce a process of 'hiving off' of new specialisms in the continual creation of chairs for new disciplines. In this creation of a new intellectual division of labour 'renaissance man' became transformed into the specialist university scientist of today. The German university system produced the main characteristics of the modern organization of universities: the interplay of teaching and research within very narrow specialisms. It is the Balkanization of knowledge producing 'experts' with limited knowledge which is associated with the rise of the empirical-analytical sciences at the expense of the historical-hermeneutic sciences. This has not only involved the massive advances of the natural sciences but also the creation of new social sciences based upon positivist approaches.

The fact that this development has occurred under the patronage of the state has meant that this part of the intellectual division of labour has not directly reflected the needs of capital. Nevertheless the specialization that has occurred has been entirely compatible with the needs of modern accumulation as we have seen. The success of the German economy from the late nineteenth century onwards was partly due to its possession of technical knowledge derived from its universities. The latter became generally accepted as the model for expanding higher education elsewhere. Hence the establishment of geography as a discipline within the German university sector was the essential prerequisite for its dissemination to the rest of the world in the twentieth century. The key breakthrough therefore was in Germany, but the nature of that success has plagued the discipline ever since.

By the end of the nineteenth century specialization was fully institutionalized in the natural sciences and medicine but was still only evolving in the social sciences. The full development of that phase of specialization was the product of the American university system and its promotion of the social science trilogy of economics, sociology and political science. This is the specialist world geography has had to adapt to but it was not the world in which it was created. These new specialisms distanced themselves from the holistic social thought of the early nineteenth century created by classical political economy and classical philosophy, 'the principal forms in which the critical and scientific tendencies of the emerging bourgeoisie received their fullest expression' (Shaw 1975, 74). The 'founding fathers' of modern geography – Alexander von Humboldt and Carl Ritter – were both contributors to this holistic knowledge and in the words of Dickinson (1969, 21) were 'imbued by the philosophy of their time'. They left a holistic legacy to geography which was at variance with the growing specialization in the natural and social sciences. This was particularly crucial in the area of social enquiry as the classical framework was abandoned in a retreat to positivism and instrumentalism. These trends were to leave holistic geography stranded in a new world of specialization.

BLOOD AND IRON ... AND GEOGRAPHY

Neither von Humboldt nor Ritter was in any direct sense the creator of modern geography as a university discipline. Their ideas were used by the creators so that they could become *symbolic* founding fathers but their direct impact was limited. Both men died in 1859 and they left no direct legacy of geography within German universities. Von Humboldt had held no university post and although Ritter had been Professor of Geography at the prestigious University of Berlin from 1820, with his death the only chair of geography in Germany disappeared. In 1859 Darwin's *Origin of Species* was published and this made the work of both men, especially Ritter's religious teleological holism, seem less relevant to the new sciences of the late nineteenth century. For this reason we can agree with the view that this 'classical period' of Humboldt and Ritter represents the culmination of a geographical exploration tradition rather than the beginning of modern geography.

Who were the creators of modern geography then? The answer to this question is not what most geographers would want to hear. The creators of modern geography are to be found in the bureaucracy of the Prussian state in the period immediately following the creation of the German Empire. In 1874 Kaiser Wilhelm I decreed that all royal universities should create chairs of geography. This political decision led to the immediate imposition of geography on universities in the Prussian state and was soon copied throughout the German universities. Hence by the end of the nineteenth century geography was an established discipline in the German university system. Figure 4.2 shows the establishment of chairs of geography in the German-speaking area in the late nineteenth century. This map is the direct predecessor of the world-wide distribution of a century later (figure 4.1) as the establishment of geography like so much else in German universities was imitated by other countries.

Geography really does seem to be unique, however. A creation of a militaristic state it is no wonder that the discipline's centenary in 1974 was totally overlooked and that Humboldt and Ritter are preferred to

98 THE FUTURE OF GEOGRAPHY

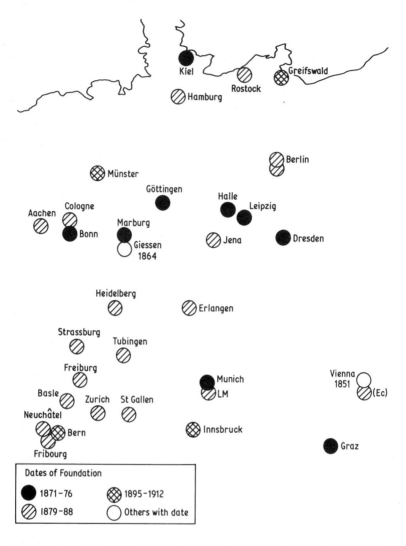

Figure 4.2 The establishment of chairs of geography in German-speaking Europe, 1871–1912 (derived from data in International Geographical Union 1975) Key: LM – Ludwigs-Maximilians University; Ec – Vienna School of Economics.

faceless Prussian bureaucrats in histories of geography. In fact the crucial decision of 1874 only warrants a few lines in the standard disciplinary histories. James (1972, 217) devotes most consideration to this event (two paragraphs under the heading 'New professorships in geography') and asks the obvious question 'Why did Prussia take this step in 1874?', but then continues, 'the answer is not entirely clear'. He attributes it to the popularity of geography after the Franco-Prussian war and specific individual lobbying. Dickinson (1969, 59) uses the lobbying argument and identifies two geographers (Wagner and Kirchoff) as the crucial lobbyists, claiming the government decision was 'largely due to their efforts'. Here we have two possible respectable alternatives to Humboldt and Ritter! But, of course, this type of explanation is unsatisfactory. The government would only have imposed geography on its universities if it was perceived that the subject would be useful for its particular purposes at that time. In short the Prussian state education bureaucrats must have thought there was some value to a geographical perspective in Germany in 1874.

What was that value? Geography seems to have been seen as useful on two counts. First there was the need to educate teachers to service the geography being taught in the expanding state school sector. Traditionally taught with history, these two subjects were seen as essential to the creation of the new German nation (Capel 1981, 48). Quite simply in history and geography Germany was treated as Germany and not merely as an example of 'modern society' as described by the new social sciences. Second Germany, as master of Europe, was now to turn to the remainder of the world to obtain her fair share of colonies. British hegemony in the world was beginning to decline and the age of imperialism was beginning. Geography was potentially useful for both commercial and military reasons in a world of imperial rivalry. In France this usefulness was expressed in the creation of many commercial geographical societies in the decade following defeat by Germany (McKay 1943). In Germany the established higher education system proved to be the chosen vehicle for the advance of geography. From being 'on the point of disappearance' (Capel 1981, 46) geography was found to be useful to German needs for nationalist and imperialist reasons and created as a university discipline accordingly.

It is not surprising that this unusual creation of the discipline produced difficulties usually expressed as the 'methodological controversies' of the late nineteenth century in the standard histories. To a university system expanding through increasing specialization a classical holistic discipline was added. The problem was that 'in the eyes of the already well-established university disciplines geography hardly qualified as a subject for real and profound research' (Bartels 1982, 25). One solution was to emphasize the physical geography aspects of the new discipline which had closer affinities to the new empirical sciences. This culminated in Gerland's proposal of 1887 to omit people from geographical study (Hartshorne 1939, 89). But this entailed cutting geography off from its 'classical' roots in Humboldt and Ritter. The debate resulted in rejection of Gerland's proposal and the encompassing of a methodology resting on the holism of the 'founding fathers'.

CYCLES OF 'PURE' AND 'APPLIED' GEOGRAPHY

The trend towards specialization and the creation of a 'holistic' geography were obviously at cross-purposes: for over half a century geography was directed away from the mainstream of intellectual change The crisis of traditional geography was indeed very slow to come. To understand the delay we must move on to consider cyclical changes in society and their effects in geography. Grano (1981) identifies two types of outside influences on the development of the discipline. Within academia geographers had to be given an intellectual foundation to satisfy intellectual peers, and outside in the wider world geography had to be justified as a useful activity on which to spend public money. These two pressures produce 'pure' and 'applied' geographies. Success in both activities is necessary for the long-term survival of the discipline.

The dichotomy of 'pure' and 'applied' in any discipline is of course a myth based on a theory of knowledge that claims a separation of knowledge from society. A 'pure' science is supposed to be an autonomous product of scientists' researches that generates theoretical knowledge. It is characterized by having no direct use for solving any particular problems. In contrast applied research is explicitly problem-orientated although it will claim to be based on the theoretical foundations set by 'pure' scientists. The point has been made many times that science is a social activity which can never be independent of the society that produces it. This is particularly clear in fields of social enquiry where the emphasis has always been more on the teaching than the research function in universities. If 'pure' research is not autonomous knowledge what is it? Quite simply it represents a set of intellectual concepts brought together to provide the theoretical foundations of a discipline. Geography is shown to be a 'natural' segment of the world of ideas. 'Pure' geographers define geography's place in the division of knowledge and promote the status of the discipline within academia. They produce the necessary core of ideas to justify the existence of geography. 'Applied' geographers accept these ideas and may use them to solve problems.

At any one time the discipline will have a mixture of pure and applied practitioners but the balance will change as the nature of the pressures on geographers changes. Outside pressures will be particularly acute in periods of economic recession when all public expenditure has to prove its worth. All disciplines will tend to emphasize their problem-solving capacity and we can expect applied geography to be in the ascendancy within our studies. In contrast in a period of expanding economies and social optimism outside pressures will diminish and academia can be expected to be under less external pressure. Geographers will be able to contemplate their discipline and feel much less guilty about this activity.

We can expect bursts of 'pure' geography to occur in these periods when pressures from academia seem more threatening than pressures from the outside. This analysis suggests that since the creation of geography as a university discipline we should expect the balance between pure and applied geography to approximate the long economic waves of the world economy. We may postulate therefore three 'applied geography' tendencies, in the late nineteenth century, between the world wars and in the current recession, separated by two 'pure' geographies flowering in the early twentieth century and the post-1945 economic boom. Let us trace these cycles of applied and pure geography.

Geography was created as an applied discipline by the Prussian state at the beginning of the late nineteenth century depression as we have seen. Although the standard discussions of this period have emphasized the methodological debates the discipline was very clearly balanced towards its applied aspects. As well as carrying out the teacher-training functions, political, commercial and colonial geographies were developed to aid diplomatic, military, trade and settlement policies. In this age of imperialism geography was a tool of nationalism and imperialism both political and economic. During the methodological debates, however, the seeds of a pure geography were germinating and culminated with the widespread acceptance of region and regional synthesis as the core of a 'new' geography in the early twentieth century. This first 'pure' geography was expressed differently in different national schools – French *pays*, German *Landschaft*, British 'natural regions', Russian 'zones' (Grano 1981) – but there was only one core geographical philosophy based on the holism of the 'founding fathers'. In fact the justification for geography was taken back as far as the late eighteenth century by reference to Immanuel Kant's division of knowledge into systematic, historical and geographical parts. Dickinson (1969, 10) describes it thus: 'Kant divided the communication of experience between persons into two branches, narrative or historical, and descriptive or geographical' and this hermeneutic element was brought into the first 'pure' geography especially in the French national school as modern humanistic geographers have discovered.

This first 'pure' geography remained the theoretical basis of geography throughout the first half of the twentieth century. But in the inter-war depression such 'pure' concerns were put aside as geography was required to show its usefulness. The most notorious example was the transformation of political geography into geopolitics in Germany. In Britain Dudley Stamp was carrying out the massive Land Utilization Survey and generally geographers were contributing to the growth of the planning movement in many countries. The emphasis here is on technical knowledge and the practical communicative aspects of the first pure geography are becoming lost. It is at this point that Hartshorne (1939) appears on the scene with his mammoth *Nature of Geography*. This impressive statement can be interpreted as a personal reaction to the neglect of the first 'pure' geography by Hartshorne's contemporaries. His work is full of references to the provincialism of English-speaking geographers, chiding them for their ignorance of the fundamental German contribution to geography. We would argue that this was not so much a language problem, however, but a symptom of the times. Hartshorne (1939, 26-9) introduces his work with an attack on those who would want to reform the nature of geography by making the subject more problem-orientated, more 'scientific' and more in tune with national interests. Hartshorne's great work stands therefore as a defence of the first 'pure' geography in a period of emphasis on applied geographies.

Geography entered the post-war era of economic prosperity and social optimism as an academic anachronism. With the social sciences fully established as specialized disciplines of social enquiry geography's holistic pretensions were becoming an intellectual liability. As late as 1951 Wooldridge and East (1951, 25-6) could still be aggressively asserting that the *raison d'être* of geography arises 'in large part from the shortcomings of the unco-ordinated intellectual world bequeathed us by the specialists' but this was merely a swan song of the first 'pure' geography. David (1957) cruelly attacked regional syntheses as empty intellectual rhetoric. Where were the grand syntheses that were the goal of geography? He concluded that geography's place in universities should be reduced. Such internal pressures led to the development of a second 'pure' geography that rejected the theory of the predecessor. Geographers at the University of Washington came to be the vanguard of the discipline's regeneration. Their motives have recently been made most explicit by Morrill (1984, 59): 'we were deeply disturbed by the fundamental question of whether geography deserved to be in a university at all, but we became determined to show that it did'. Inevitably, debate centred initially on Hartshorne's (1939) work and Schaefer (1953) came forward with an alternative programme for geography. Regional geography was dismissed as ideographic and geographers were instructed to stop contemplating regions and begin the nomothetic task of finding morphological laws. Schaefer ushered in the spatial school which involved a new theoretical structure (Bunge 1962) accompanied by the acceptance of statistical techniques in the 'quantitative revolution'. In Britain Haggett and Chorley (1967) proclaimed a 'new geography' just as Mackinder (1887) had done two generations earlier: a second 'pure' geography had been created. The 'spatial specialist' Schaefer had succeeded where the 'physical specialist' Gerland had failed and geography was dragged screaming into the twentieth century (Taylor 1976). The processes of specialization, partially held in check by the old holistic philosophy, now overwhelmed the discipline. Both human and physical geographers found space and location inadequate as a cohesive core to their researches and geography began to splinter apart in the manner predicted many times before by the older generation. By the 1970s the nightmare had arrived. Both the Institute of British Geographers and the Association of American Geographers came to have their academic activities dominated by organized groups of specialist researchers – 'study groups' and 'speciality groups' respectively. Geography had become a collection of diverse and only loosely related groups of researchers travelling in no particular direction.

In one sense the current recession has saved geography. Instead of having to face up to the fundamental problems of the legacy of the second pure geography, outside pressures have diverted attention back to applied geography. In the USA 'applied geography' has even reached the status of being the new 'paradigm' (Frazier 1978)! There has been much discussion on the role of geography in government (Anderson 1979), in the wider job market (Stutz 1980), and of its need to consider national priorities (Moriarty, 1978). Wilbanks and Libbee's (1979) solution to the dilemma they identify is to prove that geography is 'useful'. Fortunately, it has been found that the technical baggage associated with the spatial school constitutes a very useful set of equipment in showing how geography can contribute to solving problems today. Computer models, modern cartography and information systems all make geographers as spatial analysts employable in the current recession. We have, in a sense, come full circle from geographers employed by the Prussian state to help instil nationalism into its citizens to geographers employed by the American state to interpret remote-sensing images from NASA for the CIA.

The second 'pure' geography has been found, after the event, to have been 'positivist'. This label was introduced by critics in the 1970s – the term does not appear in the index of Harvey's (1969) standard methodological statement *Explanation in Geography*. The result has been the division of human geography into three broad groups; those that largely maintain the second 'pure' geography, those that have developed a humanistic critique and those of a more radical disposition (Johnston 1983). The first group continue to dominate as more and more applied geography is spawned. With rare exceptions they have not entered into the methodological debate but have concentrated on getting on with 'doing' geography. The other two groups have seriously undermined the existing 'pure' geography, however. The humanists have pointed to its mechanistic view of the individual and the resulting neglect of the truly 'human' aspects of our species. The radicals have condemned the failure to come to terms with power relations in capitalist society. The result is that the 'pure' geography upon which the current growth of applied geography is based is discredited but with only minor overall effects. The second 'pure' geography continues to underlay most activities of modern geographers who ignore the methodological controversies going on around them. It seems that there may well be parallels here with the late nineteenth-century methodological debates. From this previous period of applied geographies the political, commercial and colonial geographies are now largely forgotten but the methodological debates are remembered. This suggests that if we wish to look forward to a third 'pure' geography we should look to the current philosophical issues and not the substantive applied geographies.

A THIRD 'PURE' GEOGRAPHY?

Our previous discussion of cycles of applied and pure geography was roughly based on the identification of economic (Kondratieff) cycles. If the capitalist world-economy can sort out its current problems we can expect another upturn of economic growth in the 1990s. As pressure on public expenditure relaxes and academia can be more self-reflective the time will be ripe for a new 'pure' geography to emerge in the early twenty-first century. But things need not be that simple. The existence of 'pure' disciplines has always required the existence of a relatively autonomous and prestigious university system. Each discipline has operated as a hierarchical social system with research in the form of published papers being the 'currency' for advancement. In this process 'pure' research is generally held in higher esteem than applied research. In the current recession, however, we are beginning to see major changes in this process. The autonomy of universities is being further eroded as business schools are being imposed on universities just like geography was a century ago but on a far larger scale. Not only is government moulding academia to its own ends but the nature of the research process is changing. Research is becoming commodified as project work for outside bodies grows. Individual evaluation is beginning to be in terms of how much contract money can be attracted as selected bits of knowledge are sold in the market place. Although often carried out in explicitly 'multi-disciplinary' research institutes the research is very narrowly focused. Since these research institutions are separate from teaching functions they have no need to be within higher education and may well become part of the growing private sector of research and development. In the future therefore we may expect universities to incorporate less of the total research activity in a country and for that work to be evaluated in new ways. The problem, however, is to separate the particular pressures of the current recession from the longer-term trends. Universities, and geography within them, will survive for one of several reasons. The many new researchers of the new information-based world will require a training and education only available in the universities. The question is the degree to which this teaching function will be separated from research and thus break the concept of the university developed in nineteenth-century Germany. This is where we return to universities as the source of the theoretical foundations of disciplines which legitimize applied work. Only intellectual workers in the university sector will have the opportunity of developing 'pure' research ideas. The prestige attached to this work will return with the relaxation of economy cuts. Hence I do think that a third 'pure' geography will emerge in a relatively smaller but still prestigious university sector in the information-based world after the 1990s.

What will be the nature of this third 'pure' geography? One thing we can be sure of is that it will not be like its immediate predecessor. The theoretical basis of the second 'pure' geography had been fundamentally eroded so that there can be no return to the optimistic simplicity of the spatial school. That is not to say that many of the attributes of the recent past will not remain. The technical aspects will continue to be valued and be supported. But minimal contemplation will soon show that there is little reason why spatial analysts should not be transformed into systems analysts with little or no disturbance to their function. Spatial analysis opens up a road to the disappearance of geography and so the new generation of 'pure' geographers will be at pains to distance themselves from past excesses of 'spatial separatism'. The techniques may remain but the theory will have to be ditched.

If we cannot return to the optimism of positivism neither can we expect a rebirth of regional synthesis. And yet the latter's holistic approach is curiously closer to both the humanistic and radical positions in current debates. They both reject the spatial school's specialization and both provide a much broader view of the geographical perspective. The humanistic approach can find sympathetic examples in the former regional school, especially its French variant as we have noted, and the radical school is based on Marx's critique of classical political economy and therefore shares the holistic assumptions of early nineteenth-century social enquiry like the first 'pure' geography. But obviously geography cannot be returned realistically to philosophical positions formed before industrialization. Nevertheless we can expect a more holistic emphasis in the third 'pure' geography. And this need not be against societal trends like the earlier holistic geography. The specialization resulting from the changing division of labour in capitalism is reaching a crucial stage. The rate of change is now such that it can no longer be guaranteed that a set of specialized skills will remain useful for a working lifetime. The retraining of teachers is only the tip of a potential intellectual crisis. In this situation we can expect the attractions of specialization to diminish in the intellectual division of labour and more general training will become acceptable. Geography will be in an unique position to prosper from such a trend as an intellectually respectable alternative to *ad hoc* interdisciplinary arrangements.

Of course the two critiques of positivism will not be equally attractive as the basis of the third 'pure' geography. Although both present anti-capitalist positions in their view of the world the humanist approach is much less threatening due to its lack of a political programme. Hence we can expect the humanist critique to form the basis of a third 'pure' geography. In fact hermeneutic understanding can be viewed as an ideal training for middle management ('keeping the troops happy'), working under a top management of positivist economists maximizing profits. In the affluent 'first world' this need not appear to be at all repressive. New economic growth and shorter work-period concessions will, we are constantly being told, lead to a new 'leisure society'. Again geography will be an ideal medium this time for a new type of traveller – the educated tourist. This will, of course, be diversionary from the real nature of the society with the poverty of the Third World under-theorized. In fact watered-down elements of radical geography will have to be wheeled in to deal with the less affluent areas of the world. The third 'pure' geography will perforce consist of a dialogue between idealist and materialist philosophies. We can already see the beginning of this process in, for instance, Gould and Olsson (1982). I think their title *A Search for Common Ground* can be translated as 'a search for the third "pure" geography'.

WHITHER THE RADICAL CRITIQUE?

The radical critique is the most interesting for the future of geography because it seems to have firmly established a Marxist position within modern geography. Past radical geographers were unsuccessful (Blaut 1979). Reclus and Kropotkin were able to merge their anarchism with geography's holism but did not create a distinctive school of geography. In a later generation the more moderate Lattimore was hounded out of academia during the McCarthy witch hunts (Newman 1983). In the last generation, however, the radicalism generated in the USA by the civil rights campaign and the Vietnam war has produced a recognizable school of radical geography (Peet 1977). The nature of this geography varies among its producers. Bunge, for instance, recognizes the existence of geography as a given segment of knowledge, 'the geography' no less, and uses its concepts for his Marxist ends (Bunge 1973, 1981). In contrast Slater (1978) and Eliot Hurst (1980) insist that since geography as a discipline is a product of capitalism we cannot be expected to work within its anti-revolutionary confines. Radical students must therefore abandon geography for the comprehensive and holistic field of historical materialism. Between these two positions most radical geographers have attempted to construct a Marxist geography (Peet 1977; Soja 1980; Harvey 1984). The degree to which this should be distinct and separate from other geography has been debated with those who advocate a more eclectic approach to geography than a purely Marxist discipline (Clark and Dear 1978). It is the latter who are most likely to contribute to the third 'pure' geography, of course. The most important point, however, is the fact that the radical critique has established Marxism on the geography agenda. Marxist geography will not be the third 'pure' geography but it has a foothold in our small part of academia; it is tolerated and it will remain. It is one small part of the anti-systemic forces in the capitalist world-economy which have been growing since the middle of the nineteenth century (Wallerstein 1974). Many elements of the radical critique may be co-opted but a dissenting tradition will be kept alive in both practical and technical aspects of geography (Blaut 1979).

For a Marxist the particular form that geography takes is a political decision. The value of a geographical perspective will depend on the interpretation of the stage reached in transformation of capitalist society to socialism. For all Marxists geography's value is determined by its contribution to that transformation and to the prevention of regression to barbarism. If revolution is perceived as being imminent then the superseding of geography by historical materialism may be justified. If capitalism is expected to survive beyond this generation it seems foolish, on the other hand, to give up modern geography with its Marxist foothold. Whether we should spend much time delineating a 'Marxist geography' is problematic however. Since this will not be the next 'pure' geography the effort is probably not worthwhile. It seems to me that the role of Marxists in geography in the future is to consolidate our foothold by showing through our teaching and research the superiority of our theory (Taylor 1982). this is best achieved by 'solving' outstanding problems which contemporary theories in geography cannot adequately handle. We will win debates but winning the discipline is beyond our direct control – it depends on broader societal developments. The great advantage that the Marxist approach gives its practitioners is the ability to synthesize key aspects of the first two 'pure' geographies. I will conclude by briefly outlining such a programme.

My purpose is not to reject past geographies but to interpret them, to point out their limitations and to incorporate them into a broader synthesis. Let us begin in reverse order. The second 'pure' geography's limitations have been documented over and over again in the past decade. What we need to offer is a description of the historically specific spatial framework which the capitalist world-economy has created and continually re-creates. Different definitions of the capitalist world-economy produce alternative starting points but we can all agree on the need for a historical geography of the growth of the bounds of the world-economy and a geography of the reorganization of spaces within the world-economy. The nature of the spatial structure and the nature of the change in the spatial structure in relation to the structural determinants of the world-economy will become fundamental research questions for the study of economic, social, cultural and political topics in geography. In all studies a geographical holistic perspective will prevail so that changes in one part of the world-economy are part of a single process encompassing the world-economy as a whole. Just as every dominant class requires a dominated class so every core requires a periphery. But we will not be producing 'geographic theory'. We may contribute to the third 'pure' geography but the prime task is to provide radical political economy with a meaningful and useful geographical perspective in terms of concrete analyses of the history of spaces and locations.

The first 'pure' geography is in some ways more compatible to Marxist analyses, as I have previously noted. The geography of places will be the terrain colonized by the humanist geographers and it is here that we can expect conflict between the two critiques of positivism. In Marxist analysis the growth of the world-economy by the spread of the law of value has destroyed 'moral economies' throughout the world to produce one market. But the variety of pre-capitalist societies has not been eradicated at other levels. There is a regional geography of the world-economy with different places and their distinctive social formations being incorporated in different ways and being treated subsequently in a wide variety of ways. This uneven development is vital to understanding cultural and ideological reactions in different places. These must be understood because ultimately people can only be mobilized for socialism within their communities and regional geography will be a necessary part of our intellectual tools for that task. In the meantime this will be the intellectual arena in which the limitations of the humanist geographer's third 'pure' geography will be exposed. While they will have a classical geographical scale problem in terms of how to relate these local places to the world-economy rolling over them, in the Marxist analysis there is a geographical holism which unites regions and their experiences to the whole.

In a very small way the above programme will contribute to creating Habermas's emancipatory constituent-interest out of the current dialectic of the practical-communicative and technical-mechanical traditions. The real test, however, will be if it is to have a progressive role in the transformation to socialism. We must ensure that this is so.

REFERENCES

Anderson, J.R. (1979) 'Geographers in government', *Professional Geographer*, 31, 265–70.
Bartels, D. (1982) 'Geography: paradigmatic change or functional recovery? A view from West Germany', in P. Gould and G. Olsson (eds), *A Search for Common Ground*, London, Pion, 24–36.
Ben-David, J. and Zloczower, A. (1962) 'Universities and academic systems in modern societies', *Archives of European Sociology*, 3, 45–84.
Blaut, J.M. (1979) 'The dissenting tradition', *Annals, Association of American Geographers*, 69, 157–64.
Bunge, W. (1962) *Theoretical Geography*, Lund Studies in Geography, Series C, 1.
—— (1973) 'The geography', *Professional Geographer*, 25, 331–7.
—— (1981) *The Nuclear War Atlas*, Victoriaville, Quebec, Society for Human Exploration.
Capel, H. (1981) 'Institutionalization of geography and strategies of change', in D.R. Stoddart (ed.). *Geography, Ideology and Social Concern*, Oxford, Basil Blackwell, 37–69.
Clark, G. and Dear, M. (1978) 'The future of radical geography', *Professional Geographer*, 30, 356–9.
David, T. (1957) 'Against geography', *Universities Quarterly*, 13, 261–73.
Dickinson, R.E. (1969) *The Makers of Modern Geography*, London, Routledge & Kegan Paul.
Eliot Hurst, M.E. (1980) 'Geography, social science and society: towards a de-definition', *Australian Geographical Studies*, 18, 3–21.
Frazier, J.W. (1978) 'On the emergence of an applied geography', *Professional Geographer*, 30, 233–7.
Gould, P. and Olsson, G. (eds) (1982) *A Search for Common Ground*, London, Pion.
Grano, O. (1981) 'External influence and internal change is the development of geography', in D.R. Stoddart (ed.), *Geography, Ideology and Social Concern*, Oxford, Basil Blackwell, 17–36.
Habermas, J. (1972) *Knowlege and Human Interests*, London, Heinemann.
Haggett, P. and Chorley, R.J. (1967) 'Models, paradigms and the new geography', in R.J. Chorley and P. Haggett (eds), *Models in Geography*, London, Methuen.

Hartshorne, R. (1939) *The Nature of Geography*, Washington, DC, Association of American Geographers.
Harvey, D. (1969) *Explanation in Geography*, London, Edward Arnold.
—— (1984) 'On the history and present condition of geography: an historical materialist manifesto', *Professional Geographer*, 36, 1–10.
International Geographical Union (1975) *Orbis Geographicus*, 4th edn, Wiesbaden, Steiner.
James, P.E. (1972) *All Possible Worlds*, Indianapolis, Odyssey.
Johnston, R.J. (1983) *Philosophy and Human Geography*, London, Edward Arnold.
Jumper, S.R. (1984) 'Departmental relationships and images within the university', *Journal of Geography in Higher Education*, 8, 41–8.
McKay, D.V. (1943) 'Colonialism in the French geographical movement, 1871–81', *Geographical Review*, 33, 214–17.
Mackinder, H.J. (1887) 'On the scope and methods of geography', *Proceedings of the Royal Geographical Society*, 9, 141–74.
Mikesell, M.W. (1979) 'Current status', *Professional Geographer*, 31, 358–60.
Moriarty, B.M. (1978) 'Making employers aware of the job skills of geographers: a promotional program', *Professional Geographer*, 30, 315–18.
Morrill, R.L. (1984) 'Recollections of the "quantitative revolution's" early years: the University of Washington 1955–65', in M. Billinge, D. Gregory and R. Martin (eds), *Recollections of a Revolution*, London, Macmillan, 57–72.
Newman, R.P. (1983) 'Lattimore and his enemies', *Antipode*, 15, 12–26.
Peet, R. (1977) 'The development of radical geography in the United States', *Progress in Human Geography*, 1, 64–87.
Research Working Group (1979) 'Cyclical rhythms and secular trends of the capitalist world-economy', *Geographical Review*, 2, 483–500.
Schaefer, F.K. (1953) 'Exceptionalism in geography: a methodological examination', *Annals, Association of American Geographers*, 43, 226–49.
Shaw, M. (1975) *Marxism and Social Science*, London, Pluto Press.
Slater, D. (1978) 'The poverty of modern geographical enquiry', in R. Peet (ed.), *Radical Geography*, London, Methuen, 40–58.
Soja, E. (1980) 'The socio-spatial dialectic', *Annals, Association of American Geographers*, 70, 207–25.
Stutz, F.P. (1980) 'Applied geographic research for state and local government: problems and prospects', *Professional Geographer*, 32, 393–9.
Taylor, P.J. (1976) 'The quantification debate in British geography', *Transactions, Institute of British Geographers*, NS 1, 129–42.
—— (1982) 'The question of theory in political geography', in N. Kliot and S. Waterman (eds), *Pluralism and Political Geography*, London, Croom Helm, 9–18.
Wallerstein, I. (1974) 'The rise and future demise of the capitalist world system', *Comparative Studies in History and Sociology*, 16, 387–418.
Wilbanks, T.J. and Libbee, M. (1979) 'Avoiding the demise of geography in the United States', *Professional Geographer*, 31, 1–7.
Wooldridge, S.W. and East, W.G. (1951) *The Spirt and Purpose of Geography*, London, Hutchinson.

II
PHILOSOPHY AND METHODOLOGY

5
GEOGRAPHY AS A SCIENTIFIC ENTERPRISE
John U. Marshall

> Science, in its most fundamental definition, is a fruitful mode of inquiry, not a list of enticing conclusions. The conclusions are the consequence, not the essence. (Stephen Jay Gould 1984)

The purpose of this essay is to clarify the nature of the scientific method. The term 'scientific method' denotes the *logical structure of the process by which the search for trustworthy knowledge advances*. We are not concerned with the *content* of research but rather with the *pattern of thought* which both accompanies and facilitates the continued development of our understanding. We seek, in short, to elucidate the 'fruitful mode of inquiry' that Gould rightly identifies in our opening quotation as being the essence of the scientific enterprise.

The 'scientific enterprise' just mentioned is to be construed broadly, encompassing all disciplines whose primary task is to explain empirical phenomena. There is no need to argue that geography 'ought to be' a science. Geography simply *is* a science by virtue of the fact that it is a truth-seeking discipline whose raw material consists of empirical observations. There is no suggestion here that geography should undergo any sort of epistemological restructuring. The intention is merely to clarify certain aspects of scientific reasoning which often appear – both inside and outside geography – to be poorly understood.

The plan of the essay is as follows. The first section discusses the problem of induction, a famous (or infamous) philosophical problem which lies at the heart of all rational attempts to make sense of the world. Many philosophers believe that the problem of induction may well be insoluble, although there is at least one way, as we will see, in which it can be outflanked. The second section demonstrates that certain widely held beliefs concerning the distinctive character of the scientific approach are, when viewed in the light of our discussion of induction, properly labelled as misunderstandings or at best as half-truths. The third section, drawing chiefly on the work of Karl Popper, presents an account of scientific methodology in which inductive reasoning, though by no means eliminated, plays a role that is not problematical. In this section Popper's principle of falsifiability is given special attention. The fourth and final section shows, by means of a simple example, how the modern fallibilist approach can be put to good use in geographical research.

THE PROBLEM OF INDUCTION

A pedestrian is standing at a street intersection waiting for the traffic signal to change so that he may legally and safely cross the road. No vehicles are nearby, but a car is approaching from the right. As this car draws near, the traffic signal changes in the pedestrian's favour. Seeing the green light, the man steps confidently into the road. The car fails to stop and the man is struck down and seriously injured.

A few days later, recovering in hospital, our hero is visited by a neighbour who happens to be a lecturer in philosophy. 'I do not understand what happened,' the injured man complains. 'The light was in my favour and yet the motorist hit me.'

'The matter is very straightforward,' the philosopher replies. 'You were simply the victim of your own inductive reasoning.'

Hardly a comforting thought. But the philosopher has a point, for his injured friend evidently took it for granted, when he stepped into the path of the approaching car, that the factors which had protected him from injury in the past would serve him well on this particular occasion. He believed, among other things, that the traffic signal would not show green in both directions at the same time, that the motorist was sober and not bent on doing him an injury, and that the car's brakes were in good working order. All these beliefs, however, had been obtained *by induction*; that is, they were based on the assumption that what had been generally true of traffic signals, motorists, and automobiles in the past would remain true in the future. On the fateful day, unfortunately, this assumption of uniformity proved to be unwarranted.

In formal terms, induction is a type of reasoning in which the premisses do not logically entail the conclusion. Another way of saying this is that induction increases content, or simply that the conclusion goes beyond the available evidence. *Deduction*, in contrast, is a type of reasoning in which the conclusion follows *necessarily* from the given premisses. Deduction, in other words, does not increase content, although one may require intellectual abilities of a high order (as in mathematics) in order to trace all the steps that lead from the premisses of a deductive argument to its conclusion. In everyday speech people often mistakenly say that they 'deduced' a certain conclusion when they really mean that they reasoned inductively. A particularly glaring example of this error is provided by Sherlock Holmes, who delighted in 'deducing' people's occupations, if not their entire life histories, from details such as their dirty fingernails and the colour of their socks. Holmes's 'deductions' on these occasions were not deductions at all, but inductive guesses. One wonders whether Holmes's creator would have passed a first-year examination in logic.

Induction is important to our concern with scientific method because of the role it plays in the formulation of empirical generalizations. These generalizations may conveniently be divided into two types, namely the *summative* and the *extended*. A *summative* empirical generalization describes a property which has been confirmed by the actual observation of all the relevant cases – for example, the statement, 'All the men in this room are bald.' Since the men in question have actually been examined, the generalization does not go beyond the evidence and therefore does not involve inductive reasoning. But consider, to take a standard textbook example, the statement, 'All swans are white.' This is an example of an *extended* empirical generalization – one that goes beyond the evidence on which it is based. We may have examined thousands upon thousands of swans and found them all to be white, yet we cannot *guarantee* the truth of the general statement because it remains possible that a non-white swan may some day be found (as in fact did happen with the discovery of black swans in Australia). In short, we cannot guarantee the truth of an empirical generalization unless, as in the summative case, we can point to *all* the singular instances upon which it is based. Much of the credit for this important insight belongs to the eighteenth-century Scottish philosopher David Hume (1739; 1978 edition, 86–94, 130–42); The problem of induction is often referred to simply as 'Hume's problem' (Magee 1974, 19–22).

At this point it is useful to introduce the distinction between the *idiographic* and the *nomothetic* attitudes towards scholarly work. In brief, the idiographic attitude implies a concern with the uniqueness of individual phenomena or events, whereas the nomothetic approach implies a desire to subsume individual cases under laws or law-like statements of very general, if not universal, applicability (Hartshorne 1961, 378–97). A historian, seeking to come to grips with the causes and consequences of the French Revolution as a specific upheaval which occurred in a particular country at a particular time, may be said to have adopted an idiographic approach. A sociologist, viewing the French Revolution as one political upheaval among many, may well define his goal as that of understanding *revolutionary behaviour in general*. In this case the approach is nomothetic and the investigation is likely to be carried out in a comparative manner which involves a consideration not only of the French experience but also of other revolutions thought to be similar in kind, such as the American, the Russian, the Chinese, and the Cuban.

It is important to recognize that the terms 'idiographic' and 'nomothetic' are not antonyms. They identify attitudes which are distinct from each other but which are by no means mutually incompatible. Indeed, it can be argued that the two approaches are complementary rather than competitive. Nevertheless, many fields of enquiry tend to lean heavily one way or the other. History, for example, remains almost entirely an idiographic discipline despite recent inroads by the cadre of researchers known as cliometricians (i.e. quantitative historians). Physics, in contrast, is strongly nomothetic. In general, the nomothetic attitude tends to be dominant in precisely those fields to which the term 'science' is most commonly applied. It should be noted, however, that physicists study phenomena which, viewed as individual events, are unique, and that historians, like everyone else, are obliged to employ general concepts and categories in order to communicate. There is an important sense in which all disciplines are inevitably both nomothetic and idiographic at the same time.

The link with the problem of induction can now be made clear. General laws, by virtue of their generality, go beyond the evidence of the individual (and unique) cases upon which they are based. Such laws are not merely summative in character, they are *extended* in the sense that they are assumed to apply to specific cases which have *not* been examined. It follows that the nomothetic approach, as a result of its concern to establish general laws, necessarily involves inductive reasoning. Hence, to the extent that the scientific method embodies a commitment to the nomothetic attitude, science is guilty of induction. This conclusion is difficult to reconcile with the commonly held – but, as we will see, largely mistaken – view that the ultimate goal of science is the establishment of a body of knowledge which is both perfectly general and absolutely certain. The implications of Hume's problem for the character of scientific knowledge may be stated succinctly as follows. If induction is permitted, generality is possible but certainty is denied. If induction is forbidden, certainty is attainable but generality, other than summative generality, is lost.

WHAT SCIENCE IS NOT – OR NOT MERELY

We must now examine what may be termed the 'traditional' view of the nature of the scientific method. This traditional view has been widely held in the past and is still held today by many persons unacquainted with modern developments in epistemology (Magee 1974, 18–19). The foregoing discussion of the problem of induction will enable us to see clearly why the traditional view of science must be regarded as inadequate. Once the weaknesses have been exposed, we will be able to construct a more satisfactory account of the process by which trustworthy knowledge actually evolves.

According to the traditional view, a scientist begins an investigation by selecting a particular category of phenomena which happens to be of interest – plants, rocks, electrical phenomena, the planets, and so forth. The scientist then proceeds by carefully building up a body of accurate observations and measurements which describe the properties or 'behaviour' of the phenomena in question. With any luck, other scientists will be engaged in observing and measuring the same category of phenomena, so that the body of pertinent empirical data will grow to a considerable size as time goes by. The scientists constantly sift and sort their observational results, arranging them into classes and hoping to find a pattern of recurring relationships among measurements of different kinds. At length a regularity emerges. Whenever possible, this regularity is expressed in the form of a mathematical equation; in many cases this is a regression equation, summarizing the joint behaviour of two sets of measurements which are closely correlated with each other. The scientists then continue their empirical work with the aim of assembling such a large number of corroborating observations that the validity of the equation is no longer in doubt. At this point the established relationship is granted the status of a 'law of nature'. The desire to discover the laws of nature is seen as the fundamental motivation behind all scientific research.

Once a law of nature has been discovered and properly verified, it can be put to work (so runs the traditional account) in the formulation of deductive-nomological explanations of particular events. A deductive-nomological explanation, also known as a 'covering law' explanation, takes the following form. The premisses consist of (a) one or more laws of nature and (b) a list of specific initial conditions or circumstances which describe the setting of the event to be explained. From these premisses the occurrence of the event in question can be inferred *by a strictly deductive chain of reasoning*. In other words, given the truth of the premisses, the occurrence of the event to be explained is inevitable. On this view of the nature of explanation, the criterion of success in formulating a satisfactory explanation of an event is the ability to show, with the aid of established laws of nature, that the event in question could not possibly have failed to occur. If the explanatory argument were not deductive in form, the possibility would exist that the event might *not* have occurred, in which case the proposed explanation would have to be judged incomplete.

We noted that the premisses of a deductive-nomological explanation are of two kinds, namely (a) laws of nature and (b) specific initial conditions. In principle, the specific initial conditions can be made the target of further scientific data-gathering in the hope that they, in their turn, will be subsumed under new general laws. It is assumed that the final outcome of this process will be the discovery and verification of a set of natural laws from which *all* events can be rigorously deduced. When that point is reached, science will have achieved its ultimate goal.

Now it must be acknowledged that the traditional view of science which we have just described is not *totally* lacking in value. It provides a reasonable account of *some* of the features which characterize the day-to-day activities of the scientific community. Scientists *do* spend a lot of time making observations and measurements and they *do* look for regularities in their data. There *are* such things as deductive-nomological explanations and at least *some* scientists subscribe to the brand of determinism implicit in the belief that all events can be subsumed under natural laws. These features, however, have more to do with the outward appearance of scientific research than with its internal logical structure. Viewed as an account of the process by which knowledge advances, the traditional view of science leaves much to be desired. Its basic shortcoming is its reliance upon the inductive leap of faith which connects a mass of observations of individual cases to the statement of a general law of nature. As we saw in the preceding section, empirical generalizations which are extended beyond the singular instances on which they are based can never be known with logical certainty to be true. This being the case, we can never know whether or not a deductive-nomological explanation is correct, since the premisses of such an explanation contain general statements (laws) whose truth or falsity cannot be determined.

The same difficulty arises in connection with another type of activity that is often thought to be a hallmark of science, namely the making of predictions. In the traditional view, it is possible accurately to predict an event which has not yet taken place provided that we can specify (a) the laws of nature which govern the occurrence of events of the type in question and (b) the initial conditions which describe the environment within which the event may or may not be about to take place. If, according to purely deductive reasoning, these laws and initial conditions logically entail the event, then the occurrence may confidently be predicted.

It may be noted that the general form of the above predictive argument is identical to that of a deductive-nomological explanation. In both cases a strictly deductive conclusion is derived from premises composed of general laws and initial conditions. This identity of form lies behind the common view that prediction and explanation are essentially the same thing: the ability to predict implies the ability to explain, and vice versa. The presence of a deductive component in each case tends to create the impression that both the explanations and the predictions rest upon secure logical foundations. But the fact of the matter is, of course, that the deductive component is only one link in the complete chain of reasoning. A chain is only as strong as its weakest link. Behind each deduction lies a general law (or laws), and behind each general law lurks the *inductive* inference which brought it into existence. From the standpoint of logic it is irrelevant that many predictions made on the basis of the deductive-nomological model in various fields of science and technology have turned out to be correct. Even if *every* prediction were successful, the fact that induction involves a leap of faith means that we would still have no valid basis for justifying our belief in the truth of the relevant general laws.

The formal identity of explanation and prediction applies only to explanations of the deductive-nomological type. Other kinds of explanations exist, such as functional, dispositional, and evolutionary explanations (Nagel 1961; Passmore 1962). The general topic of the nature of explanation gives rise to a number of important philosophical issues which lie beyond the scope of this essay – notably the question of the logical status of the concept of causality. It needs to be stressed, however, that it is misleading to equate explanation with prediction, or to regard the latter as the acid test of the former. Explanation applies to events which have already taken place and whose character can therefore be apprehended in various ways. Prediction applies to events which have yet to occur and whose character is logically unknowable.

The view of knowledge outlined above is known to philosophers as *positivism*. It is a view which was favoured by such nineteenth-century thinkers as Auguste Comte and John Stuart Mill and which reached its fullest expression in the work of the Vienna Circle during the period from 1922 to 1936 (Kraft 1953; Feigl 1968; Achinstein and Barker 1969). On the metaphysical level, positivism can be contrasted with various so-called 'humanistic' approaches to understanding, such as the idealism of Collingwood and the phenomenology of Husserl (Johnston 1983, 11–86). (It can be argued that 'humanistic' is an inappropriate label here, for 'humanism' was originally a denial of the existence of the supernatural;

in this respect, positivist writers are generally more 'humanistic' than both idealists and phenomenologists.) A more significant contrast, however, and one which has attracted considerable attention among philosophers during the past twenty years, lies on the epistemological level and stems from the problem of reformulating the positivist conception of knowledge in such a way that the role of induction becomes less troublesome. The contrast in question is that between positivism and *fallibilism* (also known as 'falsificationism' and 'critical rationalism'). The fallibilist point of view is the subject of the following section.

THE FALLIBILIST APPROACH

The positivist view, described above, was the joint product of many minds. The fallibilist position, in contrast, is largely the creation of a single individual, namely Sir Karl Popper. As a young teacher in Vienna during the 1920s Popper became familiar at first hand with the ideas put forward by the members of the Vienna Circle. But he found himself disagreeing with much of the positivist doctrine, and he resolved to undertake the task of reformulation. Popper's critique of positivism developed into the philosophy of fallibilism as we know it today (Popper 1934, 1959, 1962, 1972; Magee 1974; Levinson 1982; Marshall 1982).

The starting-point of the fallibilist conception of man's search for trustworthy knowledge lies in the realization that there is an asymmetrical relationship between the logical status of a general law and that of its negation. Consider once again the classic case of the swans. As we have seen, we are unable to show conclusively, on the basis of empirical evidence, that the statement 'All swans are white' is true. However, we *can* prove the truth of the *negative* general statement, '*Not* all swans are white.' The truth of this latter statement can be demonstrated by the observation of a single non-white swan. Putting the matter another way, a general law is never conclusively verifiable but always conclusively falsifiable. Thus, viewed as strategies for assessing the validity of inductively established general statements, verification and falsification are not merely the opposite sides of a single coin. Falsification is possible but verification is not.

Popper's development of the principle of falsifiability led him to reject the traditional positivist view that science begins with the collection of observational data and proceeds inductively to the establishment of general laws. In place of this view he suggested that a scientific investigation begins *with a problem*. Once the problem has been clearly specified, the next step is the introduction of a *theory* which, if correct, will have the effect of solving the problem – that is the theory will provide a satisfactory explanation of the phenomenon under investigation. But the theory must not only be capable of solving the problem in an abstract sense; it must also, if it is to qualify as a 'scientific' theory, be *empirically testable*. This means that the theory must lead to factual propositions which are capable of being disconfirmed by observational data. If the appropriate observations do not falsify any of the factual propositions, the underlying theory can be regarded as being 'acceptable' in the guarded sense that we have no empirical basis for doubting its validity. But if any of the factual propositions turn out to be false, the theory must be modified or, in extreme cases, replaced by a different theory. The new or revised theory then becomes the basis of a new series of testable propositions, and progress thus continues towards an acceptable solution of the original problem (Marshall 1982, 92–107).

We have noted that the theory, in addition to being empirically testable, must be 'capable of solving the problem'. This means that the theory must incorporate a plausible *causal mechanism* (or a series of such mechanisms). Now causality is a non-empirical concept in the sense that the connections between causes and their effects are not observable or measurable. Causality is not a species of 'data' but an explanatory framework imposed upon the data by the observer. It follows that theories do not present themselves *spontaneously* to the investigator. They do not emerge 'naturally' from the data, as inductivists claim to believe; they have to be *invented*. As one Nobel laureate has put it: 'There is a widespread impression that scientists are people who deal only in facts, who don't exercise their imaginations. That, of course, is a quite mistaken view, because the *generative* act in science is an *imaginative* act' (Medawar 1984, 106, emphasis added). The fallibilist, in sharp contrast to the inductivist, regards theories as being logically prior to observations. Without some sort of theory, no matter how poorly articulated, to guide our approach to a problem, we have no rational basis for deciding what experiments to perform or what kinds of data to collect.

To be 'scientific', in the eyes of the fallibilist, is to recognize explicitly that knowledge is approached by means of a process of *conjectures and refutations* (Popper 1962). Our theories – those hypothetical creations that we put forward in order to explain the world – represent the conjectural side of the scientific method. We test each theory by deliberately attempting to refute it. If we fail, the theory's lease on life is extended (though never permanently). If we succeed, modification of the theory is indicated. Progress therefore depends upon the falsifiability of our theories. Note, however, that being falsifiable is not at all the same thing as actually being false. The difference between being falsifiable and being false is like the difference between being breakable and being broken. The point is simply that a theory (or hypothesis, or explanation) is not regarded as 'scientific' in character unless we can specify some conceivable empirical observations which would contradict it.

It is extremely important to realize that in actual practice we normally do not reject a theory *in toto* on the basis of a single refutation, even if the reliability of the empirical evidence seems to be unquestionable. The main reason for our hesitation is that it is extraordinarily difficult – and perhaps even, in principle, utterly impossible – to test any single theory in isolation. Theories can be separated from one another for the purposes of discussion, but in any empirical context their observable consequences occur simultaneously. We are therefore skating on thin ice if we declare unequivocally that certain empirical results are completely explained by one particular theory. Conversely, we can always evade an apparent refutation by asserting that unknown or 'external' factors have come into play and that the theory under consideration has not been given a fair trial. Operational difficulties of this kind account for the fact that, in general, refutations must *accumulate*, and must take place in a variety of settings, before a theory will be abandoned. In fact, even a theory that is widely acknowledged to be unsatisfactory will normally survive until a superior alternative theory is proposed (Lakatos 1970, 1978). Nevertheless there is general agreement among epistemologists that the fallibilist approach expresses an insight which is fundamental to an understanding of the true character of the scientific method. As Kitcher has recently remarked, 'there is surely something right in the idea that a science can succeed only if it can fail' (Kitcher 1982, 45).

We must now consider the role played by inductive reasoning in the fallibilist conception of the quest for knowledge. Putting the matter briefly, induction is involved in the development of conjectures (theories, hypotheses) but not in the search for refutations. Induction is involved in theory-building by virtue of the fact that theories incorporate (often implicitly) laws or law-like statements that extend beyond the evidence upon which they are based. For the fallibilist, however, law-like statements and the theories in which they become embedded are not the hard core of science; they are never anything more than heuristic speculations. The distinguishing feature of the scientific attitude is the willingness – indeed, the desire – to confront these speculations with pertinent empirical observations, thereby exposing them to the possibility of refutation. And a refutation, when it occurs, is a logically conclusive result in which inductive reasoning plays no part. For the fallibilist, therefore, a scientific enquiry has two distinct phases—an imaginative, hypothetical, theory-proposing phase and a critical, objective, experimental phase. Inductive inferences may be involved in the former, but they are strictly banned from the latter (Marshall 1982, 149–55).

In the traditional positivist view of science, general laws established by inductive reasoning are not regarded as speculations but as *verified scientific knowledge*. For the positivist, induction is the very foundation of science. For the fallibilist, induction leads only to conjectures, and science does not begin in earnest until these conjectures are confronted by the threat of falsification.

In bringing our all-too-brief excursion into philosophy to a close, it is appropriate to mention the fact that the very real and important differences which separate fallibilism from positivism are frequently ignored or misrepresented, even by writers who have evidently studied the relevant philosophical texts. In some cases Popper is actually classified as a positivist! One source of this troublesome error appears to be an influential article on historical understanding by Donagan (1964) in which Popper is incorrectly identified as a proponent of the idea that historians should rely on deductive-nomological explanations. Donagan's misunderstanding of Popper needs to be rectified, not only because historians have generally rejected the covering law concept as being unacceptably deterministic, but also because geographers interested in learning about historical explanation have turned to Donagan, among others, for enlightenment. This may help to explain why geographers have tended to dismiss Popper as being just another positivist. It is to be hoped that this misconception will now be corrected. For as Magee has observed:

> The truth is that Popper was never a positivist of any kind; quite the reverse, he was the decisive anti-positivist, the man who put forward from the beginning the arguments that led (after an excessively long time) to logical positivism's dissolution. (Magee 1974, 49)

A BRIEF GEOGRAPHICAL EXAMPLE

In this final section we will attempt to convey an impression of the manner in which the fallibilist approach can be used to clarify the internal logic of a specific piece of research. It is well to acknowledge immediately that the example chosen – an investigation of population density within a Canadian province – is quite elementary. But this is deliberate, since we wish to focus on the *logical structure* of the enquiry and not on its substantive content.

As indicated above, the investigation should be conceived as proceeding according to a definite series of steps, as follows:

1 statement of the *problem*;
2 formulation of an appropriate *theory*;
3 derivation of *factual propositions*;
4 carrying out of *empirical tests*; and
5 *evaluation* of results.

We will consider each of these steps in turn.

The manner in which a researcher arrives at a *problem* lies outside the scope of the scientific method as such and is strongly influenced by personal tastes and preferences. The normal practice, understandably enough, is to select a problem directly related to a current focus of interest within one's discipline. Originality is an asset, but too much originality can lead to frustration, especially if the chosen problem turns out to be highly intractable. As Medawar has noted, science is not (or should not be) an intellectual form of masochism; rather it is 'the art of the soluble' (Medawar 1967). A problem will attract little interest unless there is reason to suppose that progress can be made towards its solution.

For our working example we have selected the problem of explaining the spatial variation of population density within the province of Ontario, Canada. According to the 1981 census, Ontario contains 8.6 million inhabitants on an area of 917,000 square kilometres, giving an average population density of 9.4 persons per square kilometre for the province as a whole (Canada 1982). But the density varies considerably from one part of the province to another. Even if we exclude the two census districts which contain the Toronto metropolitan area, population density in the remaining fifty-one administrative units ranges from less than 1 to more than 300 per square kilometre. In general, densities are high in the southern part of the province and low in the north. It is this broad contrast between north and south that we seek to explain.

The next step in the logical sequence is to introduce a *theory* capable of accounting for the observed spatial pattern. Here we can draw not only upon our own imaginations but also upon the accumulated writings of the past. Indeed, we inherit so much from the past in the world of scholarship that in most instances we could not 'start from scratch' even if we wanted to. In the present case, one theory which comes readily to mind is that population density varies from place to place as a result of differences in the character of the natural environment. Drawing upon what we have been taught concerning other parts of the world, and perhaps also upon what we have observed at first hand while travelling, we put forward the conjecture that population density in Ontario is low where the natural environment is harsh and high where relatively favourable conditions prevail. This theoretical conjecture will guide the subsequent empirical portion of the research.

The *causal mechanism* in the proposed theory arises from the idea that population density is fundamentally a function of the number of opportunities available for earning a livelihood. In particular, areas capable of producing an agricultural surplus should be more densely populated than areas in which farming is difficult or impossible. The theory can be further developed by noting that areas of productive farmland also support market towns, and that these same areas, being well peopled, are more likely to attract manufacturing industries than districts where the population is sparse. The growth of towns and manufacturing will lead to an even higher level of population density within the favoured areas, thereby increasing the contrast between densely and sparsely inhabited regions.

Given the fact that the above theory attributes causal power to the natural environment, it is appropriate to note in passing that there is no intent to deny the operation of human free will. We are not, in other words, advocating environmental determinism. Indeed, a consistent, comprehensive determinism of any kind, environmental or otherwise, is by definition incapable of being falsified and therefore lies outside the domain of scientific enquiry. If one wishes to make freedom of the will explicit, the theory can be phrased in such a way that the inhabitants of Ontario are described as making free and rational choices based upon self-interest, the availability of pertinent information, the futility of farming on barren land, and so forth. The natural environment does not have causal power in the sense of being an active agent in its own right but only in the sense that it enters human consciousness and becomes a factor in the making of decisions.

It is not difficult to show that natural environmental conditions in Ontario become progressively harsher as one moves from south to north. In large part, of course, this is simply a matter of climate, but surface features also play a significant role. Most of northern Ontario lies on the Canadian Shield, an extensive area of Precambrian rock which was largely stripped bare of soil by the action of continental ice sheets during the Pleistocene glaciation. Southern Ontario, in contrast, was a region of glacial deposition rather than erosion, and here the soils are capable of supporting prosperous farms. The line separating the zone of glacial erosion from the zone of deposition corresponds closely to the southern limit of the Precambrian rocks, and this boundary has become the conventional dividing line between south and north. For residents of the south, being 'on the Shield' and being 'up north' are virtually synonymous.

From this thumbnail sketch of Ontario's physical geography we can proceed directly to the factual proposition that population density will be higher than the provincial average in those administrative areas which lie wholly or largely to the south of the southern limit of the Shield, and lower than the provincial average to the north of this line. Note that our ability to formulate this proposition depends not only upon our factual knowledge of environmental conditions in Ontario *but also upon our theory*. Indeed, since the theory, at least implicitly, is *general* in character, the testable proposition can be obtained *as a deductive consequence*, although in practice it is rare for geographers to acknowledge, or even to recognize, that their theories incorporate implicit general laws.

We may now quickly deal with the facts. Of the forty administrative areas classed as 'southern', only one – Manitoulin Island – has a population density *below* the provincial average. Conversely, of the thirteen areas classed as 'northern', only two – the resort area of Muskoka and the mining district of the Sudbury Basin – have densities *above* the provincial average. Thus, fifty out of fifty-three administrative areas are 'well behaved' in the sense that their population densities are in accord with our theoretical expectations.

There is no denying that these are encouraging results. From the fallibilist point of view, however, the significant feature is not the fact that fifty out of fifty-three areas are well behaved but the fact that the three exceptional cases *falsify the theory in its initial form*. What is now required, in principle, is the modification of the theory in such a manner that the anomalies are satisfactorily explained. In the case of Ontario this might be accomplished by taking account of relationships between population density and such factors as the presence of mining and tourism. The important point is that it is the anomalies, not the well-behaved areas, that point the way towards a deeper understanding of the topic under investigation.

In this and in many other examples which could be cited, it is apparent that knowledge evolves through the *interplay* between theory and observation. Theory on its own is little more than a form of fantasy. Observations in the absence of theory are simply disconnected facts, a 'mere heap' of unrelated bits of data. Only when theory and observation are consciously interrelated within a fallibilist framework can we begin to assemble a meaningful and coherent picture of the structure of the world.

CONCLUSION

The above example illustrates, albeit in rudimentary form, the logical structure of the fallibilist procedure as applied to a specific empirical problem. The same procedure of conjecture and refutation is present, explicitly or otherwise, in all investigations worthy of the label 'scientific'. Inductive reasoning, putatively general laws, and deductive-nomological arguments play important roles both in the study of natural phenomena and in the study of human affairs, but in *neither* case do they represent sufficient conditions for the existence of a science. In the final analysis, the essence of the scientific approach is a readiness to expose our theories to the risk of empirical disconfirmation. Only by our willingness to fail can we hope to make progress towards the truth.

REFERENCES

Achinstein, P. and Barker, S.F. (eds) (1969) *The Legacy of Logical Positivism*, Baltimore, Johns Hopkins Press.
Canada, Statistics Canada (1982) *1981 Census of Canada*, vol. 2, Provincial Series, Catalogue No. 93–906, Ontario. Ottawa, Ministry of Supply and Services.
Donagan, A. (1964) 'Historical explanation: the Popper–Hempel theory reconsidered', *History and Theory*, 4 (1), 3–26.
Feigl, H. (1968) 'The *Wiener Kreis* in America', in D. Fleming and B. Bailyn (eds), *The Intellectual Migration: Europe and America, 1930–1960*, Cambridge, Mass., Harvard University Press, 630–73.
Gould, S.J. (1984) 'Sex, drugs, disasters, and the extinction of dinosaurs', *Discover*, 5 (3), March 1984, 67–72.
Hartshorne, R. (1961) *The Nature of Geography: A Critical Survey of Current Thought in the Light of the Past*, 2nd edn, Lancaster, Pennsylvania, Association of American Geographers.
Hume, D. (1739) *A Treatise of Human Nature*, ed. L.A. Selby-Bigge, 2nd edn, 1978, London, Oxford University Press.
Johnston, R.J. (1983) *Philosophy and Human Geography: An Introduction to Contemporary Approaches*, London, Edward Arnold.
Kitcher, P. (1982) *Abusing Science: The Case Against Creationism*, Cambridge, Mass., MIT Press.
Kraft, V. (1953) *The Vienna Circle: The Origins of Neo-Positivism*, New York, Philosophical Library.
Lakatos, I. (1970) 'Falsification and the methodology of scientific research programmes', in I. Lakatos and A. Musgrave (eds), *Criticism and the Growth of Knowledge*, Cambridge, Cambridge University Press, 91–195.
—— (1978) *The Methodology of Scientific Research Programmes*, Philosophical Papers, vol. 1, Cambridge, Cambridge University Press.
Levinson, P. (ed.) (1982) *In Pursuit of Truth: Essays on the Philosophy of Karl Popper on the Occasion of His 80th Birthday*, Atlantic Highlands, NJ, Humanities Press.
Magee, B. (1974) *Popper*, London, Woburn Press.
Marshall, J.U. (1982) 'Geography and critical rationalism', in J.D. Wood (ed.), *Rethinking Geographical Inquiry*, Geographical Monographs No. 11, Toronto, Department of Geography, Atkinson College, York University, 75–171.

Medawar, P.B. (1967) *The Art of the Soluble*, London, Methuen.
—— (1984) 'Interview: Sir Peter Medawar', *Omni*, 6 (4), January 1984, 106.
Nagel, E. (1961) *The Structure of Science: Problems in the Logic of Scientific Explanation*, New York, Harcourt, Brace & World.
Passmore, J. (1962) 'Explanation in everyday life, in science, and in history', *History and Theory*, 2 (2), 105–23.
Popper, K.R. (1934) *Logic der Forschung*, Vienna, Julius Springer Verlag.
—— (1959) *The Logic of Scientific Discovery*, New York, Basic Books.
—— (1962) *Conjectures and Refutations: The Growth of Scientific Knowledge*, New York, Basic Books.
—— (1972) *Objective Knowledge: An Evolutionary Approach*, London, Oxford University Press.

SCIENTIFIC METHOD IN GEOGRAPHY[1]

Alan Hay

In writing about scientific method in geography one is addressing four groups of readers. The first consists mainly of physical geographers who believe that their disciplince is a field of natural science and therefore do not doubt that scientific method is appropriate. In the second are those human geographers who see scientific methods as being appropriate to their discipline as a social science, although they may also recognize that such an application poses certain problems not encountered in classical natural sciences like physics and chemistry. The third group consists of those, mainly human geographers, who believe that the subject matter of geography makes scientific or quasi-scientific methodology inappropriate. Most recently a fourth group has emerged which seeks to apply Marxist methods in geography and believes that such methods are scientific although not in the mould of classical natural sciences. In order to understand these differing views it is necessary to identify the main elements of scientific thinking and practice, to outline some of the philosophical problems in scientific method not always evident to scientists themselves and to examine some additional issues which arise when scientific method is applied to geography and similar disciplines. These topics constitute the first four main sections of this essay, while the final main section offers a modified version of scientific method which the author believes is, and will be, applicable in both physical and human geography.

SOME KEY ELEMENTS IN SCIENTIFIC THINKING

Scientific work is often characterized by the type of thinking used and four key elements can be identified: theory, law, logic and reduction; the meaning of these terms will now be examined.

Theory refers to an organized and coherent body of assumptions and arguments. It may be directed to the explanation of a unique phenomenon (Wegener's theory of continental drift had only one world to consider) or a whole class of phenomena (the theory of air masses). Natural sciences frequently use theories (in the plural) because two or more rival theories may offer competing explanations of a single class of phenomena, and because parallel but as yet unconnected theories may be used to account for different phenomena. There is therefore no *requirement* for a single overarching general theory but there is an underlying predisposition to such theory for two reasons. First, if such a grand theory can be proposed and existing sub-theories be seen as special cases of it the maximum explanation is achieved for the minimum of theory. Second, grand theory is attractive because in interrelating sub-theories it reflects the interrelationships of phenomena evident in the world as perceived. The theory of continental drift is a good example of a grand theory which accounts for a wide range of distinct but yet related phenomena (the matching shapes of the continents; the geological, floral and faunal distributions).

Implicit in this brief description of theory is the second key element in scientific thinking – *law*. Any fully developed scientific theory contains embedded within it certain statements about unvarying relationships (for example the power of a base to neutralize an acid, or the relationships between the pressure, volume and temperature of a mass of gas). These laws may be evident at the level of everyday experience (for example the laws of gravity) or only at the level of scientific investigation, for example by controlled experiment or microscopic investigation. As with theories there is a predisposition among scientists to seek laws which cover broad categories of phenomena: the claim that an acid will be neutralized by a base is more general than the claim that sulphuric acid is neutralized by sodium hydroxide.

There is also a preference within science for deterministic laws: that *wherever* A and B are present C *will* result. But it is recognized that some laws have a probabilistic form even if they represent a transient stage in the development of the discipline and will give way to deterministic laws as the discipline develops (Braithwaite 1953).

But *laws* are not the only type of connecting statement used in scientific theory – indeed a theory which consists entirely of laws based on experience and experiment is viewed with disfavour; it is seen as more satisfactory if most of the laws and other links in the theory can be shown to be *logically* derived from a much smaller number of fundamental assumptions and laws. Scientists have tended to use mathematics (algebra and geometry) as the language for expressing and developing this logic, but other abstract languages are also used (for example chemical equations and bonding diagrams). The attachment to such abstract languages is based on three beliefs: first that they are more powerful in that the general form of a relationship is more likely to be apparent in an abstract representation; second, the rules of algebra, geometry, chemical equations, etc., can be specified and then followed more closely than corresponding rules for statements in everyday language; third it is believed that such logics are less likely to be affected by relational shades of meaning which are common in everyday languages and create difficulties in translation between them.

The final key in much scientific thought is *reduction* – the idea that the laws and theories of a discipline can be re-expressed as special cases of the outworking of the laws of a more fundamental discipline. So for example it can be argued that meteorology is a special case of physics, weathering is an example of applied chemistry, and some authors believe all social science is reducible to psychology. Reduction is often accompanied (though this is not logically essential) by a tendency to seek explanation by small-scale studies (Hay and Johnston 1983). For example, the explanation of chemical reactions in terms of molecules and atoms, beyond that in terms of neutrons and electrons, is both reductionist and disaggregate. In geography similar reductionism leads from soil geography to soil chemistry and hence to studies of molecular behaviour and from urban geography to the behaviour of households and individuals and hence into psychology.

These four elements – theory, law, logic and reduction – are thus key parts of scientific thinking. There is however a fifth element – the *research hypothesis* – which provides a link to the area of scientific practice. In a well-developed natural science a research hypothesis predicts the outcome of an experiment or observation if the theory is correct. In this way a theory or its extensions can be tested in contexts other than those for which it was originally devised.

SOME KEY ELEMENTS IN SCIENTIFIC PRACTICE

It is difficult to identify the best term for the first such element because all words commonly used carry with them specific and contentious overtones – but the word *test* seems to be used most often. A scientific theory needs to be tested or assessed not only for its own internal consistency, but also for its consistency with the world as observed. Such tests may be focused upon assumptions (are the specified assumptions of the theory realistic?), to test laws (are the specified relationships observable?), and most often to test hypotheses (is the outcome postulated by the hypothesis observable?). In the popular conception of science these tests are most commonly carried out by laboratory experiments in which sealed vessels and pure materials ensure that effects are not suppressed or spuriously created by undetected contamination. One of the most dramatic tests of a theory occurs when it successfully predicts the outcome of an experiment or observation never before conducted (Gardner 1982). This provides the basis for the common idea that a second element – successful *prediction* – is the hallmark of science. But prediction has two other aspects: first there is a logical sense that if there is a valid explanation for C – 'because A and B then C' – it can also be used to predict 'if A and B then C'; second successful prediction can be a useful practical tool either in unconditional (the sun *will* set at 18.30 hours) or in conditional form (*if* you apply Y tonnes of fertilizer D per hectare *then* yield will increase by x per cent).

Such tests often involve a third element in scientific practice – *measurement*. This arises first in the context of description because important features of the natural world arise from differences of degree (consider climatic temperatures) and cannot be handled well by qualitative categories ('hot' and 'cold' carry less useful information than temperatures in degrees Celsius). But the problem is even more acute in testing because some laws and theories yield hypotheses which are quantitative in form. A considerable amount of scientific effort has therefore been devoted to developing methods of measuring the characteristics of phenomena, an effort which has been paralleled in geographical efforts to measure evapotranspiration, soil fertility or the level of economic development.

The development of measurement is closely linked with questions of *statistical inference*. It was recognized at an early stage that even experimental results include errors arising from the observer, from instrument calibration and from small disturbances or contaminations of the experiment. Much early statistical work was related to inference in the presence of errors. The use of statistics to infer population characteristics from a sample of observations (in which geographers have shown much interest) also became important. (The seven elements reviewed here would be recognized by most scientists as part of scientific thought and practice. One key word missing is *model*. This omission is deliberate because although the word is common in the literature of science and geography its meaning is extremely varied and vague (Chao 1962).)

Underlying these elements in scientific thought and practice there is often an implicit attitude known as *scientism*: the belief that scientific methods are *the only* methods of acquiring real knowledge. The consequences of this belief are dangerous in two ways. On one hand a devotee of scientism may attempt to apply scientific methods in inappropriate contexts and end up by redefining concepts (e.g. the beauty of a landscape) in terms of much more limited scientific measurement. The other scientistic response is to exclude such concepts from all discussion simply because they are incapable of scientific investigation (the most extreme examples of this approach appeared in the writings of the philosophers known as logical positivists). This belief in scientific method as the only route to reliable knowledge has become increasingly less tenable for two reasons: first, investigations of the history and philosophy of science have identified many gaps and inconsistencies in the popular account of scientific method, and second, as other disciplines have attempted to follow scientific method new dilemmas have become apparent.

PHILOSOPHICAL ISSUES IN SCIENTIFIC METHOD

The elements of scientific method sketched in this account were developed by practising scientists without much explicit reference to philosophical issues, but when historians and philosophers turned their attention to science a number of problems were identified. Central to all these is the distinction between the logic of scientific methods and the practice of science. Some accounts, for example the work of Karl Popper (1959), have stressed logic, but others, for example Kuhn (1970a), have stressed the history, and yet others, including Mulkay (1975), have stressed the sociology of scientific ideas. Even where scientists are agreed as to how, in a logical sense, they ought to develop and test their theories it is clear that these rules are not always followed and some very important scientific discoveries occurred despite breaches of 'the rules' of scientific method.

One of the first issues to be tackled was the initial derivation of laws and theories. Many scientists had claimed, or appeared to claim, that laws or theories were derived from the observation of repeated regularities (the rising of the sun in the east is a simple example of such a regularity in nature). This method – often referred to as *induction* – was criticized on two grounds. First it was evident that in many cases the observations were themselves made with preconceptions (right or wrong) as to what constituted characteristics worthy of observation and recording. Similarly the exact form of the law-like statement derived was seldom free from theoretical presuppositions and *a priori* definitions. It seemed that pure induction seldom occurred. Even more damningly it was recognized that the logical justification of induction (that it has worked successfully in the past) itself relies upon the inductive method. These conclusions are important not only for an inductivist theory of scientific method but for any other empiricist school of thought which believes that facts should and can be allowed to 'speak for themselves'.

If the principle of induction is discarded a new reliance is put on the approach through theory, research hypothesis and testing. For many years it was assumed that the function of such tests was to prove the 'truth' of the hypothesis and by inference the truth of the theory from which it had been derived. This approach is referred to as *verification*. So, for example, in the 250 years following the publication of Newton's *Principia* a wide range of observations verified Newton's theory. The twentieth century however saw the whole fabric of Newtonian physics overthrown by modern physics (especially relativity theory). Popper, as Kuhn (1970b) acknowledges, was the first philosopher to explore fully the implication of this. Newtonian physics had never been proved true, it had survived because it had not been proved false. Popper (1959) introduced falsification as a concept to replace verification: under this new concept all theories were deemed to be provisional – coherent systems of not yet falsified hypotheses about the nature of phenomena. The falsificationist position had two other implications. In the first place it required that all scientific statements should have the logical possibility of being proved false: in this sense the claim that 'All swans are white' is incapable of being proved conclusively true (the next swan observed might be green) but could be proved false (by the observation of one truly green swan). The second consequence was for the form of empirical enquiries: no longer was it necessary to find evidence in support for a hypothesis, it became part of the scientific enterprise to find evidence which disproved it.

But the falsificationist position had its own logical flaws, especially in that it assumes that the observation is correct and that the hypothesis allows the crucial falsehood in a theory to be identified. As an example of the first point Chalmers (1978) refers to the observation that the moon appears larger when it is near the horizon: at an early stage of European scientific thought this was believed to be conclusive evidence against the theory that the moon's orbit is circular, but it is now known that this effect is illusory and its falsifying implications can be ignored. In a geographic case the presence of striations might be seen as falsifying the claim that a highland area had not been glaciated in the Pleistocene, but the presence and the date of the alleged striations might be called in question.

On the other hand suppose a complex and well-tested theory predicts an effect which fails to occur. This failure may be due to a fundamental falsity of the theory or to some low level error of logic in deriving the research hypothesis. The falsification of the hypothesis requires a search for the false step, not the immediate abandonment of the theory as a whole (Chalmers 1978). If, for example, a limestone region fails to exhibit any of the characteristic solution forms this does not falsify the general ideas about limestone landforms but leads to a question as to what other variables (climate, solution-inhibiting chemical composition of the rock) have rendered those ideas inapplicable.

A final problem in this area is posed by the existence of probability statements (the probability of tossing a six with this die is 0.1666) for logically they can never be conclusively proved or disproved and yet they have an important even if transient role in science (Braithwaite 1953). It is interesting to note that in discussing this issue Popper (1959, 209) uses one of his few geographical examples – the flow of water over a waterfall. It can be argued that these hypotheses can only be tested by using statistical inference because a probability statement can never be proved incontrovertibly false (the next run of observations may make the observed frequency approach that hypothesized by the probability statement).

The debate between verification and falsification is further confused because neither prescription seems to fit the observed behaviour of practising scientists and the scientific community at large. In some cases there is a great reluctance to abandon theories held for some time, so rather than obey the falsificationist logic scientists have adjusted their theories in an attempt to cover a contradictory result. If that stratagem failed the contradictory results were dismissed as due to freak conditions or experimental error. Koestler's (1959) account of the astronomical discoveries of Kepler, Copernicus and Galileo gives clear examples of both these stratagems. On the other hand some theories have been superseded before being falsified – the new theory being judged superior as an integration of existing ideas and observational or experimental results.

These considerations have led some authors to a more historical concept of scientific change. One of the best known of these is Kuhn (1970a) who introduced the ideas of *paradigm* and *normal science*. Kuhn's belief is that at any one time most practitioners of a discipline are working within a framework of theories and methods (paradigms) which together constitute normal science. Under this argument a normal science is overturned when a crisis occurs: this crisis may be evident from contradictory results (as in falsification) or from the inability of existing methods to solve central problems of the discipline. The crisis is resolved by a paradigm shift and a new period of normal science ensues. But such crises are not frequent – a particular paradigm can last for decades or even centuries – nor are they easy to identify at the time and the crucial observations or experiments may only be identified with hindsight. A similar argument (but with a stronger emphasis on the logic of the crisis and its resolution) is offered by Lakatos (1970) in his theory of *research programmes*.

Both these authors recognize that while rival paradigms or research programmes co-exist within a discipline it is difficult or impossible to arrive at clear-cut judgements as to which is superior; a word used to describe this dilemma is *incommensurability*. The dilemma arises because the two different programmes pose different questions, define different variables, hypothesize in different forms and require different tests. It is only rarely that two programmes make contrasting and directly comparable predictions about the same observation or experiment.

It is also worthwhile to note that some geographers (Taylor (1976); and, though he finally rejects it, Johnston (1979)) have adopted Kuhn's terminology in their discussions of shifts within geography, but Kuhn's analysis is concerned about changes within a single tradition about the nature of knowledge: it is not clear that Kuhn intended it to refer also to the shifts between quite different theories of knowledge which have been the basis of recent changes in human geography.

But there are other issues too on which normal scientific methods have been subjected to criticism, especially the issues of *reductionism* and *objectivity*.

In discussing the key elements of scientific thinking reductionism was defined as the idea that the laws and theories of one discipline can, and should, be reformulated as special cases of the outworking of a more fundamental discipline, often linked to a change of scale of enquiry. But such a reduction may be fallacious in two ways. In some systems the investigator may mistakenly suppose that micro behaviour controls behaviour at the macro level when 'in reality' the micro behaviour is only a reflection or outworking of macro-level relationships. For example, is a stream's flow pattern really reducible to no more than the laws which govern the paths of individual molecules, or is the buying pattern of a community no more than the aggregation of individual consumer preferences? Even if the micro elements can be shown to act independently of the macro system there may be middle order laws which describe how such micro elements behave in aggregations of certain sizes (those are often called *laws of composition*). For example the behaviour of individual persons may well change according to the size of the community in which they find themselves, and thus the behaviour of the group is more than the aggregation of individual behaviours (Hay and Johnston 1983).

The second alleged fallacy of scientific thinking is the belief in the *objectivity* of scientific knowledge. It is sometimes asserted that whereas the humanities have important areas which are matters of personal subjective value judgement, science is independent of such personal judgement, an independence assured by its use of abstract logics (like mathematics) and neutral measurements. But this view is difficult to maintain when it is recognized that the type of question investigated, the theory used and the observations conducted all depend on the research paradigm adopted, and that the paradigm in turn reflects the value system of the society and its philosophical and scientific presuppositions. If objectivity does exist it is not the objectivity of the individual scientist but a relative objectivity of the knowledge itself because it has been tested and corrected by many individuals working in different contexts (Popper 1972).

THE GEOGRAPHICAL APPLICATION OF SCIENTIFIC METHOD

The last twenty-five years have seen continuing debate on the application of scientific method in geography. One side argues that scientific method should be introduced (even if in modified form) into both physical and human geography. On the other hand some geographers claim that the subject is in some sense an exceptional discipline which may be excused (if not completely excluded) from the constraints of scientific method. This debate has origins in the nineteenth century which are summarized by Hartshorne (1939, 102–29). Despite the counter-arguments the period from 1960 saw a vigorous expansion of geographical research using quasi-scientific methods. The philosophical and methodological base for this was described by Harvey (1969) and a number of text-books in both human and physical geography laid out the need for theory, laws, hypotheses, measurement and statistical testing (e.g. Abler, Adams and Gould 1971; Young 1972). But the enthusiastic practitioners and protagonists were often unaware of the problems inherent in the scientific approach, and did not clearly identify the additional problems posed by its geographical use. Most of these problems stem from the twin facts that geography as a whole deals with multi-variable open systems and that human geography deals with knowing subjects.

The first problem has been recogized by geographers over many years under the term *uniqueness*. Important geographical phenomena are often complex in character and causation. An elementary text-book on soil geography, for example, outlines the fifteen characteristics of soil necessary for a full description, and the causal processes invoked include physical, chemical and microbiological mechanisms. With all these variables it is seldom possible to identify two completely identical soil sites – but even if they appear identical they will differ in their location in geographical space. The conclusion drawn is that geography deals with unique events, and generalization in the form of laws and theories is doomed to failure (Hartshorne 1939, 378–97).

This position certainly provides a powerful argument against inductive methods in geography (but more fundamental reasons for abandoning induction have already been given): it is therefore strange that many geographers who state the uniqueness case nevertheless argue for an inductive approach. It is less clear that uniqueness is a valid objection to a theoretically based hypothesis-testing approach, because a collection of unique (in the values taken by the variables) cases may nevertheless confirm or reject a hypothesized relationship: uniqueness is only an obstacle to that if it can be shown (not just asserted) that the causal relationships are themselves unique to each instance and change inconsistently from place to place, and from time to time. It is also arguable that uniqueness with respect to some trivial property (and for some geographic phenomena even location *may* be a trivial property) or some peripheral relationship is insufficient reason to invalidate scientific method.

A second consequence of geographical systems being large open systems is the difficulty of carrying out experimental tests. The sheer size of a geographical system (the atmosphere, a river basin, a city) makes the *laboratory experiment* impossible. Scaling down the system may alter its properties in unknown ways. Even if the system can be reproduced in the laboratory there is no assurance that all the variables relevant in reality have been included in the laboratory version. The second-best solution in scientific terms is the *field experiment,* but it is difficult to ensure that the only variables allowed to vary are those being investigated and certain experiments in human geography would be politically or morally unacceptable. So in *field experiments* and certainly in field *data collection* much of the control of extraneous variables is achieved by purely statistical means which in theory allow the isolation of a two-variable relationship when 'all other variables are held constant'. Yet even such methods can only 'hold constant' recorded variables; there is no way of controlling for the possible effects of unrecognized and unrecorded additional variables.

A third consequence of the multi-variable nature of geographic systems concerns the use of theory from other disciplines. This may be applied and synthetic (the attempt to bring theory from other disciplines to bear on a geographic problem) or reductionist (the attempt to interpret geographical relationships as special cases of more general theory in other disciplines). Such borrowing is especially difficult if more than one other discipline is involved, each with a scale of analysis, a conceptual framework and definitions which may not be compatible with each other or with the geographical terms of reference. A common geographical solution to this is to adopt one discipline as the source of a central theoretical framework (an *economic* approach to urban geography, the *physics* of slope development) and to use theory from other disciplines as modifying the central theory. A problem remains however that the best current theory in geography requires some knowledge of many natural and social sciences (Davidson 1978). It is this, one suspects, that makes most geographers reluctant (if not unable) to pursue a wide slice of the discipline at the highest level.

A quite different problem which arises in applying scientific method in geography is the interference by the observer with the phenomenon observed. This problem is encountered in laboratory sciences but it is usually possible to design the experiment to minimize the effect. In a trivial sense it occurs in physical geography – the art of placing a measuring device in a stream disturbs the water flow, the act of digging a pit to place tablets of calcium carbonate may increase or decrease the rate of their subsequent solution. In human geography (as in all social sciences) the same problem occurs in two more acute forms. First, if the presence of an observer is known to the actors (either in the role of observer or as a simple stranger) it may lead to a short-run change in behaviour, conscious or unconscious; the results of observation will thus be untypical of normal behaviour. Second, the interaction between observer and observed (at the time of observation or later by the publication of research findings) may produce long-run changes which would not otherwise have occurred. If such changes are towards a research hypothesis a false hypothesis may be spuriously confirmed, if counter to the hypothesis a correct hypothesis may be mistakenly rejected.

SCIENTIFIC GEOGRAPHY IN THE FUTURE

The main focus of this essay has been scientific method and its problems as seen by those who, in broad terms, defend it and its application to geography as a way of obtaining useful and reliable knowledge. It is not part of this essay to review the criticisms offered by those who are opposed to scientific method in geography (especially in human geography). This last section is written in the belief that despite these criticisms, and rightly or wrongly, scientific methods will be retained in both physical and human geography for three reasons. First, although scientism is a mistaken and dangerous ideology, the scientific method does have the ability to provide coherent and testable theories about the nature of geographical phenomena. Second, the scientific method remains appealing because it is in many respects a codified and logically connected extension of thought structures developed in everyday life, including the willingness to correct theories or hypotheses in the light of experience. Finally, and partly as a consequence of these two points, knowledge of a scientific type is required by society for its purposes of managing social and natural systems (and if geography fails to provide such knowledge some other disciplines will develop to fill the gaps).

But the scientific geography of the future cannot be untouched by all criticisms, and the elements which are retained will be modified. As a result theories and laws in geography will have two aspects. Many of them will be *derivative* in the sense that they specify the application to geographic phenomena of theories which have been established in cognate disciplines (in the natural and social sciences). Indeed the ability or inability of such theories to provide the basis for geographical explanations may be a test of their overall value as research programmes. But in addition to such derivative theories there may be a need for specifically geographical theories or laws which are essentially *laws of composition,* specifying the ways in which these derivative laws interact to produce the multi-faceted phenomena which geographers seek to understand. The level at which these laws of composition operate may be much greater than the scale at which the derivative laws and theories operate. It is possible to identify a number of examples of derivative theory already used in geography. Economic location theory uses the concepts of economics (demand curves, long-run average cost curves) and seeks to apply these in geographic space. In a similar way geographical studies of weathering apply the chemistry of ions and cations to the specific chemical composition of parent rocks under stated conditions of temperature and humidity. The examples suggest that although some geographical problems require 'spatial laws' not all derived theory in geography will be concerned with spatial relationships (Sack 1972). It is less easy to identify laws of composition in current geographical work but some recent applications of choice theory to the choice of destination and mode of travel bring together economic and non-economic variables in a choice calculus which seems more convincing than explanation in economic terms alone.

The use of both these types of theory and law implies an openness to *reductionism* – a willingness to accept reduction but not an assumption that all geographical problems can and must be solved in reductionist terms. It also implies an open approach to the *types of logics* adopted. The fact that many of the cognate disciplines (physics, chemistry, economics) themselves depend on abstract logic, and that abstract languages have proved extremely powerful makes it likely that such languages will be used in geography. The work of Alan Wilson (1974) is important precisely because he has shown how an abstract mathematical language is capable of integrating quite different parts of urban and regional systems. But the openness to abstract logics (like mathematics) must not be allowed to exclude from geographic theory those variables and concepts (e.g. the quality of a landscape) which are not capable of representation. By the same token geography must remain open to the introduction of new languages which may prove more flexible. Atkin (1974) has argued that many problems in applying mathematics to human affairs have arisen because of an obsession with nineteenth-century mathematics of differentiable functions and too little attention to the language of sets and topology (see Gatrell 1983).

At the level of practice geography will need to retain most of the elements of scientific method. The *research hypothesis,* the hypothesis *test* and *prediction* as a testing device will be needed. But the idea that any one set of observations conclusively proves or definitively falsifies a theory cannot be retained. It is the accumulation of such results (positive and negative) which will lead to the advance and decline of rival research programmes.

It also seems inescapable that scientific geography in the future will require the retention, development and refinement of *measuring devices,* guided by the emergent geographical theories and by the cognate disciplines from which they are derived. This too requires an openness as to what levels and forms of measurement are most appropriate for a given study. The collection of measurement data will carry with it a continuing need for *statistical analysis* although again an openness to new techniques (e.g. Bayesian statistics, Tukey's exploratory data analysis) will be essential (Cox 1978).

These proposals for a future scientific geography are not intended to preclude all other approaches to research and teaching in the discipline. Even the most robust adherents of a scientific method would accept the possibility that humanistic or phenomenological approaches may yield valuable new insights into the nature of geographical phenomena (Nagel 1953). Many others still believe that methodological heterodoxy (allowing the co-existence of radically different approaches within the discipline) is desirable (Johnston 1980). It is however intended to defend a place for scientific method in the subject, for without it, the author believes, geography will cease to offer a convincing interpretation of the earth's surface and the activities of individuals upon it.

NOTE

1 This essay includes material which was presented to seminars in the Departments of Geography at Queen Mary College London and Bristol University. I am grateful to participants in those seminars, and to Stan Gregory, who read a draft of this paper, for their comments and suggestions.

REFERENCES

Abler, R., Adams, J.S. and Gould, P.R. (1971) *Spatial Organization: The Geographer's View of the World,* Englewood Cliffs, NJ, Prentice-Hall.

Atkin, R.H. (1974) *Mathematical Structure in Human Affairs*, New York, Crane Russak.
Braithwaite, R.B. (1953) *Scientific Explanation*, Cambridge, Cambridge University Press.
Chalmers, A.D. (1978) *What Is This Thing Called Science?*, Milton Keynes, Open University Press.
Chao, Y.R. (1962) 'Models in linguistics and models in general', in E. Nagel, P. Suppes and A. Tarski (eds), *Logic, Methodology, and Philosophy of Science*, Menlo Park, Stanford University Press, 558–66.
Cox, N.J. (1978) 'Exploratory data analysis for geographers', *Journal of Geography in Higher Education*, 2 (2), 51–4.
Davidson, D.A. (1978) *Science for Physical Geographers*, London, Edward Arnold.
Gardner, M. (1982) 'Predicting novel facts', *British Journal for the Philosophy of Science*, 33 (1), 1–15.
Gatrell, A.C. (1983) *Distance and Space; A Geographical Perspective*, London, Oxford University Press.
Hartshorne, R. (1939) *The Nature of Geography*, Lancaster, Pennsylvania, Association of American Geographers.
Harvey D. (1969) *Explanation in Geography*, London, Edward Arnold.
Hay, A.M. and Johnston, R.J. (1983) 'The study of process in quantitative human geography,' *L'Espace géographique*, 12, 69–76.
Johnston R.J. (1979) *Geography and Geographers*, London, Edward Arnold.
—— (1980) 'On the nature of explanation in human geography', *Transactions, Institute of British Geographers*, NS 5, 402–12.
Koestler, A. (1959) *The Sleepwalkers: A History of Man's Changing Vision of the Universe*, London, Hutchinson.
Kuhn, T.S. (1970a) *The Structure of Scientific Revolutions*, 2nd edn, Chicago, University of Chicago Press.
—— (1970b) 'Reflections on my critics', in I. Lakatos and A. Musgrave (eds), *Criticism and the Growth of Knowledge*, Cambridge, Cambridge University Press, 231–78.
Lakatos, I. (1970) 'Falsification and the methodology of scientific research programmes', in I. Lakatos and A. Musgrave (eds), *Criticism and the Growth of Knowledge*, Cambridge, Cambridge University Press, 91–196.
Mulkay, M.J. (1975) 'Three models of scientific development', *Sociological Review*, 23, 509–26.
Nagel, E. (1953) 'On the method of *verstehen* as the sole method in philosophy', *Journal of Philosophy*, 50, 154–7.
Popper, K.R. (1959) *The Logic of Scientific Discovery*, London, Hutchinson.
—— (1972) *Objective Knowledge*, Oxford, Clarendon Press.
Sack, R.D. (1972) 'Geography, geometry and explanation', *Annals, Association of American Geographers*, 62, 61–78.
Taylor, P.J. (1976) 'An interpretation of the quantification debate in British geography', *Transactions, Institute of British Geographers*, NS 2, 129–42.
Wilson, A.G. (1974) *Urban and regional models in geography and planning*, London, John Wiley.
Young, A. (1972) *Slopes*, Edinburgh, Oliver & Boyd.

ARGUMENTS FOR A HUMANISTIC GEOGRAPHY
Stephen Daniels

Since its first use by Yi-Fu Tuan (1976) the term 'humanistic geography' has become a keyword in geographic thought. Perhaps inevitably its meaning has dispersed with its use. For Tuan humanistic geography was a perspective that disclosed the complexity and ambiguity of relations between people and place, qualities eclipsed by the positivist perspective of much human geography. In Tuan's writings a humanistic perspective is not a methodology nor even a philosophy but a pattern of ironic observations on familiar and exotic forms of geographic knowledge and experience. Subsequently humanistic geography has been formulated as a substantial programme with distinctive questions and ways of answering them. The term has also been used in weaker senses, sometimes merely to label any human geography which emphasizes human agency and awareness. In this essay I will first review the main arguments of self-consciously humanistic geographers from a perspective that emphasizes the historical dimensions of geographic knowledge. I will then suggest a prospect for humanistic geography that, in terms of ideas and methods, is grounded in historical understanding.

HUMANISM AND POSITIVISM

What unites humanistic geographers is their disenchantment with the writings of positivist human geographers. Criticisms that positivist assumptions and procedures do not adequately explain human issues are intellectual and moral, sometimes inextricably so. For David Ley (1980) spatial analysis is so abstracted from the world it claims to explain as to lose sight entirely of its subject, a 'geography without man' that is at once intellectually deficient and morally blind. All reasoning implies abstraction of some kind and degree. Humanistic geographers propose that reasoning in humanistic geography should conserve contact with the world of everyday experience and recognize, if not celebrate, the human potential for creativity. The scope of this brief is broad in terms of both subject matter and approach. It includes the populist essays on social geography in *Humanistic Geography: Prospects and Problems* (Ley and Samuels 1978) and the more elitist essays on novelists and poets in *Humanistic Geography and Literature* (Pocock 1981b).

It is customary to describe the positivist reformulation of human geography as a 'revolution'; Ley and Samuels (1978) describe their humanistic reformulation as a 'reawakening', a sign of their moral evangelism and their conservative claim to restore human geography to its rightful role, 'the study of the earth as the home of man' (Ley 1980, 3). Despite or because of their disaffiliation with much post-war geography, positivist revolutionaries were intent on constructing an ancestry made up of scholars from many sciences, to legitimate their undertaking (Taylor 1976, 36). Humanistic geographers have been no less intent on creating a tradition to disinherit positivist geography and legitimate theirs. Because of their interests in landscape and culture eminent geographers like Paul Vidal de la Blache and Carl Sauer have been retrospectively converted to humanistic geography. If alive, they would probably protest that their scientific persuasions had been overlooked. Humanistic social geographers appeal to a non-positivist, sometimes explicitly humanist, tradition in sociology. They have in the process disentombed the non-positivist persuasions of scholars like Robert Park and Alfred Weber and reclaimed them from the ancestry of spatial analysis (Smith 1981). Tracing a tradition beyond modern academic disciplines is both easier and less meaningful because the term humanist has been used to describe almost any thinking which addresses human issues undogmatically (Ley and Samuels 1978, 4–10).

Moral criticisms of positivist geography are usually situated in and draw authority from a more general questioning of ways of reasoning and acting in modern western culture, paricularly the powerful combination of science, technology and capitalism which conventionally defines rationality. Relph (1976) sees logical connections between various narrowly conceived, efficient and reiterative constructions: intellectual constructions like Central Place Theory, material constructions like suburban housing and perceptual constructions like tourist guides. Relph envisages more 'authentic', more wholesomely human ways of understanding, creating and perceiving the world and finds examples in the lives of various pre- or non-industrial peoples. Relph is not alone in his conspiratorial view of modern western culture nor in his romanticism of other cultures but it is worth examining his argument more closely because it reveals the historical myopia that afflicts much humanistic geography. As an example of an authentic place, an expression of an 'I–thou relationship between man and god', Relph (1976, 75) includes a photograph of Tintern Abbey, or more precisely its ruins. The problem here is the historical complexity of both the subject and its representation. To be sure, Cistercians were pious but they were also unromantically mindful of their worldly interests as the peasants who were cleared from their estates and the monarch who ruined their abbeys realized. Relph (1976, 67) maintains his photographs are intended to 'demonstrate' not just 'illustrate' his argument but his pictorial interpretation of Tintern Abbey actually conflicts with his writtten interpretation. The view of Tintern Abbey Relph photographed has a histroy no less than the abbey itself. The view was first constructed by eighteenth-century tourists and artists according to a conventional picturesque formula that still informs tourist perceptions of England as a series of beauty spots and the photographs they take of them. Relph thus reproduces a tourist's way of seeing that he is at pains a few pages later in his writing to deplore: 'this is inauthenticity at its most explicit; the guided tour to see works of art and architecture that someone else has decided are worth seeing ... such guides stress the picturesque and the monumental' (Relph 1976, 85).

CONCEPTS

Much humanistic geography concentrates on developing a set of concepts to articulate issues of value and meaning ignored or misconstrued within positivist human geography. These concepts have, or are given, a social, sometimes psychological reference as well as a spatial one. Concepts like 'landscape' and 'region' have been retrieved from the periphery or from beyond the pale of positivist geography. In the process they have been infused with a greater sense of perception and experience than during the heyday of regional and landscape geography (Tuan 1979). Humanists reject the reduction of space and place to geometrical concepts of surface and point; humanistic conceptions of space and place are thick with human meanings and values (Entrikin 1976, 623–5).

Place is a key concept in humanistic geography; much humanistic writing is devoted to illustrating and clarifying it. From a humanistic perspective the meaning of a place is inseparable from the consciousness of those who inhabit it. The scope of place as a concept varies according to the extension of the thoughts, feelings and experiences that make up the consciousness of inhabitants. For any one person a favourite armchair may be a place, also a room, street, city, nation, perhaps the whole earth (Relph 1976, 141; Tuan 1977, 149). From a humanistic perspective place is not so much a location as a setting, less a thing than a relationship.

Most humanistic geographers acknowledge the implication of process in place – a sense of place is something that develops through time – but few explore or explain it (Cosgrove 1978, 70). This is not just because some see place as 'an essentially static concept' (Tuan 1977, 179) but because the form of writing they use to explicate place cannot adequately account for issues of process and development. Much humanistic commentary on place is cast in a categorical form and conducted in the present tense. Categories and sub-categories of place and place-consciousness are usually illustrated by a variety of examples from various historical periods but it is seldom clear how particular meanings of place develop in particular settings. Meanings are noted, seldom narrated.

The concept of place itself has a complex history. Like 'landscape', 'nature' and 'community' it resonates with ideological implications but these are scarcely explored. Place is a positive word. To be displaced, out of place or placeless is a negative, even 'unnatural', condition. Common to many uses of the word place, including those by humanistic geographers, are connotations of stability, belonging and propriety. What is seldom noted is how these connotations are informed by a strict, even oppressive, sense of social order and control, explicit in the expectation that the poor should know their place. Place is a predominantly conservative notion, one that can perhaps help us understand, even resist, destructive changes but one that threatens to oppose change itself. As Buttimer (1981) recognizes, intellectual appreciations of a sense of place among the underprivileged can amount to a tacit condoning of conditions like poverty and injustice.

In developing its concepts humanistic geography assumes its philosophical tone. This is less an echo of the analytical tradition that prevails in the philosophy departments of Britain and North America than the more speculative philosophical tradition of continental Europe. The style and subject matter of existentialism and phenomenology have proved alluring to humanistic geographers. Analytical philosophers are more reticent about dramatic thoughts or experiences and more modest about what philosophy can do, often arguing it can at best reduce obscurity that is largely linguistic in origin and can do so best by adopting a restrained style. The claims and language of existentialists and phenomenologists are characteristically less restrained. This is not to say their writings are irrational but the humanistic geographers who invoke their authority have done remarkably little reasoning themselves. Philosophy in humanistic geography is less reasoning than reporting. It often amounts to quoting passages with a geographical flavour from this or that existentialist or phenomenologist or just classifying incidents in life or literature in terms of technical concepts like 'lifeworld' or 'insideness' (Buttimer 1976; Samuels 1978; Seamon 1981a). Such concepts are occasionally modified in translation to English and to geography but are seldom themselves the subject of philosophical scrutiny. In a study of 'insideness' and 'outsideness' literature is seen as a 'testing ground to confirm and amplify exising phenomenological claims' (Seamon 1981a, 85). If we accept that reason should be at least potentially critical this seems not just a rejection of positivist criteria of reason, of hypothesis testing, but a rejection of reason itself.

METHODS

There is less reticence about how humanistic geography should be conceived than about how it should be done. Ley and Samuels (1978, 121) suggest that conceptual clarification is itself a method, a way of achieving a clearer or deeper understanding of an issue. There is some agreement that humanistic methods cannot be defined as a set of formal procedures or techniques. One informal procedure appealed to is phenomenological reduction, the suspending of conventional ways of knowing, such as ordinary language or academic expertise, to achieve a primordial intuition of the world (Seamon 1981a). It is not hard to see why this procedure has not been satisfactorily used or explained. It rests on the fallacy that a convention like language is an impediment to understanding. Some languages or terminologies may occlude certain qualities. For example, an urban geography constructed around concepts like 'housing' and 'land use' may well not capture the values and experiences of people living in particular 'homes' and 'neighbourhoods' (Buttimer 1981). But it is only by failing to discriminate some things that a language can discriminate at all. We may argue about the kind of understanding particular languages offer but language itself is a convention that makes understanding possible. Seamon (1981b) speculates that phenomenological reduction might disclose essential 'time-space routines' to neighbourly places that underlie or transcend the particularities of history, culture and personality. His calling such essences 'place ballets' shows that if they do exist language, or some other form of communication, is essential to their recognition. In Seamon's account such essences could only be identified by someone who knew what a 'ballet' was.

Most humanistic geographers forgo a phenomenological search for essences for more existential accounts of experiences of particular people in particular places. This they argue implies reducing the detachment and authority of the geographer. 'An existentially aware geographer,' announces Buttimer (1974, 24), 'is less interested in establishing intellectual control over man through preconceived analytical models than he is in encountering people and situations in an open, intersubjective manner.' Methods of participant observation developed in sociology and anthropology provide an example for humanistic social geographers in how to account for how those under study understand their world (Smith 1981). Some, less social, humanistic geographers (Rowles 1978) notionally blur the distinction between investigator and investigated, a distinction that is inevitably sharpened when an investigator writes up his research for academic purposes or even conceives of what he or she does as research in the first place.

The incapacity of standard statistical techniques to describe the fluency of human life and in some cases their capacity to distort it out of all recognition has provoked a cautious if not hostile attitude among humanists to quantification. While some settle for a conversational language to capture the nuances of everyday experience others adopt a more self-consciously literary style to express more profound experiences or to explicate the profound in the apparently everyday. Olsson (1978, 110) sees quantitative geography and humanistic geography as essentially opposed, the former aspiring to 'formal reasoning', the latter to 'creative writing'. Such a creative humanistic geography should, Olsson maintains, celebrate ambiguity and allusiveness. What is striking about humanistic aspirations to creativity is the equation of creativity with experiment. In his essay in *Humanistic Geography and Literature* Olsson (1981, 126) asks,

> What does it *now* mean to yearn for home? Jokingly serious, seriously joking?
> Joker trumps the trumps! Follow suit!
> To yearn for home is to experience double bind. It is to be torn between irreconcilable identities, sometimes enjoying the illusory freedom of singing with the wind, sometimes missing the real subjugation of being tied to the ground.

It is not difficult to see what Olsson is getting at here but hard not to suggest that it could have been expressed more staightforwardly, if less playfully, without the various conceits. You don't have to be a linguistic puritan to conclude that the literary styles of humanistic geographers are no less gratuitous than the quantitative or algebraic flourishes of some positivist geographers (Billinge 1983).

HUMANISTIC GEOGRAPHY AND THE HUMANITIES

Some of the more puzzling humanistic writings are those about literature. This is not so much because of their style but because few of their authors have been reading literature at all. In discussions of D.H. Lawrence (Cook 1981), George Eliot (Middleton 1981), Doris Lessing (Seamon 1981a), Mary Webb (Paterson and Paterson 1981) and various English novelists of 'place' (Pocock 1981a) there is little or no recognition of the literary conventions these novelists employ, for example their methods of narration or description. It is no defence to say (Pocock 1981b, 9–10) that these are matters for the literary critic not the geographer. The issue of place and consciousness of place cannot be isolated from the issue of how these are, can or perhaps cannot be represented (Barrell 1983). When analysis of how authors communicate is evaded it is easy as well as convenient to substitute phrases like 'heightened perception' and 'mysterious intuition' (Pocock 1981b, 11–12). The mystery here is of the geographer's making. Isolated from the issue of literary convention the arguments of humanistic geographers, and through them the writings of novelists, slide into soft focus. For Pocock (1981b, 15) 'the starting point' of a geographical reading of literature is not language but 'the artist's perceptive insight'; 'literature simply is perception', and 'in the last analysis ... it is perhaps not so much what the poet, novelist or playwright says, but what he *does* to us that matters'. For a humanist to abstract the perceptive insight of an artist from the language in which it is embodied is paradoxically to diminish an artist's humanity. From this viewpoint authors appear not to work, as they do and very humanly, with the possibilities and restraints of artistic form and language, or, in a larger context, with those of the society in which they work. They become instead vehicles for transcendent truths. Literature becomes 'literary revelation' (Pocock 1981b, 9).

While a number of human geographers have used literary evidence few have used what seems a more obvious source for a highly visual discipline: the visual arts. Rees (1973) has noted the potential of landscape painting for studying questions like national consciousness and attitudes to nature. In the quest for a tradition in which the sciences and humanities were not so strictly demarcated he has suggested parallels between the ways geographers and painters know and represent the world. It is useful to regard the visual arts as ways of knowing as well as ways of seeing. But the potential of paintings can only be realized if geographers analyse more carefully, as they do in map interpretation, the form this knowledge takes; how paintings represent the world through various conventions such as medium, perspective, framing, symbolism. We might then appreciate for example that many eighteenth-century

English landscape paintings are not so much tender-hearted renderings of an open countryside as tough-minded assertions of landed property. Oil paint and perspective combine to assemble a visible world for private consumption in pictures that were originally hung in the country houses of their owners, sometimes next to maps of their property. This is not to reduce their meaning to ideology or to deny their beauty, only to emphasize that art is not invariably the humane thing some humanistic geographers assume it to be (Daniels 1982b). Relations between beauty and morality are complex, sometimes contradictory (Daniels 1982a; Cosgrove and Thornes 1981).

The naive approach of most humanistic geographers to the arts, especially their neglect of artistic form, reveals their poor working knowledge of the humanities, especially literary and art criticism and history. Scholars in the humanities are presently looking to the social sciences for a more adequate account of meaning and value, one that places the work of an author or artist in a wider social and economic context. Those humanistic geographers who turn their backs on social science yet are blind to the humanities may, or may not, sense their isolation.

OBJECTIVITY AND SUBJECTIVITY

A dualism that has disabled much geographic thought is that between objective and subjective knowledge. In their readiness to renounce positivist notions of objectivity such as the detachment and neutrality of the observer, some humanists have settled for, even welcomed, positivist notions of subjectivity, especially the reduction of values and perceptions to private feelings and experiences. This limited notion of subjectivity has influenced conceptions of subject matter and ways of studying it. A discussion of 'experiential field work' in the study of the perceptions of old people concludes 'if I am able to come to know Marie [an old person] well, I can gain a special sensitivity to her geographical experience. Together we may develop a shared awareness' (Rowles 1978, 175). The point of such studies as this seems less to make verifiable claims about values than to report feelings and experiences. When humanistic geographers do on occasion address the critical issue of how the claims of a humanistic study to be true are to be assessed their accounts are scarcely persuasive. For Pocock (1983, 356) verification in humanistic geography is a form of intuition: 'Does the description ring true with my own and others' experience?' By ruling out the pre-conditions of rational argument Pocock logically cannot demonstrate the truth of his own assertion. Argument is here reduced to swopping experiences and verification to matching them.

The reduction of subjectivity to feelings which are, for better or worse, beyond rational scrutiny is a general historical development. It dates from the nineteenth century (Williams 1976, 259-64) and a bourgeois conception of life in which a subjective, private world of intimate feelings, symbolized by the suburban home, was separated from and compensated for an objective, impersonal world of hard facts symbolized by the workplace. Despite, or because of, their disenchantment with commercial values, this distinction has been reproduced by humanistic geographers. While sensitive to the more personal qualities of relations between people and between people and place they have proved reluctant to address the economic dimension of these relations (Sayer 1979, 33). There is, to my knowledge, no humanistic geography of work relations or the workplace and little emphasis on the economic implications of the home or private life. This is curious because to neglect the economic is to neglect a dimension of life that is *experienced* forcefully, especially when money or resources are scarce. It is as if some humanistic geographers are not so much criticizing positivist geography as compensating for it, accepting a division of labour which delegates some to go out and analyse the objective world and others to stay at home and cultivate the subjective.

Humanistic geographers with a stronger sense of the social dimension of human life and knowledge (Gibson 1978; Ley 1978) have recognized that values are not a matter of inner feelings nor are all facts as apparently clean as facts like the population or latitude of a town. It is a fact that every day in English towns people pay for goods. This is a more informative and explanatory description of these actions than stating that 'every day people exchange pieces of paper or metal for various items' because its meaning derives from and implies certain commercial values that underpin modern English urban society. There are other possible descriptions of those actions, or a sub-set of them, for example 'keeping up with the Joneses' or 'sustaining the capitalist mode of production' and each of these implies a particular system of values, in the latter case a system perhaps not shared by most of the participants. Such descriptions are more controversial. These controversies, which abound in social science, are not settled by intuition but by making rational inferences from factual evidence. It is the same with disputes explicitly about values such as justice. To move from facts to values is not necessarily to slip into solipsism. On this account objective knowledge is not value-free, but a blend of evaluative and non-evaluative description (Graham 1982).

HISTORICAL UNDERSTANDING

How may objective knowledge which is evaluative as well as explanatory be acquired and expressed? Cole Harris (1978) describes a procedure that characterizes study in history, a discipline which has for long, and unselfconsciously, dealt with issues of value and meaning that humanistic geographers, generally with little grounding in history, have rediscovered. This historical procedure is 'hardly an explicit methodology, it is better thought of as a habit of mind'. The 'historical mind' as Harris calls it is 'contextual not law finding', more a matter of incremental learning and judgement than technique. The product of immersion in the sources for a period, it 'is open, eclectic and curiously undefinable'. Harris suspects it is not solely relevant to the study of the past: 'as a form of explanation it is part of everyday experience'. In the remainder of this essay I want to clarify and amplify this procedure, or at least what I see as its main form. In the process I hope to show that it is as much a method as a mentality. It is a method that, despite the censure of some positivists, is still used in geography and deserves affirming: narrative.

Consider this extract from a well-known text on the human geography of North America (Paterson 1979, 175).

> The semi-feudal conditions of land tenure in some of the early settlements, and the pressure on the land of an increasing immigrant population soon produced a drift westward toward the mountains. Here in the rougher terrain of the upper Piedmont and later of the Appalachian valleys, independent farmers carved out their holdings, accepting the handicaps of infertility and remoteness in exchange for liberty of action. To the eighteenth-century farmer the exchange seemed a reasonable one; his twentieth century descendant, occupying the same hill farm, suffers the handicaps without the same compensation.

This is the kind of account which positivists regarded at best as a loose, weakly explanatory combination of factual description and implicit law-like generalization. In positivist terms it might be reformulated as an explanation sketch of pioneer settlement, the significance of some facts about eighteenth-century pioneers indicated by the law-like first sentence about migration. Such a sketch would require filling out by specifying the necessary and sufficient conditions of the law, or at least their statistical probability, and, in greater and stricter detail, the facts which that law explained (Harvey 1969, 421–2). Such arguments have persuaded some human geographers to strive for a positive form, or at least style, of temporal explanation. It is not unusual for developments to be described in the language of systems theory or the symbolism of flow diagrams, as processes with discrete stages connected by lines of causal linkage. If we set aside positivist criteria we can see a different explanatory form to the above passage. Without pretending that it is full or complete we may appreciate that its explanatory force is stronger and more rigorous than positivist criteria suggest.

The explanation is primarily contextual not causal. The point of the passage is not to demonstrate the necessity of pioneer settlement but to establish what westward pioneering meant in eighteenth-century North America. This is done by situating it in an evaluative context, that of the liberty of the pioneers. This context is constructed of a set of overlapping descriptions. Some are more or less straight ('rough terrain', 'infertility'); some more emotive ('suffers the handicaps'); some more technical ('the semi-feudal conditions of land tenure'). One description of pioneering suggestive of its hardship, 'carving out holdings', is directly overlapped by another, 'accepting the handicaps of infertility and remoteness in exchange for liberty of action', which enlarges its meaning by showing how the pioneers themselves conceived their action. The passage shows how narratives can negotiate the perspective of those who lived through the past and that of the modern observer. The advantages of hindsight are exploited. The key concept 'semi-feudal' was unavailable to an eighteenth-century pioneer, or at least not understood in the way it is by modern readers. The final description of his twentieth-century descendant, trapped in a region that once promised opportunity, amplifies through irony the values of the 'drift westward towards the mountains'. The opening causal sentence is a key part of the explanation but subordinate to an overall contextual account that incorporates descriptions of varying evaluative force. The account is both particular and thematic. It is also objective. It is open to amplification or to question by further empirical research which might indicate a more appropriate context for events, one perhaps that establishes a less ideological, more economic meaning.

If I have burdened this brief extract with interpretation it is to show why narrative might appeal to humanistic geographers. It explicates meaning through context and in the process mediates, or should mediate, the views of the 'outsider', the narrator, and those of the 'insider', the participants in the history the narrator constructs. The language of even the most scholarly narratives is mostly conversational, suggesting that it is less abstracted than many academic forms of explanation from the meaning of everyday life. Narratives conserve a more seamless sense of the fluency of relations between people and between people and place than do systems or structural modes of temporal explanation. But they do not reproduce the flow of lived experience. Narratives are not lived but told. Narrative is an essentially retrospective mode of understanding but it is not confined to studying the dead nor is it necessarily scholarly. For a variety of continuing processes, for example urbanization, farming, race relations, a life, a career, it can explicate what seemed to participants and observers at the time of occurrence insignificant or meaningless. A historically minded scholar might make sense of his or her own life in a way similar to the way they explain the lives of those they study. In other words historical understanding is not essentially an empirical notion, one restricted to those pursuits we call 'history', but a philosophical one.

In narrative the essential relationship between events is not antecedent to consequent, but part to whole. Having followed, and thus understood, a narrative, the element of time diminishes as a complex array of happenings – sudden events, steady pressures, long campaigns, impulsive actions – are seen together as parts of a single, if often complex, configuration: 'a landscape of events' (Mink 1969–70, 549). Humanistic geography is sometimes characterized as a broader or clearer perception of the world. An appreciation of narrative form takes some of the mystery out of this description and gives it some methodological muscle.

Although philosophical in tone humanistic geography is rarely if ever theoretical, indeed is often anti-theoretical. This is understandable given the tendency of both positivist and radical geographers to reify theoretical concepts and over-privilege their explanatory status. But to be on principle anti-theoretical is to relinquish the potential of explanatory power. Narrative can conserve theoretical concepts as it were in solution, preventing them precipitating into discrete structural categories. It is perhaps not surprising that Marxist narratives, informed, often implicitly, by a historical theory, provide the clearest examples. An example of a Marxist geographical narrative is Harvey's (1979) study of the Basilica of Sacre Coeur in Paris.

Harvey narrates the conception and building of the Basilica (a process that was neither linear nor smooth) at varying levels, some theoretical, some concrete. He recounts episodes of often violent conflicts and experiences in Paris during and after the Franco-Prussian War, many on the hill of Montmartre. He then recapitulates these episodes in terms of broader conceptual relations: town and country, church and state, bourgeoisie and proletariat, economy and society. As I read it the point of his narrative is not to test or exemplify these concepts in an ahistorical positivist manner. Nor is it merely to chronicle certain events in Paris in a historically empty antiquarian manner. Rather it is to explicate through descriptions of varying scope, from the shooting of Communards to the development of a religious cult, the meaning of a landmark that was intended to mythologize the history it commemorates and which remains one of the most striking and poorly understood in Paris.

> The building hides its secrets in sepulchral silence. Only the living cognizant of this history . . . can truly disinter the mysteries that lie entombed there and thereby rescue that rich experience from the deathly silence of the tomb and transform it into the noisy beginning of the cradle. (Harvey 1979, 381)

It is not necessary to make revolutionary inferences from historical analysis to see how narrating landscapes can revitalize them. Narrative as a form of historical interrogation can recover the conflicts and hardships which so often constitute the making of landscapes and which the conventional idea of landscape, with its implications of harmony and peace, seems to deny.

My emphasis on the formal characteristics of narrative should not give the impression that narrative is a formula which can be learned independently of particular subject matter and applied to it. Narratives are explicated from the evidence of specific times and places. This process, a dialectic of discovery and construction, can involve complex and delicate adjudications: between the perspectives of participants and observers, between particular incidents and general themes, between competing theories. Interpretation and judgement are ingredients to narrative, not operations performed before or after the evidence has been collected and processed. Pointing out the formal characteristics of narrative does not make the writing of good narratives any easier than it ever was. But it may help show why the effort is worthwhile.

CONCLUSION

If humanistic geography is to be more than a criticism of positivist geography it needs a more thoroughly reasoned philosophical base, a closer understanding of the conventions through which human meanings are expressed, a more adequate account of what humanistic methods are or might be, and above all a greater historical understanding. Humanists might extend their claims into areas such as economic geography presently dominated by positivists and radicals. In the process they might be forced to modify their claims through negotiation with other geographical perspectives. In consequence humanistic geography might lose a sense of separate identity. If human geography as a whole assimilated its more persuasive arguments there might be no need for a group of human geographers to proclaim their study emphatically human. Many humanistic geographers would surely welcome this.

REFERENCES

Barrell, J. (1983) Review of Pocock, D.C.D. (1981) *Humanistic Geography and Literature*, in *Journal of Historical Geography*, 9 (1), 95–7.
Billinge, M. (1983) 'The mandarin dialect: an essay on style in contemporary geographic writing', *Transactions, Institute of British Geographers*, NS 8, 400–20.
Buttimer, A. (1974) *Values in Geography*, Washington, DC: Association of American Geographers Research Paper No. 24.
—— (1976) 'Grasping the dynamism of the lifeworld', *Annals, Association of American Geographers*, 66, 277–92.
—— (1981) 'Home, reach and the sense of place', in A. Buttimer and D. Seamon (eds), *The Human Experience of Space and Place*, London, Croom Helm, 166–87.
Cook, I.G. (1981) 'Consciousness and the novel: fact or fiction in the works of D.H. Lawrence', in D.C.D. Pocock (ed.), *Humanistic Geography and Literature*, London, Croom Helm, 66–84.
Cosgrove, D. (1978) 'Place, landscape and the dialectics of cultural geography', *Canadian Geographer*, 22, 66–71.
—— and Thornes, J.E. (1981) 'Of truth and clouds: John Ruskin and the moral order in landscape', in D.C.D. Pocock (ed.), *Humanistic Geography and Literature*, London, Croom Helm, 20–46.
Daniels, S. (1982a) 'Humphry Repton and the morality of landscape', in J.R. Gold and J. Burgess (eds), *Valued Environments*, London, Allen & Unwin, 124–44.
—— (1982b) 'Ideology and English landscape art', in D.E. Cosgrove (ed.), *Geography and the Humanities*, Loughborough University, Department of Geography Occasional Paper No. 5, 6–13.
Entrikin, J.N. (1976) 'Contemporary humanism in geography', *Annals, Association of American Geographers*, 66, 615–32.
Gibson, E. (1978) 'Understanding the subjective meaning of places', in D. Ley and M. Samuels (eds), *Humanistic Geography: Prospects and Problems*, London, Croom Helm, 138–54.
Graham, E. (1982) 'Objectivity, values and bias in human geography', paper read at annual conference of the Institute of British Geographers, Edinburgh.
Harris, C. (1978) 'The historical mind and the practice of geography', in D. Ley and M. Samuels (eds), *Humanistic Geography: Prospects and Problems*, London, Croom Helm, 123–37.
Harvey, D. (1969) *Explanation in Geography*, London, Edward Arnold.
—— (1979) 'Monument and myth', *Annals, Association of American Geographers*, 69, 362–81.
Ley, D. (1978) 'Social geography and social action', in D. Ley and M. Samuels

(eds), *Humanistic Geography: Prospects and Problems*, London, Croom Helm, 41–57.

—— (1980) *Geography without Man: A Humanistic Critique*, University of Oxford, School of Geography Research Paper No. 24.

—— and Samuels, M. (1978) 'Introduction: contexts of modern humanism in geography', in D. Ley and M. Samuels (eds), *Humanistic Geography: Prospects and Problems*, London, Croom Helm, 1–17.

Middleton, C.A. (1981) 'Roots and rootlessness in the life and novels of George Eliot', in D.C.D. Pocock (ed.), *Humanistic Geography and Literature*, London, Croom Helm, 101–20.

Mink, L.O. (1969–70) 'History and fiction as modes of comprehension', *New Literary History*, 1, 541–58.

Olsson, G. (1978) 'Of ambiguity or far cries from a memorialising mamafesta', in D. Ley and M. Samuels (eds), *Humanistic Geography: Prospects and Problems*, London, Croom Helm, 109–22.

—— (1981) 'On yearning for home: an epistemological view of ontological transformations', in D.C.D. Pocock (ed.), *Humanistic Geography and Literature*, London, Croom Helm, 121–9.

Paterson, J.H. (1979) *North America*, 6th edn, New York, Oxford University Press.

—— and Paterson, E. (1981) 'Shropshire: reality and symbol in the work of Mary Webb', in D.C.D. Pocock (ed.), *Humanistic Geography and Literature*, London, Croom Helm, 209–20.

Pocock, D.C.D. (1981a) 'Place and the novelist', *Transactions, Institute of British Geographers*, NS 6 (3), 337–47.

—— (1981b) 'Introduction: imaginative literature and the geographer', in D.C.D. Pocock (ed.), *Humanistic Geography and Literature*, London, Croom Helm, 9–19.

—— (1983) 'The paradox of humanistic geography', *Area*, 15 (4), 355–8.

Rees, R. (1973) 'Geography and landscape painting: an introduction to a neglected field', *Scottish Geographical Magazine*, 89, 148–57.

Relph, E. (1976) *Place and Placelessness*, London, Pion.

Rowles, G.D. (1978) 'Reflections on experiential fieldwork', in D. Ley and M. Samuels (eds), *Humanistic Geography: Prospects and Problems*, London, Croom Helm, 173–93.

Samuels, M. (1978) 'Existentialism and human geography', in D. Ley and M. Samuels (eds), *Humanistic Geography: Prospects and Problems*, London, Croom Helm, 22–40.

Sayer, A. (1979) 'Epistemology and conceptions of people and nature in geography', *Geoforum*, 10, 19–43.

Seamon, D. (1981a) 'Newcomers, existential outsiders and insiders: their portrayal in two books by Doris Lessing', in D.C.D. Pocock (ed.), *Humanistic Geography and Literature*, London, Croom Helm, 85–100.

—— 'Body, subject, time-space routines and place ballets', in A. Buttimer and D. Seamon (eds), *The Human Experience of Space and Place*, London, Croom Helm, 148–65.

Smith, S.J. (1981) 'Humanistic method in contemporary social geography', *Area*,

13 (4), 293–8.
Taylor, P.J. (1976) 'An interpretation of the quantitative debate in human geography', *Transactions, Institute of British Geographers*, NS 1, 129–42.
Tuan, Yi-Fu (1976) 'Humanistic geography', *Annals, Association of American Geographers*, 66, 266–76.
—— (1977) *Space and Place: the Perspective of Experience*, London, Edward Arnold.
—— (1979) *Landscapes of Fear*, Oxford, Basil Blackwell.
Williams, R. (1976) *Keywords: A Vocabulary of Culture and Society*, London, Fontana.

8
REALISM AND GEOGRAPHY
Andrew Sayer

In the last three decades, geographers have repeatedly cast around in the literature on the philosophy and methodology of science for ideas which could give their research greater purchase on the world. In so doing they have shifted position unusually frequently. Recently, some geographers have begun to consider realist philosophy as a possible source of guidance and in this chapter I shall try to explain why.[1]

It's easiest to understand this interest in realism if we first recall some of the methodological positions that went before. The traditional regional geography which preceded the quantitative revolution combined a distaste for theory and conceptual analysis with a treatment of geographical phenomena as *unique* and as not susceptible to explanation by reference to general laws. In other words it was both *atheoretical and idiographic*. In retrospect, geography at this stage seems to have been in an academic backwater, blissfully ignorant of the methodological and philosophical issues which enlivened other disciplines. The elders of the discipline were given to anodyne talk of the 'craft' of geography, which always seemed to elude definition, though whatever it was, we were given to believe that the profs had it. Consequently – and this is relevant for the discussion of realism – they gave idiographic studies a bad name.

Probably the most refreshing and lasting contribution of the quantitative revolution was in breaking out of this complacent isolationism and mediocrity. In place of the study of the unique (idiographic method), it pushed the search for order, or regularity, i.e. a nomothetic method, which it held to be characteristic of 'science'. And 'science' easily won the battle for the minds of the younger geographers against the vapid claims of 'craft'.

It also celebrated *theory* as the means and end of geographical enquiry. But whatever was said about theory in the methodological manifestos of 'the new geography' (e.g. Chorley and Haggett 1967; Harvey 1969), in practice a rather narrow conception of theories as *ordering frameworks* came through. 'Theories' shaded into 'models', understood as formalizations of enduring regularities between variables. Once these invariances had been discovered and formalized, data could be 'plugged' into the model and used for prediction, or so it was hoped. Sometimes the ordering frameworks were derived deductively from first principles (e.g. central place models from principles of supply and demand); sometimes they were discovered inductively by fitting them to data (e.g. diffusion curves from data on rates of adoption of innovations).

There are two important points to note about this approach. First, while it was possible, by the deductive method, to generate idealized landscapes that were characterized by regularities, these turned out to bear little resemblance to actual ones. This then raised the problem of deciding whether the differences indicated errors in the model or the presence of interfering processes, or both. Where regularities were discovered inductively they turned out to be only very approximate, as well as variable across space and time, so that the models had to be re-fitted anew to each and every case. The goal of universally applicable models which could be used for the prediction of events in a variety of situations was not realized: ironically, the regularities and the values of the parameters that represented them turned out to be unique to each application. Second, in *practice*, the theories used by quantitative geographers said remarkably little about how we should conceptualize objects of interest and hence about the nature of the latter. The term 'data' was understood literally as 'given things' not as something problematic. Consequently, by default, this 'scientific' geography rested upon the unacknowledged and unexamined concepts of common sense.

Post-quantitative geography has slowly drifted away from these methods: the search for regularities has gradually weakened, and concern with conceptualization has grown. Contrast, for example, quantitative geography's modernization studies with radical geography's analyses of Third World development, in particular the former's astonishing neglect of the problem of the meaning of 'development' with the latter's near obsession with this question. Other interesting changes are a (possibly rather tentative) re-emergence of apparently idiographic regional studies, albeit of a relatively theoretical kind. Also history is coming back into geographical explanations after having been banished by quantitative geography as unscientific. Once again it is becoming respectable to be a specialist on a particular region whereas the nomothetic geographers would have found this incompatible with the search for general regularities in regional systems.

In my view, although few recognize them as such, all the above post-quantitative trends are moves towards a realist approach. The changes are clearest in radical and Marxist research in geography (sometimes rather unfortunately labelled 'structuralist'). However, this association of realism and radical geography is not a necessary one: some radical work has been done using a nomothetic deductive method (e.g. Scott 1980) and acceptance of realist philosophy does not entail acceptance of a radical theory of society – the latter must be justified by other means. Some aspects of humanistic geography might also be taken to be convergent with realism and occasionally even mainstream geographers stumble across realist methods, if only temporarily and unknowingly.

Having set the context in this possibly tendentious way, let us switch to a closer and more sober analysis of some of the issues at stake. I will begin by comparing and evaluating two basic methodological principles – one characteristic of the 'new', nomothetic or quantitative geography, the other of realism. I will then try to demonstrate the significance of these issues by reference to an extended example from geographical research. This will be followed by a discussion of the nature of theory. In conclusion, I will outline some of the methodological problems peculiar to geography and discuss whether realist geography is idiographic or nomothetic.

TWO CONTRASTING METHODOLOGICAL PRINCIPLES

When stated baldly, methodological prescriptions have a habit of appearing bland, innocuous, or too obvious for words, but when we follow through their implications, their particularity and critical significance becomes clear. Consider the following pair of injunctions, the first broadly 'positivist'[2] (and nomothetic), the second realist.

1 If we are to explain processes we must discover the regularities or universal laws governing their behaviour. Hence the thrust of research must be towards the discovery of order.
2 If we are to explain why things behave as they do we must understand their structure and the properties which enable them to produce or suffer particular kinds of change.

Both of these probably sound unexceptionable, but there are two problems with 1. First, although this strategy appears to work in some natural sciences – particularly physics – it doesn't or hasn't in human geography; regularities in the latter tend to be approximate and temporally and spatially specific, or, as we have already noted, unique. There is a convenient excuse for this state of affairs, however, namely that nomothetic human geography and other social sciences are as yet immature and that if we only persist with this allegedly 'scientific' approach, order will be discovered. But even if we accept that social science is more youthful than natural science (not an unquestionable assumption) there is still the danger that such an argument is merely a protective device – rather like saying that if prayer doesn't work its because you haven't been praying hard enough. At the very least, it seems unreasonable to refuse to consider the possibility that the search for regularities may not be *appropriate* given the different nature of objects studied by disciplines such as human geography. The second problem with 1 is that even where regularities do exist they don't provide explanations. That two or more variables behave in a regular, predictable fashion doesn't tell us why this is so, or, to put it another way, it tells us nothing about what *makes* it happen, what produces it. (This is why the conclusions of empirical research into regularities are so inconclusive!) Note that this is not simply the old problem of 'spurious correlation', familiar though that may be. Even if correlations or other indices of association and regularity concern variables which might reasonably be expected to be causally related, the mere existence of the (non-spurious) regularity says nothing about what produces it. Nor, for that matter, does the possession of a hypothesis or theory that they are related provide evidence for the regularity being causal. For example, we may succeed in fitting some sort of spatial interaction model to some migration data, or more particularly to certain regularities between variables hypothesized to be causally related to migration, such as the ratio of job vacancies in the regions of origin and destination. It's a safe bet that the parameters of the model will have to be adjusted for other data sets. Even if we feel able to call the relationship a 'regularity' and even if it fits with our expectations we have not shown what actually produces it. Note that this does not mean that regularities can never be causal, but rather that whether they are can only be determined by a different kind of analysis.

Before saying what this different kind of analysis would be like, let us stop to ask a crucial question: why is the search for regularities seen as so central if it does not provide explanation? The answer is that 1 (above) is misguided for the simple reason that *what causes something to happen has nothing to do with the matter of the number of times it has happened or been observed to happen and hence with whether it constitutes a regularity.* Unique events are caused no less than repeated ones, irregularities no less than regularities. Even showing that an individual event is an instance of a universal regularity or law fails to explain what produces it. In failing to grasp this simple point the whole debate of the 1950s and 1960s between the advocates of idiographic and nomothetic conceptions of geography was ill-founded.

We will return to the issue of regularities later, but we can now turn to what is in my view a more satisfactory route to explanation – the realist route expressed in 2. Discovering causes involves finding out what *produces* change, what *makes* things happen, what *allows* or *forces* change. The italicized verbs are all causal terms and they all allude to mechanisms or ways of acting of objects. These mechanisms exist in virtue of the structure of the objects that possess them. For example, people have the power to build houses or program computers in virtue of their anatomical and mental structures, as modified by experience and education. This is not a tautology, for there is a difference between such powers and the structures which 'ground' them. Sometimes powers which we might initially attribute to individuals (whether people or other objects) turn out to be grounded in structures of which those individuals are just a part. For example, the powers of the Prime Minister (*as* Prime Minister and not merely as Margaret Thatcher or whoever) derive not from the individual occupying that position but from the wider political structures on which the existence of that position is dependent. Some mechanisms are thoroughly mundane and already familiar in everyday discourse; for example, when we talk about the *cultivation* of land we are referring to a familiar causal mechanism. Many other mechanisms, and especially social structures, are more elusive and their identification may be a central concern of research, both theoretical and empirical; for example the debates over whether the mechanisms which produce a segregation of low income households into poor quality housing are simply grounded in the actions of autonomous individuals such as housing officers or in the wider financial structures which support the (re)production of the built environment in capitalist society.

In explaining processes it is not enough merely to posit such mechanisms and structures: careful observation and description are needed to check whether the objects to which the properties are attributed do in fact have them and to eliminate other possible contenders which might equally be responsible for the effects we want to explain. Much of the business of research is concerned with precisely such questions. Now it might be thought that the standard methods of mainstream geography, including, centrally, the search for regularities, might be appropriate for this task; if certain conditions B are causally necessary for the existence of A, then whenever A occurs, so must B, thereby constituting a regularity. The latter is true, except for the qualification that obviously if the number of As is very small then the relationship will hardly constitute a 'regularity'. But the converse is not true: that is that B always accompanies A does not establish whether the connection is necessary. However, in many cases, the regularities produced by necessary conditions or relationships are not very interesting, either because they are already known (e.g. the relationship between brains and the power to think) or more generally because the regularity itself does not disclose its own status.

It might also be thought that causal mechanisms would produce regular effects, but this depends on circumstances. If the object possessing the mechanism and/or the conditions in which it operates change, then we shouldn't expect a regular outcome. For example, if we were studying a case of land use conflict, we wouldn't expect the actions of a particular pressure group to produce regular effects if either (a) its internal structure was changing (e.g. through internal disunity) or (b) the external political and legal environment were changing. So we cannot 'read off' effects simply from a knowledge of the existence of certain mechanisms; we must find out what conditions the mechanisms are operating in as well as discover the state of the mechanisms.

We can now answer the question posed earlier concerning why some disciplines, such as physics, seem to be able to progress by searching for regularities, while others, such as geography, do not. We have just seen that the production of regularities by causal mechanisms depends on two conditions – that the internal structure of the mechanisms and their external conditions must be constant. A system which satisfies these criteria is called a *closed* system; those which don't are termed *open*. Physicists find that many of the objects they study exist spontaneously in closed systems and that where they don't they can often be produced artificially, through the scientific experiment. Geographers – and also other social scientists, and some natural scientists (e.g. geologists) – deal with open systems. Indeed the distinctive quality of human beings that they can *learn* to understand and respond to their circumstances in novel ways guarantees that social systems are open. So the success of physics has little to do with its maturity and plenty to do with the nature of what it studies, with obvious implications for the lack of 'success' of the social sciences – particularly those that follow methodological prescription 1, above. Yet, even in physics, although the availability of closed systems and hence regularities facilitates prediction, physicists still have to switch to realist methods for *explaining* regularities. (In some cases, however, they rest content with non-explanatory prediction.)

This argument may at first sight encourage a certain gloom and despondency about the possibility of progress in open system subjects, but there are compensations. In the case of *social* systems, we have an advantage over our counterparts in natural science in that actions and many other social phenomena have an intrinsic meaning which we can *understand*, and for which no equivalent exists in natural science. Atoms don't act towards one another on the basis of any understanding: people usually do, even if it isn't a correct understanding. More particularly, people's *reasons* for doing things can often be taken as the *cause* of their actions (not always because they may not always know their reasons). So through already knowing or being able to ask people about these matters we have an 'internal access' to our objects of study which, although fallible, is simply not open to natural scientists. And this is also a reason why realists can and must accept a crucial element of the manifesto of humanistic geography: viz. we must understand what people understand, mean or intend by their actions and not merely rely on measurements of their overt physical behaviour.

THE EXAMPLE OF RESEARCH ON 'MANUFACTURING SHIFT'

At this point, having come through some rather compressed and abstract methodological argument, it may help to go through a more extended substantive example drawn from geographical work, and one which can demonstrate the consequences of following either of our two methodological principles. It concerns some recent research on 'manufacturing shift' in Britain; that is the tendency for metropolitan areas and large towns to suffer major losses of manufacturing employment while small towns and rural areas enjoy gains.

In this case there is a striking, if approximate and transient 'regularity': the larger the settlement the greater the job loss, with the smallest actually showing an increase, albeit small in absolute terms. The 'messy' and transitory nature of the 'regularity' reminds us that it is hardly a closed system – indeed what else could one expect of something like employment change?! Most researchers have sought explanations through strategy 1 (searching for regularities). Accordingly various hypotheses were tested, such as the possibility that the pattern reflected (i.e. was determined by) whether places had Assisted or Non-Assisted Area status. In this particular case, the hypothesis was rejected, for Non-Assisted Areas such as East Anglia and the South Coast, as well as parts of Assisted Areas (e.g. Northumberland), experienced employment growth. Conversely, the larger settlements within Assisted Areas seemed to suffer decline no less than those in Non-Assisted Areas. But let's just reconsider whether this really was a refutation of the hypothesis. Now most researchers cast the hypothesis in the form of an expectation of an *empirical regularity*: that employment growth or decline would vary according to whether the area was Assisted or Non-Assisted. But there is another way of casting such a hypothesis – one which refers to a mechanism and which is therefore more in keeping with a realist approach. We could specify how we thought regional development grants and other incentives might influence firms – not just by positing a simple regularity between incentives and employment change but by thinking through what *kinds* of firms would find them an attraction and why (e.g. in relation to financial resources, capital intensity and to competing influences). This recasting of the hypothesis would require a different (and possibly more time-consuming) style of research to check its empirical adequacy. Interviews with firms, taking into account their differing circumstances, might be required. And it would probably produce complex answers, hedged round with qualifications about intervening, mediating mechanisms – but then this is what decision-making is usually like. We might find that even though, at an aggregate level, there is no firm correlation between Assisted Area status and employment decline, there were nevertheless cases where location and *in situ* investment decisions were influenced by regional incentives. However, this need not always translate into employment increases, for grants can be used to replace workers by machines! This latter kind of mediating factor is very common in this field of study, but a realist approach is more likely to pick it up than one which concentrates on regularities. So rather than reject the hypothesis because the expected regularity did not occur it would be more reasonable to say that it is appropriate (albeit with certain qualifications) for some cases but not for the majority.

Now this may seem a rather pedestrian conclusion, but behind such issues lies a confusion which is endemic in mainstream methodology. If we ask what determines the location of investment the answer will probably be that several different mechanisms either do or could do. We might then try to find out which ones do, and *how they work* and produce certain effects when they do. These might be called 'intensive' research questions and answering them would require looking at some instances to see how the mechanism works. But discovering how a mechanism works does not tell us how widespread that mechanism is; for example in how many instances were regional incentives influential? The latter can be called an extensive research question: instead of asking how a mechanism works it enquires into the *extent* of certain phenomena, be they events, mechanisms or (as is usually the case) regularities. The basic confusion of so much of mainstream methodology is that this second kind of question regarding distribution and extent is expected to answer the first, regarding causation. And in an open system, where mechanisms generally do not stand in a fixed relation to their effects, this expectation is especially disastrous. Yet it is virtually built into what are widely taught as 'scientific methods' – particularly in the cavalier transfer of methods of statistical inference often originally developed for the study of largely controlled and closed systems, such as experiments on plants.

Research on manufacturing shift has replicated these problems; most of it has been based on hypotheses which predict regularities and it has repeatedly confused the intensive and extensive research questions by expecting the latter to do the job of the former. One exception was research on the spatial consequences of industrial restructuring, the best-known example being that of Massey and Meegan on the electrical engineering industry (1979). This research did not put forward hypotheses, at least not in the form of expectations of empirical regularities among sequences and patterns of events. Rather it sought to research how a certain kind of mechanism – restructuring – worked and how its effects were mediated by various contingent conditions. Only subsequently, and through covering most of the instances in a small population of firms, was it able to offer answers to the extensive research question regarding the *extent* of the effects of restructuring. Conducting the research in this order was especially necessary given that often the same mechanism could produce different effects (e.g. the introduction of new technology can lead to additional or reduced employment, depending on conditions) and conversely, the same effect could result from different causes (e.g. job loss due to loss of markets or automation).

One response to the pressure to restructure was found to be the search for cheaper and more productive sources of labour, and so many firms relocated to take advantage of female labour. Later, another researcher, David Keeble, translated the restructuring thesis into a hypothesis about a possible regularity, an association between the availability of female labour and manufacturing shift (Keeble 1980). The evidence for such an association turned out to be weaker than some of the competing hypotheses concerning manufacturing shift. But this was hardly fair for (a) it failed to note that the original analysis did not claim that a shift to locations with abundant female labour was the only possible outcome of restructuring; and (b), and related to this, it misidentified what was primarily an intensive research project as one concerned with and assessable by extensive questions (Sayer 1982).

Keeble also noted a particularly strong association between manufacturing employment change and what he termed 'rurality', which led him to conclude that 'rurality' was a 'causal factor'. But in what sense could 'rurality' be said to be a causal factor, or in other words how does 'rurality' 'produce effects on' or 'operate or work upon' manufacturing? The obvious question (and a typically realist one) is: what is it *about* rurality that could produce such a change in manufacturing? Clearly the concept of 'rurality' needs 'unpacking' in order to see if it covers any relevant constituent mechanisms and conditions, for otherwise its statistical association with manufacturing shift could be interpreted as an accidental one.

But while many, perhaps most, mainstream researchers would ask our realist question, their standard methodology of seeking further regularities between more specific variables would not provide an answer, for it would merely re-pose the underlying methodological problem: that regularities do not explain themselves.

So to summarize the key points of the discussion so far: although 'regularities' are often interesting, causation is not a matter of regularities in events but rather of the mechanisms which produce them. Likewise what is interesting about regularities is (a) simply their pervasiveness and (b) their determinants. Mechanisms will only produce the precise, enduring regularities that are the goal of quantitative geography's 'search for order' in closed systems and these are not available in human geography's object of study (except perhaps in so far as machines are included in the latter). In social systems, which are open, the activation and effects of causal mechanisms will depend on the nature of the objects which have them and the conditions in which they happen to exist. Given this openness, while we may find that the search for regularities and generalizations may contribute answers to questions regarding the quantities and distributions of phenomena (extensive questions) it will throw little light on what produces them (intensive questions).

A CLOSER LOOK AT 'THEORY'

Having battled through the complexities of causal explanation we can now return to the question of the changing nature of *theory* in geography; as we shall see the two main competing conceptions of theory complement the two alternative routes to explanation explored above. This will also help to clarify how the existence of less familiar causal mechanisms and structures is established.

The ordering-framework conception of theory clearly fits neatly with the search for regularities as a route to explanation. On this view description and conceptualization of data is an unimportant preliminary to the 'real' business of 'science', the attempt to discover a framework into which data can be 'plugged' with a view to predicting certain unknowns. Accordingly, the bulk of what passes as theorizing is concerned with the development and refinement of such frameworks, most prestigiously in the form of mathematical models. If the categorizations in which the data are given are ever questioned it's usually in terms of their accuracy (e.g. susceptibility to measurement error) and only rarely in terms of their *conceptual* adequacy. This can readily be checked by scanning the pages of a few mainstream journals, for example *Environment and Planning A, Geographical Analysis*, particularly for quantitative deductive approaches, *Regional Studies, Urban Studies*, particularly for inductive approaches. (*Annals of the Association of American Geographers* and *Transactions of the Institute of British Geographers* tend to have a more eclectic mix.)

These priorities are inverted in a realist view of 'theory' where, as we noted earlier, conceptualization takes priority over the construction of ordering-frameworks; indeed if the latter are proposed at all then they are not intended as a predictive device. The connection between this emphasis and the realist view of causation lies in the fact that causation is seen as deriving from the properties of particular objects and that if these are to be grasped adequately then considerable care is needed in conceptualizing those objects. Without such care we are chronically vulnerable to the problem of attributing powers to the wrong objects or aspects of objects, in fact this is a central, perhaps *the* central, substantive problem which any enquiry into the nature of the world must face. To what should underdevelopment be attributed? We have now learned that earlier geographers repeatedly attributed to nature what was in fact due to the organization of society in the explanation of underdevelopment. But then we must ask, as we did for 'rurality' before, what is it *about* the structure of certain societies that makes them vulnerable to stagnation and decline? Is it a 'lack of capital' or a blockage of the masses' access to the means of production or what? And again in the case of industrial re-location to exploit female labour, we might ask what it is about women and society which makes them more productive (often) and cheaper to employ. Is it something inherent in their biology or a consequence of their socialization in a patriarchal society? And if the latter what aspects of socialization in particular?

The point to be underlined here is that little progress can be made in answering such questions without exhaustive scrutiny of our concepts of such objects. At first sight many mainstream geographers may be tempted to react dismissively to this by claiming that they already do it. But we should judge them by their actual research and compare it with that of radical researchers following a broadly realist approach. The contrast between the treatment of the question of the meaning of development in modernization studies and radical research is salutary. So a typical realist theoretical question would be, 'What do we mean by development/gentrification/urbanism/collective consumption/class/rent/ community, or whatever?'

And in empirical work, whereas mainstream researchers tend to quickly 'collect' data and do some sort of quantitative analysis on it, realists would emphasize the scrutiny of the categories in which the data are set, *possibly* following up with some quantitative analysis. Empirical research is not just a question of determining magnitudes and distributions once the categories have been finalized; it is also important for checking the adequacy of our categorizations. Indeed, to a realist, description should never warrant the pejorative prefix 'mere', for the adequacy of our descriptions – our applications of concepts – dictates the adequacy of our explanations. Terms like 'service employment' or 'interest group' may look quite innocuous, and in many contexts they are, but as soon as we put some 'explanatory weight' upon them they may lead us astray. The danger of 'services' is that it covers such a heterogeneous scatter of activities that it's absurd to attribute to services certain unitary powers (e.g. static productivity). The danger of the term 'interest group' when applied to groups such as trade unions, tenants' groups, pressure groups, employers' and professional associations is that it tends to imply that each group stands in an equivalent position within the power structures of society so that when we use the term to explain, say, a land use conflict, we overlook crucial imbalances and asymmetries and differences in constraints between such groups.

It is for these reasons that 'abstraction' and 'unpacking' are 'buzzwords' in realist analyses. We try to reduce a complex entity into its component parts, abstracting them out one by one in order to consider their properties. To a limited extent we do this intuitively, but the favoured procedures of mainstream methodology divert attention away from this healthy habit towards the search for empirical regularities among whatever data – however categorized – are available, in other words among phenomena whose conceptualization is seen as unproblematic. By contrast, realism attempts to formalize and develop abstraction.

One way of doing this is by distinguishing between those properties of objects which they possess independently of their relationships to other objects and those which only exist through them. For example part of my behaviour as an owner-occupier is dependent on my relationship to the building society which has granted me a mortgage. Purely individualistic explanations of my actions would therefore be inadequate; my insertion into a specific structure of social relations (in this case concerning indebtedness) would have to be acknowledged as a necessary condition of my behaviour. By contrast, my recreational activity may be independent of my being indebted to a building society. Simply by asking what are the necessary conditions for the existence of objects of interest, we can make considerable progress in working out the (causal) structure of the systems in which they exist. And where necessary relations are discovered, we can make strong theoretical claims about them, for example the existence of A *presupposes* B. Empirical work will then be needed both to check such claims and to discover the configuration (including the spatial form) of the many relationships which are contingent (i.e. neither necessary nor impossible). For example, it is one thing to note that the survival of capital presupposes the reproduction of a labour force (among other things) but the nationality of that workforce is a contingent matter only discoverable through empirical research.

PROBLEMS OF GEOGRAPHY AS SYNTHESIS

I now want to mention some realist ideas concerning the difficulty of doing geography. Prior to the quantitative revolution it was widely thought, following Kant, that geography and history were special subjects in that they were concerned with *synthesis*, in space and time respectively. The quantitative geographers put paid to that idea with their assertions that method was not determined or constrained by subject matter and that any discipline could embrace a nomothetic method: geography was not an exception. This was a bold thesis and it led to some interesting methodological experiments, but it has failed. It is not *only* that geographers deal with open systems but that they place more emphasis than many other disciplines on synthesizing the interactions of diverse systems. *Regions*, for example, are enormously complex and heterogeneous aggregates of diverse processes, some interdependent, others quite independent, and yet we frequently try to grasp the whole. Apart from history, other social sciences tend to limit themselves to particular types of object or system, be they political groupings, markets or interpersonal relations. Now I don't want to exaggerate the contrast, for it is one of degree, and in many cases other social scientists find themselves driven towards broad levels of synthesis; for example, sociologists studying working-class culture face a dauntingly large and complex set of issues. ('Political economy' is perhaps in a different category, at least in its empirical work attempting a very high level of synthesis.) At the very least, geography is far removed in character from those natural sciences which can objectify their abstractions in closed systems, isolating their objects of interest from the contexts in which they spontaneously occur. Physicists may protest that their objects are enormously complex too, but I doubt if they are *in the same sense*.

Consequently, given the heterogeneity and complexity of objects like regions, we cannot reasonably expect works of geographical or historical synthesis to provide exhaustive intensive studies on all aspects of a region. The best we can normally manage is an incomplete picture consisting of a combination of descriptive generalizations at an aggregate level (e.g. on changes in population and standards of living), some abstract theory concerning the nature of basic structures and mechanisms (e.g. concerning modes of production), and a handful of case studies involving intensive research showing how in a few, probably not very representative, cases these structures and mechanisms combine to produce concrete events. Look at any historical geography of the industrial revolution of twentieth-century Britain and you will find such combinations, although many will simply ignore abstract theories and rely on common-sense ideas. Certainly, *any* research is bound to be incomplete, because there are always more things to discover and better ways of conceptualizing what we think we already know; but works of synthesis such as these, which are particularly prominent in geography and history, are also incomplete in a different sense, which has more to do with the practicalities of this kind of research.

There are some kinds of research problems which we can't expect new kinds of philosophy and methodology to solve, indeed it is better that the latter disclose rather than conceal their existence as quantitative nomothetic geography tended to do.

CONCLUSION

A realist geography must be deeply concerned with theory, even in the conduct of empirical research, though in a different sense of the term 'theory' from that understood by mainstream geographers. In so far as it succeeds in discovering necessary connections in the world, its theoretical claims will have some generality, although a realist approach won't expect to find enduring regularities at the level of empirical events. It is not entirely clear which kind of generality a 'nomothetic' approach is concerned with, although most geographers seem to have assumed it is the second. If so, then realist geography might more properly be called idiographic. Given its concern with open systems and its preoccupation with synthesis an attempted nomothetic approach in the second sense can only lead to failure. Finally, if we are to call it idiographic, a realist geography would, in contrast to traditional regional geography, have every right to claim to be theoretical, explanatory and, if the word means anything, 'scientific'.

NOTES

1 For fuller, and I hope more rigorous accounts of realism see my (1984) *Method in Social Science: A Realist Approach* (London, Hutchinson); and in

relation to geography, my (1982) 'Explanation in economic geography', *Progress in Human Geography*, 6, 68–88 (and the references therein).

2   The meaning of the term 'positivist' is notoriously contested and those who are attacked under this heading frequently capitalize on the confusion by arguing about its meaning, thereby evading the thrust of the criticisms directed at them. I only use the label here because it may be familiar to some readers. What matters is not its 'correct meaning' but whether the actual practices – *whatever they may be called* – which I attack are defensible.

REFERENCES

Chorley, R.J. and Haggett, P. (1967) *Models in Geography*, London, Methuen.

Harvey, D. (1969) *Explanation in Geography*, London, Edward Arnold.

Keeble, D.E. (1980) 'Industrial decline, regional policy and the urban–rural manufacturing shift in the United Kingdom', *Environment and Planning A*, 12, 945–62.

Massey, D. and Meegan, R.A. (1979) 'The geography of industrial reorganization', *Progress in Planning*, 10 (3), 155–237.

Sayer, A. (1982) 'Explaining manufacturing shift: a reply to Keeble', *Environment and Planning A*, 14, 119–25.

Scott, A. (1980) *The Urban Land Nexus*, London, Pion.

9
INDIVIDUAL ACTION AND POLITICAL POWER: A STRUCTURATION PERSPECTIVE[1]
James S. Duncan

INTRODUCTION

Given the inescapable interdependence of geography and the other social sciences, one of the central issues that has been debated by geographers during the late 1970s and 1980s is what types of social theory and what modes of explanation are the most fruitful for geographers to borrow from related disciplines. Indeed this is not a one-way street; as geographers are becoming more critical and conscious of issues within social theory they have increasingly joined interdisciplinary debates. In some areas such as urban theory and research geographers such as David Harvey have been influential well beyond geography.

One of the central questions in the social sciences that geographers are becoming more and more attentive to is, 'What is the status of the individual actor within society?' In order to answer this question I will attempt to situate it within the long-standing debate concerning the relationship between structures and human action. I will briefly rehearse the by now well-known arguments against structuralism and individualism and then offer a reconstructed view that avoids the impoverished notion of the individual human actor that is found in these antinomies.

This reconstruction is based on a type of action theory which while acknowledging the central role of actors places them within an ongoing dialectical process by which individuals and structures are mutually constitutive. It follows Giddens in his 'structuration' view of the relationship between individuals and structures which claims that 'social structures are both constituted "by" human agency, and yet at the same time are the very "medium" of this constitution' (Giddens 1976). Elements of the model I will present are also drawn from such social theorists as Clifford Geertz, Antonio Gramsci, Alain Touraine, Victor Turner, and Dominick La Capra.

It will be argued that central to this structuration view of individual action is the concept of ideology which plays a key mediating role between various actors and the social system of which they are both products and producers.

A common problem with discussions of social theory in geography and elsewhere is that concrete examples which would help to flesh out and illustrate theoretical categories are often lacking.[2] In order to avoid this problem I will illustrate the reconstructed approach to structure and action presented here with some data I collected in Sri Lanka on the interrelationships between various actors: the citizens, state officials and the ideological context and 'texts' of political legitimation. I will elaborate on the notion of texts below, however first I wish to proceed with the discussion of the structuralist and individualist positions which I wish to counter.

CRITIQUE

Several years ago John Agnew and I, recognizing the derivative nature of our discipline, outlined a framework that could be used to evaluate theories that were borrowed from other social sciences (Agnew and Duncan 1981). We argued that all types of social scientists, whether they are aware of it or not, necessarily adopt a model of human action when they engage in the explanation of social phenomena. All such explanations contain often implicit, sometimes explicit assumptions about the relationship between individuals and social, economic and political structures. We suggested that it is important for geographers to be critical of these types of assumption. Therefore I will now turn to a critique of this set of antinomies with particular reference to theories utilized by geographers.

THE STRUCTURE–ACTION DEBATE

This debate centers on the issue of whether or not explanation in social science should be sought in terms of autonomous or quasi-autonomous structures or whether it is to be found in the actions of individuals. Structuralists in support of their position often point to the fact that there are regularities in the behavior of groups of people which enable us to talk about a group having a stable structure in spite of its fluctuating membership. They claim that rather than discussing facts about individuals it is better to examine the social roles that are filled by different people at different times. This line of reasoning often, although not always, leads into a functionalist mode of explanation. If within groups there are regularities of behavior not accounted for by the intentions of individuals and yet these regularities seem to serve the purpose of maintaining the group as a whole, then explanation is framed in terms of social wholes. Within the structural-functional tradition many of the research problems involve identifying the ways in which various elements in a social system are functional or dysfunctional to the working of the system. Thus in this type of explanation there is an absence of individual social psychology.

There have been a number of objections raised against this type of approach to explanation. The first of these centers on the problem of how these groups and organizations are maintained if not by the will of the individuals who participate in them. The second revolves around the fact that individuals are represented as if they were 'cultural dopes' who somehow mysteriously conform to the will of institutions. Functional explanations which are predicated upon the 'needs' of the social whole such as the group, the organization, or capitalism, raise a related set of objections. Can one ascribe needs, intentions, and goals to social wholes apart from those of the individuals that compose them? Surely these social wholes cannot be said to possess these needs and intentions unless one accepts a teleological position – a belief perhaps in some supernatural purpose, in the hand of God, social evolution, or the like. I have developed elsewhere some of the difficulties in attempting to sustain such a position (Duncan 1980; Duncan and Ley 1982).

Within geography an example of structural explanation can be found in the work of certain Marxists. This particular version of Marxism, and it is important to note that it is only *one* variant of Marxism, can be found in some of the writings of Harvey (1975), Peet (1979), and Santos (1977). It envisages the social structures of advanced capitalism being driven forward toward some preordained end. These structural Marxists see 'the logic of capitalism', 'social formations', and 'the capitalist mode of production' as autonomous and self-determining structures which display regularities that cannot be accounted for in terms of individual or collective action. Individuals in this view are essentially passive agents who simply act out their roles as agents of their class within the grand social evolutionary schema. The rhetoric of structural Marxism may sound radical but the model of human action is ultra-conservative, portraying actors as prisoners of fate caught in slowly grinding structures not of their own making, waiting for the evolution of the social structures through contradictions to move them on to a higher stage in the evolutionary process.[3] In general, the structural approach in social science appears to create an imbalance, that is, as Geertz (1973a) has pointed out, its sociology is too 'muscular' and its psychology is too 'anemic'.

Institutions in this perspective, too, strongly determine action, while consciousness is dismissed as a product of ideology and hence false. The structural approach is deeply, some would claim fatally, flawed by its inability to demonstrate adequately the ontological relationship between autonomous social structures and individual action and by its teleology. It is perhaps for this reason that structural functionalism declined rapidly in American sociology after 1970 (see Gouldner 1971) and why Castells (1978), one of the foremost structural Marxists of the 1970s, has moved away from this position.

In opposition to this view is a model based wholly upon individual attitudes and intentions. Here the regularities that appear in social life are explained by reference to individual and collective action rather than by autonomous and reified structures. Emphasis is placed in this view on intentionality. The problem with much of this analysis, which is commonly found in humanistic, behavioral, and economic geography, is that it is overly individualistic and idealistic, focusing exclusively upon the ideas and behaviors of individuals with little regard to the socio-cultural and institutional structures of which their individual acts are an integral aspect.

Within geography let us consider for a moment as an example of an overly individualistic action theory the work in the neoclassical economic and behavioral traditions. Explanations in both these traditions are based upon the assumption that social explanation must ultimately be based upon individual attitudes and desires. Scholars working within these traditions pay virtually no attention to either the role of the state or social and economic institutions, and consequently the socialized nature of individual actors within an institutional framework is ignored. They tend to employ a model of society that is classless and based upon social harmony. Individuals are seen as having free choice and the social origins of their preferences are not investigated by the researcher. As a result of these assumptions the individual is seen in generic terms and the social context of action ignored. If the structural model overemphasizes the context into which individuals are dissolved, then the individualistic variant of action theory does the reverse. The irony is that this form of hyper-individualistic action theory is based on nearly as impoverished a social psychology as the structural model. To posit the actor as a free decision-maker may have appeal; however it fails to reveal the richness and complexity of the origins of social action and individuality.[4]

RECONSTRUCTION

The difficulty in constructing a workable theory of action is to avoid on the one hand the determinism of the structural view and on the other the idealism and hyper-individualism of some non-structural approaches. One of the most influential social theorists at present among social geographers is Anthony Giddens (1979; 1981); for an example in geography see Pred (1984). His notion of structuration is an attempt to mediate between structure and consciousness. For Giddens structuration means that the structural properties of a social system are at the same time the medium and the outcome of the practices that constitute that social system. He differs from many earlier theorists in that he considers structures to be enabling as well as constraining. They are integrally involved in the process of social action. He argues that the structural conditions in a society are an invaluable resource to be used by those members of a society who seek power. There exists, in other words, a dialectical relationship between structure and action. Structure is not merely context or background, it is a mode of action, a property of social interaction. These structuring properties are composed of rules and resources through which social systems are reproduced. Structures then are processes, and as such are continually being modified as the action which constitutes them changes.

It would however be a great mistake to think that every individual has an equal ability to modify structures. A person's ability to intervene in the social process depends upon his position in the society. Politicians, for example, while subject to many structural influences are apt to contribute to the structuration process to a greater extent than is a factory worker. Since the former are able to manipulate structures on a larger scale they may be less likely to take these structures for granted or see them as fixed elements in the cultural or political environment. And yet ordinary individuals collectively probably have a greater influence on the reproduction and transformation of structures than do those in positions of power when considered individually; the 'great man' theory of history therefore tends to exaggerate the importance of powerful individuals. Ordinary individuals are a force not only because of their vast numbers but also precisely because they are alienated from the structuration process due to the fact that their actions, from the point of view of the individual actor, have an imperceptible effect upon it. This often unquestioning acceptance leads, for better or worse, to the reproduction and institutionalization of the situation (N.G. Duncan 1984).

The concept of ideology is related to structuration in a number of important ways. It can be seen as a set of values, ideas, and taken-for-granted cultural prescriptions which mediate between individuals and the structural properties of social systems. If not clearly defined the concept may be confusing, however, because of the many different, often pejorative, connotations associated with it. For conservative thinkers such as Daniel Bell (1962) and Edward Shils (1958) it is associated with what they consider to be politically distorted but formal doctrines such as Fascism and Bolshevism. For theorists influenced by Marxism such as Habermas (1972), Althusser (1969), and Lefebvre (1976), it is a set of ideals including everyday assumptions, which distort reality thereby serving the purpose of justifying class interest and domination. On the other hand, Clifford Geertz (1973b) argues for a non-pejorative notion of ideology, expanding the concept to mean a belief system. My use of ideology is closer to that of Geertz, although somewhat narrower, in that I will use it in a non-evaluative sense to mean a body of interrelated ideas which include taken-for-granted assumptions as well as more formal doctrines. Examples of ideologies might be, for example, equality of opportunity, possessive individualism, the right to national self-determination, and the divine right of kings. They are, in other words, beliefs about what constitute natural or just social arrangements. A non-evaluative concept of ideology is useful because it is less open to a priori theorizing and functional interpretations. This has the advantage of leaving open to investigation what the effect on society of certain beliefs may be. While some ideologies might have the effect of serving the interests of certain dominant classes and may serve to strengthen their political or economic position in society, other ideologies may have a quite opposite effect. When an ideology which presents social arrangements that favor the interests of a dominant group as being in the interests of the community as a whole becomes widely accepted its effect may be said to be hegemonic. Only at this point does the concept of ideology take an evaluative connotation. Such ideas may be perpetuated through the conscious manipulation of popular channels of communication or they may be cultural values held by the great majority of the population including members of the dominant class, who may or may not accept them unquestioningly in reified terms. In either case ideology is best described as an outcome of a structuration process whereby individual consciousness both shapes and is shaped by ideology. Ideologies are either reproduced or transformed by the acceptance or nonacceptance of individuals.

Perhaps one of the best-known treatments of hegemony is found in the writings of Antonio Gramsci (1971). He believed that open class conflict was largely avoided because the dominant culture becomes diffused throughout society, pervading both its institutional and private domains. But Gramsci, as Femia (1975) points out, did not conceive of the underclasses as unthinking types who were totally duped by the dominant group. He placed a great deal of emphasis upon the fragility and instability of the consensus. There exists, he claims, a contradictory consciousness in the ordinary individual reflecting the incongruity between an abstract ideological perspective and the objective reality of his or her own life situation.

Alain Touraine (1981) adds a slightly different twist to this argument and it is one that I wish to pursue here. He claims that at present the principal form of domination is not by the bourgeoisie over the proletariat (slippery social categories in today's complex societies), but by the state apparatus over the citizens and that this domination is perhaps greatest in the Marxist states. Put slightly differently and in terms more suitable to an action approach, increasingly the agents of domination are government officials, members of an intelligentsia who in the name of one ideology or another are gathering increasing control over people's lives.

Not all ideologies serve a political purpose; however some ideologies play a central role in legitimating the central governing institutions in a society. Ideology, then, is intimately connected to the classical problem of legitimacy – how is it that some people come to be credited with the right to rule over others? Ideology is an important facet of action theory precisely because it is a socio-psychological concept. It is a mechanism which links institutions to actors. If the ideology is accepted by the actors then it makes the central institutions seem proper and an appropriate extension of the individual actors. In other words it provides them with legitimacy. If on the other hand, as sometimes happens, some of the actors fail to accept the ideology then the legitimacy of the institutions is questioned and government officials may need to use force in order to maintain the central institutions.

In order to be widely understood and acceptable, political ideologies tend to draw more or less explicitly on the broader cultural tradition of a society. Ideologies, seen from a structuration perspective, are not static, they are products of ongoing action and are therefore in a process of transformation. They are created out of the general cultural context or drawn from an outside set of ideas that have been grafted onto the cultural background. This ideological base then becomes modified as politicians utilize it to justify their policies or to legitimate their power. It may be further transformed as citizens react positively or negatively to it or to particular aspects of it. Ideologies should not simply be conceived of as 'lies' that government officials tell citizens. Ideologies serve the purpose of legitimation precisely because they are able to capture and summarize a part of the citizens' experiences or yearnings. Political leaders, whether they are defenders of the political establishment or reformers, often fervently believe in the righteousness of the ideologies they are espousing above and beyond the knowledge that this will serve the purpose of rallying support to their cause. Politicans it must be remembered do not stand outside of the society's cultural context; like the rest of the citizens, their consciousness is formed within this cultural matrix. Ideology then is a process. It is a form of interaction between political systems supported by officials and citizens, usually characterized by the domination of the former. But that domination is fragile, as Gramsci has argued, for ideology is a socio-psychological process and the citizens can and do evaluate, accept, reject, or modify the ideology.

Ideologies have an important symbolic component (Geertz 1973b). The notion of symbol is used here in a general sense to refer to something which signifies something else. As such, symbols are simplifying devices used to communicate complex sets of ideas. The symbol stands for the idea; it is something that people can easily grasp and understand. The promulgation of a successful ideology, therefore, is dependent upon the adoption of a set of symbols which can summarize and communicate it effectively. Given the tremendous importance of symbolism not only in political life but in cultural transmission in general, it is ironic that the term 'symbolism' is used in popular parlance to mean that which is somehow unreal or lacking in substance, that is the leader's action is seen as being merely symbolic. Action is always symbolic but never merely so. It serves the purpose of communicating a set of ideas.

Symbols which stand for important ideas in a society are not simply strewn randomly across the intellectual landscape. They are ordered, stitched together into a coherent pattern, formed into texts which have a narrative structure woven through them. The term 'text', which has come into increasing prominence of late within the humanities and social sciences (see La Capra, 1983; Geertz 1973a), derives from *textere*, to weave or compose. For social scientists it is often used metaphorically not only to capture the interpretive role of actors within a social structure but also the interpretive role of the social scientist *vis-à-vis* the data. I would argue that three types of social text can be identified: written texts composed of offical documents which structure action, oral texts which shape to a large degree the popular consciousness of individuals, and landscape texts which represent the transformation of the former two into the medium of the concrete.

Ideology, then, is communicated through the narrative structure of these three types of text. Although the texts structure action they are also in turn structured by ongoing action, for they are open to interpretation and hence change. In this respect, of course, the notion of text is remarkably like that of structuration. The concept of text, however, has a particular usefulness as an interpretive tool as it allows one to conceptualize the symbols that compose ideologies as a series of transformations of a communicated code.

Landscape texts should be of great interest to geographers who traditionally have been interested in the man–environment relationship.[5] Such texts can, in turn, be subdivided into objects, structures, and places. Examples of objects whose primary purpose is symbolic communication would be flags, royal regalia, and statues of leaders, past or present. On a larger scale there are symbolic structures such as Lenin's tomb, the wailing wall, and Sacre Coeur cathedral, and finally there are places of symbolic importance such as Peterloo to the English, Selma, Alabama to Americans, and the city of Kandy which serves for Sri Lankans as a symbol of Ceylonese resistance to British imperialism. Not uncommonly individuals who have been killed in the name of some political cause have been memorialized in the landscape on all these various scales. One such individual is Miguel Hidalgo, the parish priest at Dolores who started the Mexican Revolution of Independence in 1810.

> The cultural landscape itself is signed with Hidalgo's name. An entire state and many towns, suburbs, parks, and streets bear that name, while every year at midnight on September 15 the president of the republic repeats from the main balcony of the National Palace in Zócalo of Mexico City the supposed words of his *Grito* at Dolores, literally, his 'Cry', his proclamation, 'Mexicanos, viva Mexico!' Statues of Hidalgo abound throughout the land in plazas and parks. (Turner 1974, 99)

The landscape serves as a vast repository out of which symbols of order and social relationships, that is ideology, can be fashioned. Such symbols, as I discussed earlier, are social, that is they arise and are sustained within the public arena. The landscape thus should be viewed as a text that can be interpreted by those who know the language of built form. The interpretation of symbols in the landscape may be a conscious and articulated activity; to use Giddens's (1979) term it may constitute 'discursive knowledge', or it may more commonly be a form of 'practical knowledge', a kind of tacit understanding among participants in a cultural system. The concreteness of the landscape makes it seem more real, more unquestioned, more easily graspable both intellectually and emotionally precisely because it is a concrete object or place that can be visited, touched, venerated, or otherwise celebrated. It is for precisely this reason that statues, buildings, and places are used to symbolize belief systems.

A STRUCTURATION APPROACH TO POLITICAL LEGITIMATION: A CASE STUDY

In the preceding section I argued that ideology served to mediate between institutions manned by state officials and the citizens. Although this relationship is generally one of domination, citizens can and do influence the ideology; its use is constrained by the cultural-historical background of the ideas and by expectations of citizens' reactions.

THE IDEOLOGICAL FRAMEWORK

In Sri Lanka there are three principal ideologies of authority, two of which are long-standing and closely interwoven while the third is a relatively recent import from the west. The first concerns the role of the ruler as a supporter of Buddhism, the second focuses upon the national integrity of the island, and the third embodies English notions about parliamentary democracy. Each of these ideologies is composed of a set of 'written texts' as well as 'oral texts' and each in turn has been transformed into an accompanying 'landscape text'.

The ruler as a supporter of Buddhism

Buddhism has been closely associated with kingship in Sri Lanka since its arrival there in 250 BC. Relics have played an important role in the religion as symbols of the Buddha, and therefore when a tooth relic of the Buddha arrived on the island in the fourth century AD it became the palladium of kings, and battles were fought by rival claimants to the throne to take possession of this symbol of sovereignty. The king came to be seen as a potential future incarnation of the Buddha, one of his principal reasons for ruling being to protect and enhance the religion. This ideology, which was elaborated in such 'written texts' as the *Mahavamsa* (sixth century AD) and elaborated over the centuries in 'oral texts', has formed a part of the popular culture and is inscribed in the urban fabric of Kandy in the relationship between the palace and the various temples. This ideology was kept alive during the period of British domination (1815–1947) by the priesthood and certain nationalist leaders, in part, at least, to deny the British the legitimacy they sought. Since independence political leaders have had to deal with this ideology or ignore it at their own risk.

Nationalism

The issue of nationalism is inextricably intertwined with the religion. Again, the *Mahavamsa* records that the Buddha himself ordained that the island of Lanka was to be the land where Buddhism would flourish. Sri Lanka, in other words, has a national destiny to fulfill the wish of the Buddha, a destiny which can only be fulfilled by Buddhist rule over the whole of the island. Perhaps the most popular part of the *Mahavamsa* deals with how Duthagamini (101–77 BC), the militant Buddhist king, killed his Tamil rival and established Buddhist control over the whole country. The *Mahavamsa* was invoked to support the nationalist cause against British domination during the colonial period and at present against western and Japanese cultural and economic imperialism. It is also interesting that this reasoning is still used by many Buddhists to argue against the Tamil nationalists who today support the idea of a separate Hindu state in the north of the island.

Democracy

This third strand in the ideological framework is of more recent origin and possibly less secure than the other two strands. It is embodied in 'written text' in the form of a constitution and in an 'oral text' in popular expectations about the duties, and rights, of citizens and political leaders. But these texts are not accorded a sacred character like those of the other two strands of the ideology and as a result both the constitution and peoples' expectations are not infrequently revised.

THE MUTUALLY CONSTITUTIVE RELATIONS BETWEEN POLITICAL STRUCTURES AND CITIZENS' ATTITUDES

In this section I will consider the manner in which ideology serves to legitimate the political order. Particular attention will be focused on the interaction between state officials who propagate the ideology through manipulation of 'environmental texts' and citizens in the formation of this order. One can also see that the influence of past actors, those many actors who produced and reproduced the religious and political 'texts' over the centuries, particularly those who kept such 'texts' alive during the British rule, impinges heavily on the present structuration process.

Traditional ideology as enabler and constraint

Political officials such as the president of the country in seeking political legitimation need not devise *de novo* a form of legitimacy. In this sense the traditional ideology serves, from their point of view, as an enabling structure. But it also at the same time acts as a constraint for it places bounds upon the types of action that are considered appropriate. These can be traversed but only at the risk of delegitimizing one's power. The president, therefore, does not create the structures of legitimation, he operates within them. It is an historic creation, fashioned over the centuries by kings, priests, and commoners and maintained and elaborated over that time in the popular consciousness through written, oral, and environmental texts. The democratic strand in the ideology, although a more recent import, is well entrenched and therefore also serves the same enabling and constraining function. It must not be forgotten however that the president and other officials not only use ideologies, they contribute to their reconstruction and transformation as I will show below.

From ideology to political action: exploring the limits

The first two strands of the ideology (the ruler as a protector of Buddhism, and nationalism) are problematic for contemporary officials in that both were elaborated centuries ago under a different political system (quasi-divine kingship), for they must therefore be made consonant with the third thread (modern democracy). This, for the present president, involves a conscious reaching back to the days of the kings for symbols to be tested out in the contemporary political arena. I will focus here upon some of the environmental components of this process. The 'environmental text', as stated earlier, is but a transformation of the written and oral texts, and yet it is terribly important in that it concretizes and makes immanent these texts. A recognition of the importance of the environmental text is even to be found within the society's 'written texts', ' I speak thus [in images] because of the frailty of the intelligence of the tender children of men' (Lankavatara Sutra 1972).

In his attempt to appear to be an 'ideal Buddhist ruler' and nationalist to the people, the president has modeled his actions with respect to the symbolic environment upon those of the kings of Kandy. He spends much of his time visiting Buddhist temples throughout the country as did the kings. His first act after his election was to come with his cabinet to the Temple in Kandy to worship the relic of the Buddha, the palladium of kings, and he is the first modern head of state to address the nation from the Octagon of the Temple in Kandy where the king addressed his people. The president, in these and many other ways, acts out the traditional model of the Buddhist king and yet while the rhetoric is about Buddhist antimaterialism and cultural nationalism, since 1977 his administration has opened the country to a flood of western commodities and cultural values that seem to directly contradict the ideology that he propounds. Many Sri Lankans wonder how long this 'Janus-faced' president will be able to orient his ceremonial political actions to a nationalist past and his economic ones to an international present. Perhaps the disjuncture will lead to his political undoing or it may be the very thing that will allow him to carry out his western-oriented economic program.

Citizen reaction: a fragile consensus

The attempt to draw upon traditional sources of legitimacy entails a risk for functionaries for it is no simple matter to draw upon an ideology which was shaped under a past social system, in this instance based on quasi-divine kingship, and apply it to a very different one based on parliamentary democracy. Success depends upon the subtlety with which the symbols are blended. With this in mind I conducted with the help of assistants 170 interviews in an attempt to gauge citizen reaction to the president's use of traditional symbols of political legitimacy. An analysis of these interviews reveals that whereas the majority of the citizens accept his use of the symbols, a sizeable number do not, and that these skeptics tend to be either more highly educated or supporters of the opposition party in parliament. The consensus on his use of the traditional symbols of authority is, in other words, a fragile one. A number of things worried people about the president's use of the symbolism of the Buddhist kings. That he should model himself upon the ideal of the righteous, religious king was seen as admirable but it was felt that he often went too far in his use of the symbols of kingship. For example, going with his cabinet to worship at the Temple of the Relic was seen as highly appropriate, as was visiting the temples on a regular basis, although many thought that his visits were designed primarily to get publicity and drum up political support among the influential clergy. Much more controversial was his emulating the last king of Kandy in addressing the nation from the Octagon of the Temple. In this instance many people felt that he had gone too far and was starting to think of himself as if he really *were a king*. The Temple of the Tooth in Kandy is so charged with symbolic meaning that its use makes a powerful political statement. The difficulty, of course, in adopting king-like symbols lies with the democratic strand of the ideology. How literally were the president's symbolic gestures to be taken? Was there any connection between this symbolism and his tampering with the constitution, or with his cancellation of the last election in favor of a referendum on his rule? Were the first two strands of the ideology of legitimation to overwhelm the third one, and was some form of dictatorship going to be the result?

The use of ideology in legitimating power is dialectical. The official draws upon a traditional source which is an historic creation and then attempts to shape the elements of this ideology to conform to the political and social realities of the present. The citizenry reacts to this manipulation of ideology and this reaction shapes the future use of it. In the case of Sri Lanka the jury is still out on the effectiveness of this ideology, in part because of the present dominance of the issue of Tamil separatism.

CONCLUSION

The role of the individual actor in the social process has been hotly debated by academics for generations, although not infrequently the debate has produced more heat than light. In the hope of adding to the latter rather than the former I have argued that traditional approaches have either erred on the side of overstressing the power of structures, as in structural Marxism, or conversely have granted near autonomy to the individual, as in neoclassical economic and behavioral geography. Both of these perspectives, although radically opposed in many respects, share a common failing; they do not capture the richness and the complexity of social action because they are underpinned by an impoverished social psychology which does not adequately relate individuals to social structure.

I therefore suggested that a reconstructed view of social action could be fashioned from the work of a number of theorists. First, Giddens's notion of structuration provides us with a basic building block to construct a bridge between the individual and social structure. To this I added the concept of ideology, a key socio-psychological mechanism underlying the structuration process. Gramsci, Geertz, and Touraine provide us with different facets of the concept of ideology but share the view that it is a symbolic system. Finally from La Capra I drew the notion that symbol systems could usefully be conceived of as texts. In this view ideologies are a type of social narrative which undergo transformations into written, oral, and landscape texts. As such they make manifest the ideology and tie the actor in his everyday use of the environment into the structuration process.

Finally, since discussions of this nature are notoriously and, suspiciously, I might add long on theory and short on supporting data I illustrate my position with data which I collected on legitimation of power in Sri Lanka. The bare bones of the theory are fleshed out by demonstrating the manner in which citizens and officials participate in the structuration process, both shaping and being shaped by the ideological framework of the society.

NOTES

1. My research in Sri Lanka was funded by the Social Sciences and Humanities Research Council of Canada. I am grateful to Professor Gerald Pieris, Head of the Department of Geography, University of Peradeniya, who invited me to be a visiting research fellow at the University of Peradeniya during 1983. I am indebted to my field assistants Mr Shanta Hennayake and Ms Nalini Meulilatha, and to Nancy Duncan and Shelagh Lindsey who shared their ideas on structuration and texts with me.
2. Among the few attempts by geographers to utilize a 'structuration' perspective in the analysis of empirical research are Gregory (1982) and N.G. Duncan (1984).
3. In other words, following this logic, any type of social movement to bring about change would be ineffective.
4. One can say that individuality is social as well as genetic in origin because of the unique set of *social* relations and experiences shaping each individual's life history.
5. For examples of landscape as text see Darnton (1984); Cosgrove (1982); Ley (1984).

REFERENCES

Agnew, J.A. and Duncan, J.S. (1981) 'The transfer of ideas into Anglo-American human geography', *Progress in Human Geography*, 5 (1), 42–57.
Althusser, L. (1969) *For Marx*, Harmondsworth, Penguin.
Bell, D. (1962) *The End of Ideology*, New York, Free Press.
Castells, M. (1978) *City, Class and Power*, London, Macmillan.
Cosgrove, D. (1982) 'The myth and the stones of Venice: an historical geography of a symbolic landscape', *Journal of Historical Geography*, 8 (2), 145–69.
Darnton, R. (1984) 'A bourgeois puts his world in order: the city as text', in *The Great Cat Massacre and Other Episodes in French Cultural History*, New York, Basic Books, 107–44.
Duncan, J.S. (1980) 'The superorganic in American cultural geography', *Annals, Association of American Geographers*, 70 (2), 181–98.
—— and Ley, D.F. (1982) 'Structural Marxism and human geography: a critical reassessment', *Annals, Association of American Geographers*, 72 (1), 30–59.
Duncan, N.G. (1984) 'The politics of suburban landscapes: a structuration perspective', Syracuse University, unpublished doctoral thesis.
Femia, J. (1975) 'Hegemony and consciousness in the thought of Antonio Gramsci', *Political Studies*, 23 (1).
Geertz, C. (1973a) 'Thick description: toward an interpretive theory of culture', *The Interpretation of Cultures*, New York, Basic Books, 3–32.
—— (1973b) 'Ideology as a cultural system', *The Interpretation of Cultures*, New York, Basic Books, 193–233.
Giddens, A. (1976) *New Rules of Sociological Method*, New York, Basic Books.

—— (1979) *Central Problems in Social Theory*, London, Macmillan.
—— (1981) *A Contemporary Critique of Historical Materialism*, 1, London, Macmillan.
Gouldner, A. (1971) *The Coming Crisis in Western Sociology*, New York, Basic Books.
Gramsci, A. (1971) *Prison Notebooks*, London, Lawrence & Wishart.
Gregory, D. (1982) *Regional Transformation and Industrial Revolution*, London, Macmillan.
Habermas, J. (1972) *Knowledge and Human Interests*, London, Heinemann.
Harvey, D. (1975) 'The geography of capitalist accumulation: a reconstruction of Marxian theory', *Antipode*, 7 (2), 9–21.
La Capra, D. (1983) *Rethinking Intellectual History: Texts, Contexts, Language*, Ithaca, Cornell University Press.
Lankavatara Sutra, II, 112, 114 (1972), quoted in A.K. Coomaraswamy, *Elements of Buddhist Iconography*, New Delhi, Munshiram Manoharlal.
Lefebvre, H. (1976) *The Survival of Capitalism*, London, Allison & Busby.
Ley, D. (1984) 'Styles of the times: reading the texts in inner Vancouver', unpublished manuscript.
Peet, R. (1979) 'Societal contradictions and Marxist theory', *Annals, Association of American Geographers*, 69 (1), 164–9.
Pred, Allan (1984) 'Place as historically contingent process: structuration and the time-geography of becoming places', *Annals, Association of American Geographers*, 74 (2), 279–97.
Santos, M. (1977) 'Society and space: social formation as theory and method', *Antipode*, 9 (1), 3–13.
Shils, E. (1958) 'Ideology and civility: on the politics of the intellectual', *Sewanee Review*, 66, 450–80.
Touraine, A. (1981) *The Voice and the Eye: An Analysis of Social Movements*, New York, Cambridge University Press.
Turner, V. (1974) 'Hidalgo: history as social drama', *Dramas, Fields and Metaphors: Symbolic Action in Human Society*, Ithaca, Cornell University Press, 98–155.

10
ANY SPACE FOR SPATIAL ANALYSIS?
Anthony C. Gatrell

PREAMBLE

I intend to exploit an ambiguity in the question posed in the title of this essay, using it to structure the essay in two parts. Look again at the title and place an emphasis on the word *space*; the question asks if there is space, or room in the discipline, for a spatial analytical perspective. I shall answer this question in the affirmative. In the second half of the essay I shall ask you to underline the word *any*; what sort, or concept, of space is appropriate for spatial analysis?

It is abundantly clear from skimming the contributions made elsewhere in this volume that geographers (like other social scientists) cannot agree on a single, unifying approach to their subject matter. Even if we focus on a particular perspective such as spatial analysis it is clear that there is no consensus as to what it involves. In *Introductory Spatial Analysis* (Unwin 1981), for example, the emphasis is on the statistical analysis of map pattern, while the collection of essays *European Progress in Spatial Analysis* (Bennett 1981) is much more eclectic, with essays on, for instance, the multivariate analysis of spatial data, spatial interaction modelling, and automated cartography.

None the less, let me suggest that we can pick out certain common strands in spatial analytical work, themes which lie at the heart of several syntheses produced in the 1960s and 1970s. Browse through the contents of Bunge (1962; revised in 1966), Haggett (1965; see the revision by Haggett *et al.* 1977), Abler *et al.* (1971) and Morrill (1970). To a greater or lesser extent they are concerned with:

a   the geometric properties of objects as arranged in space;
b   the role of distance as the major constraint on the organization of human activity on the earth's surface;
c   the efficiency of locational arrangements and the need to optimize such arrangements.

I want initially to draw on the first of these themes and to use the standard typology of objects in spatial analysis to structure my consideration of some important problems. The selection of problems is obviously partial and personal, but most of them are current problems in spatial analytic research and the 'future geography' will continue to explore them.

What is this standard typology? At an appropriate scale some features of interest to the geographer (such as nucleated settlements) may be treated as *points* located on the earth's surface. There may be *flows* (movement or spatial interaction) between people or activities occupying such point locations. The flows may take place along well-defined *lines* (the set of lines forming a network), while differentiation of the nodes in terms of their size or status may reflect a *hierarchical structure*. But while some variables of interest to the geographer refer to points or lines, others, such as population density, may be properties of *areas* and yet others (such as temperature) distributed continuously over space as *surfaces*. My own examples focus on some point, area and surface problems and represent a tiny fraction of the rich spatial analytical tradition. The other themes I identified above, concerning distance as a crucial variable and the efficiency of spatial arrangements, impinge on some of these problems.

WHERE'S THE POINT?

A classic problem in spatial analysis is the Weberian location problem. Originally conceived by Alfred Weber this posed the question of how to locate a manufacturing plant so as to minimize the total costs of transporting to it a set of raw materials and from it a finished product to the market. But since the mid-1960s rather more interest has centred on finding an optimal location for one or more public facilities, such as hospitals, fire stations, schools and so on, for which there is some demand and which need to be accessible to the population they are designed to serve. This is not abstract 'spatial science'; such location-allocation models are useful planning tools. Consider an example (Springer 1980). In Salford and Manchester in the late 1970s there were twelve fire stations serving the two districts. A proposal was made to close three and locate four new ones; where ought these to be located, and how do the optimal locations compare with the proposed ones? Springer divided the study area into fifty-six zones and estimated 'demand' for fire cover as the total floorspace in each zone. Keeping nine stations fixed, he then located four more so as to minimize the total cost of travel by fire engines to zones (see Rushton *et al.* (1973) for computer programs to solve this problem). Note (figure 10.1) that while the original proposals are clustered near the city centre, the optimal locations yield a more dispersed arrangement. Of course, many other considerations impinge on these location decisions, and other issues are raised (notably the fairness or 'equity' of such optimal arrangements) but as spatial analysts we can contribute productively to the debate. And, to give another application, what geographical research could be more satisfying than suggesting optimal locations for rural health centres in the Third World (Fisher and Rushton 1979)?

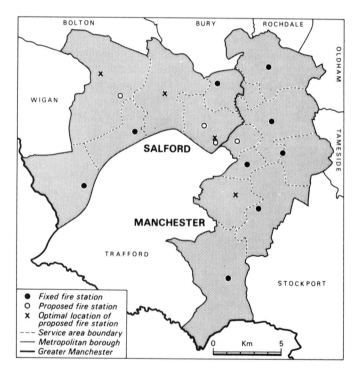

Figure 10.1   The optimal locations of fire stations in Manchester and Salford (after Springer 1980)

Suppose instead that you wanted to monitor rainfall in drought-stricken Africa. You have data from a sparse, unevenly distributed set of recording stations (point locations) and wish to use this limited data to predict rainfall at a point for which no data are available. In the process, you are also asked to recommend where additional recording stations might be set up should funds become available to permit this. One way to proceed is to borrow from geologists a method they use to predict the ore content of rocks from that of surrounding samples; this is known as 'kriging'. If we have n nearby points at which the annual rainfall is measured (call these values $z_1, z_2, \ldots z_n$) then our missing value, $\hat{z}_i$, can be obtained as a linear weighted sum of these known values:

$$\hat{z}_i = a_1 z_1 + a_2 z_2 + \ldots + a_n z_n = \sum_{j=1}^{n} a_j z_j$$

But how do we obtain the weights $\{a_j\}$?

To answer this, we need to know something about how similar or dissimilar are the rainfall measurements at pairs of stations separated by various distances. This information is provided by correlating rainfall values for stations separated by, for example, about 5 km, 10 km, 15 km and so on. We can call these distances lag 1, lag 2, lag 3, etc. This yields what is called an autocorrelation function (figure 10.2). Alternatively, and this is commonly used in kriging, we can calculate a 'variogram', which plots the mean square of the differences in data values against lag. Suppose there is a high correlation in rainfall values at lag 1. This means that stations about 5 km from that whose rainfall we want to predict give a lot of information about its value. So we can use the autocorrelation

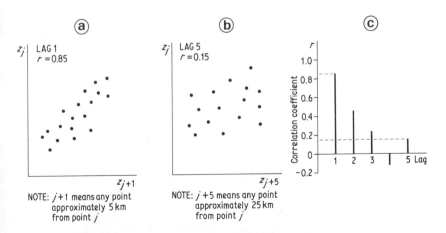

Figure 10.2 Construction of a hypothetical autocorrelation function

structure in the data to derive the weights we need (to see precisely how this is done, look at Nipper and Streit 1982). If we repeat this procedure for a large number of points we can construct an isoline map of 'kriged' values, representing our estimate of the spatial variation in rainfall.

An example comes from work in the Münster region of West Germany (Nipper and Streit 1982). There are 150 recording stations in the study area, distributed unevenly with a higher density in the south-west (figure 10.3a). The kriged map (figure 10.3b) shows a belt of above-average rainfall, running from south-east to north-west. But, as a by-product, kriging also yields a map of standard errors of the estimates, showing in which parts of the map we must be a little cautious about over-hasty acceptance of the estimates (figure 10.3c). Notice the coincidence between such areas and those parts of the study area relatively remote from recording stations. These are locations which merit an additional monitoring station. And while cover in the Münster region appears quite satisfactory, in the Third World where such monitoring stations are much more thinly spread and resources for new stations are modest, it is clearly essential to locate such a station where it is of maximum benefit to hydrologists.

AREAL DISPLAY

Many data of interest to geographers reach us as counts for a set of areal units. Much social and economic data is of this form; for instance, the 1981 British Census yields a wealth of data that can be mapped and analysed at a district, ward, or Enumeration District (ED) level. At least three issues arise as a result. First, in what form should the data be analysed and mapped? Second, do the data display any spatial pattern? Third, to what extent do the results depend on the (perhaps arbitrary) areal units chosen? Let us consider each in turn.

Suppose we wanted to describe the spatial distribution of households comprising single-parent families in Crescent Ward, Salford (Greater Manchester). There are 29 EDs, in which the number of households varies from 12 to 217. The number of households with single parents varies from 0 to 40. The conventional way of standardizing for such size variation is to calculate percentages. If we do this we will find two EDs with near-identical percentages (13.4 and 12.5), the former calculated from 22 single-parent households out of a total of 164, while the second has only 16 households of which 2 comprise single-parent families. The map of such percentages (figure 10.4a) may be misleading in this respect, as presumably the former area is more deserving of resources (whatever form these might take) than is the second. In addition, the EDs differ

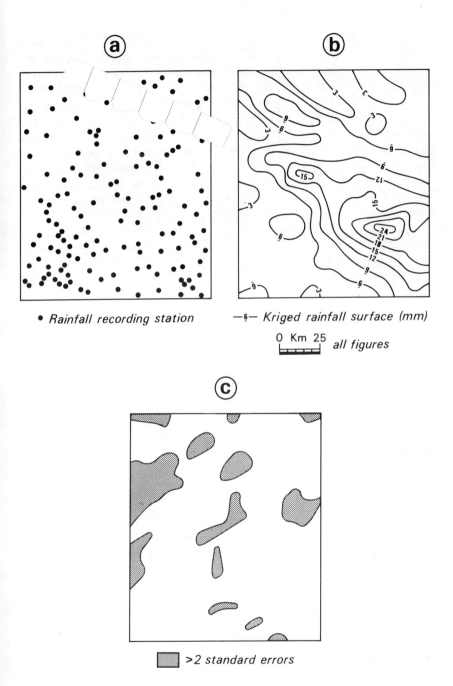

Figure 10.3 Kriging rainfall in Münster, West Germany (modified from Streit 1981)

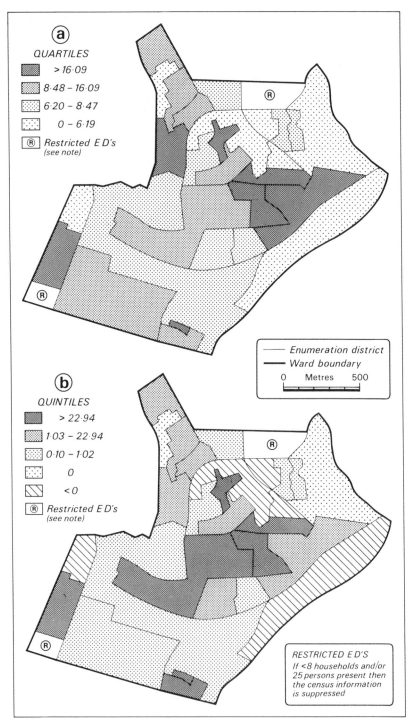

Figure 10.4 Single-parent families in Crescent Ward, Salford

substantially in physical size and the choropleth map is misleading in this respect too.

Clearly, we need a measure that makes some allowance for the variation in absolute numbers. Chi-square values (Jones and Kirby 1980; Gatrell and Cole 1983) meet this requirement. We obtain these by subtracting expected counts from observed, squaring the differences and dividing by the expected counts, attaching a negative sign to the chi-square value if the observed is less than the expected count. The expected counts are computed by multiplying the total number of households by the proportion of single-parent households in the country (or county, or metropolitan district) as a whole. The map (figure 10.4b) of chi-square values is based on expected counts obtained using the county proportion. Although the two maps are broadly similar, the chi-square map highlights particular concentrations of single-parent families. Given that census data are widely used to inform policy-makers, it is essential to beware the pitfalls of using percentage data.

One of the geographer's prime tasks is to detect spatial pattern. In choropleth maps this means asking whether an observed spatial distribution displays significant clustering of similar values. Again, this calls for a consideration of spatial autocorrelation and although the statistical theory underlying this can be demanding there are very clear pedagogic accounts available (Unwin 1981). An examination of the autocorrelation structure of a set of data assists in checking whether the assumptions of inferential statistical tests are met and whether diffusion processes are at work, and as we have seen, in interpolation (Haggett *et al.*, 1977, 353–77). Most important of all, it can prevent us from reading too much into our maps, as the following example shows.

Consider a map of Salford, showing a variable distributed over the set of twenty-two wards (figure 10.5). Suppose you are told that the variable is an index of socio-economic status. As geographers we are accustomed to searching for spatial pattern: 'Ah yes,' we say, 'this is an inner city region, and over there is a high status region.' We produce our own cognitive regionalization of the city. But let me reveal that the map is fictitious, and was created by asking twenty-two students to toss a coin and code the wards independently 'above or below average' according to whether the throw was heads or tails. There *seems* to be some spatial pattern; this is what we look for. In describing real distributions we need an autocorrelation statistic to tell us whether apparent clustering is any more than we would expect as a result of such a chance process. Moreover, if the clustering *is* significant at, say, the 95 per cent level, we should not forget that there is a 5 per cent probability that it arose by chance.

My discussion of areal patterns is couched in terms of census data.

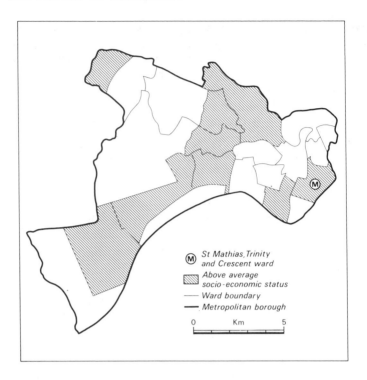

Figure 10.5 'Socio-economic status' in Salford

Such data are collected from individual households but the most detailed data that geographers can usually hope for are available only at the ED level. Such EDs are aggregations of households; on average about 170, although this number can vary widely as noted above. To what extent would our picture of the socio-economic landscape differ if we aggregated households into a new set of EDs? Can you contemplate the (literally) vast number of ways this might be done? Different aggregations yield different results; this is the 'aggregation problem'. But suppose we wanted a much smaller set of areal units? Almost inevitably this will give us a different picture; this is the 'scale problem'. Let me illustrate using an example from Openshaw and Taylor (1979).

If we measure, for each of the 99 counties of Iowa, the percentage of population aged over 60 in 1970 and the percentage vote for Republican candidates in the 1968 congressional elections we obtain a correlation between the two of 0.35. If we correlate these variables over a set of six Republican-proposed congressional districts we get $r = 0.48$. The correlation for six Democrat-proposed districts is $r = 0.63$. The correlation increases as scale becomes finer, but different results are obtained from different arrangement of zones. Indeed, Openshaw and Taylor show that, taking six zones, we can generate any correlation we like from −1.0 to +1.0! Look at Openshaw's recent pedagogic monograph to see how pervasive the problem is and what can be done about it (Openshaw 1984).

ON THE SURFACE

If we map a variable using isolines we are projecting onto a plane a three-dimensional surface. We learn, at an early stage in our geographical education, how to translate from a topographic contour map into a three-dimensional view of a landscape. The surface is continuous; height above sea level can, in principle, be measured at any location. This is true too of air pressure, temperature and other physical variables. Such surfaces may show complex spatial variation, but there may be simpler 'trends' in the data which may be described by linear, quadratic, or higher-order polynomial functions of latitude and longitude. This is what trend surface analysis seeks to do (Unwin 1981).

In the social domain it is harder to justify this assumption of continuity, but this has not prevented some authors applying such analysis to some data sets which ought never to have been handled so. Further, in the social domain such variables will frequently be categorical rather than measured on a ratio scale. To surmount this problem Wrigley (1977) devised an extension of trend surface analysis to categorical data, probability surface analysis. Wrigley illustrates this using data on reactions to aircraft noise in south Manchester. There are three categories of response: highly annoyed, moderately annoyed and only a little annoyed.

Given three categories there will be three probability maps for each order of surface. Wrigley selects the third-order surfaces as best-fit descriptions of the data, and the predicted probabilities of being highly annoyed are shown in figure 10.6. The ridge of high probabilities running south-west to north-east coincides with the major air traffic route. Other 'nuisance fields' may be described as surfaces, indicating in graphical form that noxious facilities (be they sewage plants, nuclear reprocessing plants, or whatever) have negative impacts (externality effects) which spill over into a surrounding area. Those living close to a major British football ground may regard it as having undesirable effects. Recent work in Southampton (Humphreys et al. 1983) illustrates this and several examples of nuisance surfaces are portrayed, reflecting the

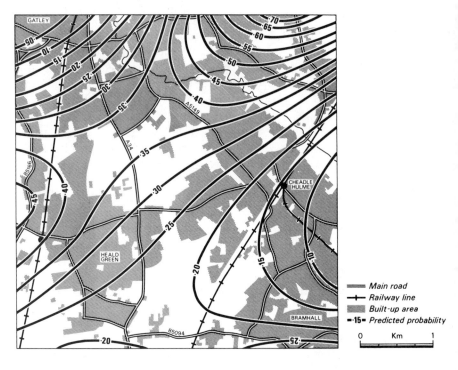

Figure 10.6 A third-order probability surface: probabilities of noise annoyance in south Manchester (modified from Wrigley 1977)

differential impact on noise levels, car parking availability, traffic flow, and so on.

But while we may want to place some distance between ourselves and such facilities in other cases we may wish to minimize distance and maximize accessibility. In recent years the spatial analysis literature has given some attention to the problem of measuring accessibility and many empirical illustrations of accessibility surfaces are available. Most of the applications are to a single class of facility; for instance, primary health care (general practitioners), libraries or shops. Yet while the importance of the concept of accessibility is taken for granted in spatial analysis there is no consensus as to how to measure it (Guy 1983). One widely used measure is based on the notion of potential, itself arising out of the elementary Newtonian gravity model. Suppose, for instance, that we are interested in how accessible to the population is a set of public libraries (Cole and Gatrell 1984). We can imagine that readers will not necessarily visit the nearest library but may trade off the distances travelled to all possible destinations against the likely benefits or utility of each library, where this might be measured in terms of book stock. Denoting the size (stock) of the j'th library as S_j and distance from the i'th neighbourhood as d_{ij} we could write

$$A_i = \sum_j \frac{S_j}{d_{ij}^\beta}$$

where A_i is the accessibility and β is some parameter which reflects how sensitive users are to the deterrence of distance. If we compute this measure for a sufficiently large number of neighbourhoods (and if these are well distributed throughout the study area) we can legitimately map accessibility as a continuous surface.

But this raises at least two issues. First, the way we draw a boundary around our study area affects the results. Unless the facilities all lie in the periphery we will tend to get low accessibility scores in this region; yet residents in such areas may patronize libraries just outside the arbitrarily defined boundary. Second, do users really evaluate all possible libraries – or are they concerned perhaps only with the two or three nearest to their houses? As an alternative we might contour accessibility values based on only the nearest facility, but experience suggests that this simply highlights the locations of the libraries. Finally, how sensitive are the results to the choice of β?

There seem to be some fundamental issues involved in this problem area (Guy 1983), which future spatial analysis will perhaps resolve. In the meantime, this kind of work helps to build a bridge between those working in a spatial analytic tradition and those concerned with the distribution of 'real income' (involving access to all society's scarce resources as well as payments for labour).

BRIEF INTERLUDE

I hope I have succeeded so far in arguing that there is space for spatial analysis in geography, and that part of its future lies in researching some of the as yet unresolved problems. I want now to examine critically the twin concepts of distance and space, since they clearly underpin the spatial analysis tradition and yet are also implicated in other research traditions. I do this because I think that part of the onslaught against spatial science reflects its (at times) naive conception of spatial relations; its emphasis on straight-line (Euclidean) distance as the primary spatial relation. And I think that a way forward in spatial analysis is to widen our conception of what distance and space entail. Space prohibits me from developing this theme as fully as I would like and the reader may care to look at more detailed arguments (Gatrell 1983a, 1983b).

WHAT IS DISTANCE?

Imagine three landmarks (shops, churches, public monuments, or whatever) in your home area. Let us call them (rather unimaginatively – but I do not know your home area) a, b and c. We can refer to the location of each as (x_a, y_a), (x_b, y_b), and (x_c, y_c). We learn to measure the distance between any pair by squaring the x and y differences, adding them together and taking the square root. For instance,

$$d_E(a, b) = [(x_a - x_b)^2 + (y_a - y_b)^2]^{0.5}.$$

This 'crow-flight' distance is called the Euclidean metric. But in some areas, notably cities with a grid-like street pattern, we cannot behave like crows and we must traverse city blocks. The appropriate metric to describe this is the Manhattan or 'taxicab' metric:

$$d_T(a, b) = |x_a - x_b| + |y_a - y_b|.$$

Suppose the three landmarks are major supermarkets, all of approximately equal size, and that we want to construct theoretical market areas around them. Using the Euclidean metric we bisect the distances between all pairs of supermarkets and join these mid-points to form Thiessen polygons (figure 10.7). But measuring distance by counting city blocks, and again finding and joining mid-points, yields a different set of market area boundaries. Note how our Euclidean perception seems to suggest that the 'taxicab' boundary between A and C is 'pushed' towards C (figure 10.7). This configuration of market areas is entirely a function of geometry, not differential attractiveness of stores.

In any metric several conditions must hold, two of which I want to highlight here. First, distances are symmetric: $d(a, b) = d(b, a)$. Second, distances obey what is called the 'triangle inequality': $d(a, b) \leq d(a, c) + d(b, c)$. By definition, these conditions hold for each of the two metrics considered above, and for any other metric. But travel time, cost or cognitive estimates, all of which we use to describe spatial separation, do not necessarily share these metric properties. Moreover, while there is a *single* distance from a to b, there may be several modes of travel available, each with perhaps a quite different journey time or cost.

WHAT SPACE FOR COGNITIVE MAPS?

Let me elaborate on and illustrate some of these issues with reference to cognitive mapping, the processes of acquiring, storing and representing information about our environment. Interestingly, the early monograph

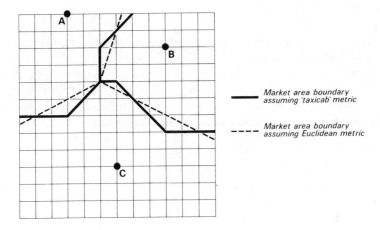

Figure 10.7 Hypothetical market areas

by Lynch (1960) categorized elements of the urban landscape recognized by residents as 'nodes', 'paths', 'districts', 'edges' and 'landmarks', a categorization that partly anticipated the standard spatial analytic typology. Much of the analytic work on cognitive mapping has started by defining a set of landmarks and eliciting from respondents (using a variety of experiments or tasks, the choice of which affects the results) estimates of distance between all pairs. We have available then a matrix of distance estimates; how do we analyse this?

Two classes of approach have generally been adopted. The first is to relate these cognitive distances to the Euclidean distances and to describe this relation using regression methods; this approach is reviewed in Gatrell (1983a, 63–73). I want to comment here on the second approach, which uses the matrix to produce a map showing the configuration of landmarks in what might be called a cognitive space. This may be accomplished using a method of exploratory data analysis called multidimensional scaling (MDS) which, briefly, tries to locate points (landmarks) in a space so that the distances in that space match as closely as possible the input 'distances' (known as 'dissimilarities'). There is a superb introduction to MDS by Kruskal and Wish (1978) and a geographical perspective in Gatrell (1983a, chapter 4). A substantial volume of work on this topic has been generated by Golledge and his students, in particular on the city of Columbus, Ohio; some of this work is brought together in Golledge and Rayner (1983).

An illustration is provided by some research done for an undergraduate dissertation on spatial cognition in Humberside (Page 1979). The MDS

204 THE FUTURE OF GEOGRAPHY

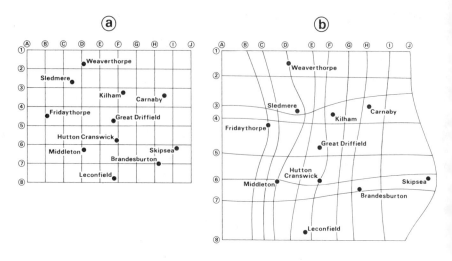

Figure 10.8 A cognitive transformation of rural Humberside (after Page 1979)

configuration (figure 10.8), derived by averaging distance estimates for a set of respondents, reveals some cognitive distortion in a north–south direction but rather more in the east–west direction; Skipsea, on the coast, is 'pulled away' from other villages while to the west cognitive distances seem to be underestimated. Deviations of cognitive locations from real locations may be explained using regression methods (Golledge and Spector 1978), while we might explore how the newcomer's cognitive configuration develops as he or she explores a different environment.

Scaling the distance estimates by using a taxicab metric is also possible and there is some evidence (Richardson 1981) that this gives a better description of the data. But, in general, the assumption that a metric space is appropriate for portraying cognitive relations has gone unquestioned. There is empirical evidence to suggest that, when the order of a pair of landmarks is reversed, distance estimates differ; yet there is only one distance shown on the MDS configuration. Further, distance estimates frequently show evidence of the violation of the triangle inequality, which again calls into question the imposition of a metric space on data that may display non-metric properties. Other methods for exploring the structure of cognitive relations should be used. We may have to forfeit the ability to produce a map-like representation of relations, such as MDS provides, but this is the price to pay for a more faithful description of our data.

WHAT SPACE FOR INNOVATION DIFFUSION?

Let me develop this argument further, with reference to a classic theme in spatial analysis, the diffusion of innovations. Much of the traditional research suggests that geographical distance constrains social interaction and thereby inhibits innovation diffusion. This is enshrined in the concept of a 'neighbourhood effect' that lies at the heart of Hägerstrand's highly original and imaginative work, now over thirty years old (Hägerstrand 1967). The analytical tool that is used to embody this effect and to simulate the diffusion process is the 'mean information field', in which the probability of an adopter contacting a potential adopter declines sharply with geographical distance.

Hägerstrand evaluated his simulation results by comparing them visually with maps showing the actual adoption of innovations. Subsequent analysis of his results by others questioned whether the geographical neighbourhood effect really existed in Hägerstrand's data (on agricultural diffusion in rural Sweden during the 1930s). In any event, perhaps we might consider whether it is always useful to regard geographical space as the space within which diffusion unfolds. Let me explain this with reference to more recent work on agricultural innovation diffusion, research carried out in the Cova da Beira, a valley in eastern Portugal (Gaspar and Gould 1982).

The majority of farmers in the region live in small villages and travel each day to their fragmented plots. The farmers meet socially in the village bars or squares and while some have a wide circle of friends and acquaintances others have a much smaller social circle. Not all have the same potential as information transmitters. Consequently, if we want to understand information flow we need to look at the structure of the social network or 'social space' within which such flow might take place.

The authors asked each member of a carefully defined set of 250 farmers to list, in order of importance, five others from whom he would seek agricultural advice. Suppose we focus on the three most important sources of advice and information. We may then construct a matrix with 250 rows and 250 columns, the rows representing the farmers as 'information seekers', the columns as 'information givers'. We place a 1 in row i, column j if farmer i seeks advice from farmer j. There will be three 1s in each row but a varying number of 1s down each column; columns with a large number of 1s represent highly sought-after farmers, while those with few are rarely consulted.

Each information-giver may in principle be depicted as a geometrical figure or polyhedron, the vertices of which are the farmers who solicit his advice. For instance, a farmer who gives advice to three colleagues could be represented as a triangle (of wholly arbitrary size and shape) while one who dispenses advice to only two could be shown as a line, the end-points of which are his two contacts. These polyhedra are called 'simplexes'. But the polyhedra may of course be 'glued' together because, for instance, two or more farmers are consulted by one or more of their colleagues. The connected structure forms a topological space called a 'simplicial complex,' and although it may be difficult to represent graphically it is none the less open to algebraic description, using 'Q-analysis' (Gatrell 1983a; Beaumont and Gatrell 1982).

In the real example considered by Gaspar and Gould the social network is very fragmented, with relatively little connectivity among the Portuguese farmers. None the less, it represents the structure (a 'spatial' structure – but topological rather than geographical) within which we might expect the spread of information about innovations to take place. The authors hypothesize that large polyhedra (farmers whose advice is sought by many) are early innovators who transmit information about the item (a new tractor) to those who visit them for advice. We might, further, expect a sequential ordering of dates of adoption, with farmers some 'distance' (in social terms) from early adopters taking up the innovation some years later. In fact, there seems to be little relationship between date of adoption and social distance from the dominant simplexes. This suggests that the authors failed in their attempt to find the *relevant* space or 'backcloth' on which unfolds the diffusion process. The adoption process will be determined by economic considerations as much as by social ones and the authors were not able to capture these financial constraints adequately. None the less, despite the complexity of the diffusion process, Gaspar and Gould have succeeded in pointing out the need to find and describe the real structures that govern diffusion. We can now look beyond geographical diffusion theory to a search for the genuinely 'spatial' structures that permit or forbid diffusion to occur.

CONCLUDING COMMENT

Q-analysis is an expression of a wider concern in the discipline with the analysis of qualitative data and provides one perspective, a non-statistical one, on such analysis. It also serves to free us from what some have felt to be the excesses of narrowly conceived 'spatial science', offering as it does an approach to describing the structure of any number of different spaces that we believe to be relevant to our enquiries. Such spaces, as we have seen, arise from defining sets and relations. Geographers who would not wish to be identified as spatial analysts cannot avoid, albeit implicitly perhaps, defining sets and looking at relations on and between such sets. As a result we have, in Q-analysis and related approaches, a methodological perspective that can appeal to geographers of quite different philosophical persuasion. And anything that diminishes the fragmentation and separation of research traditions in geography is surely to be welcomed.

REFERENCES

Abler, R., Adams, J.S. and Gould, P.R. (1971) *Spatial Organization: The Geographer's View of the World*, Englewood Cliffs, NJ, Prentice-Hall.
Beaumont, J.R. and Gatrell, A.C. (1982) *An Introduction to Q-analysis*, Norwich, CATMOG No. 34, Geo Abstracts.
Bennett, R. (ed.) (1981) *European Progress in Spatial Analysis*, London, Pion.
Bunge, W. (1962), *Theoretical Geography*, Lund University Studies in Geography, Series C, 1, Lund, Sweden.
Cole, K. and Gatrell, A.C. (1984), 'Access to libraries in Salford: explorations using 1981 Census data', paper presented at IBG Annual Conference, Durham.
Fisher, H.B. and Rushton, G. (1979) 'Spatial efficiency of service locations and the regional development process', *Papers of the Regional Science Association*, 42, 83–97.
Gaspar, J. and Gould, P.R. (1982) 'The Cova da Beira: an applied structural analysis of agriculture and communication', in *Space and Time in Geography: Essays dedicated to Torsten Hägerstrand*, A. Pred (ed.), Lund, C.W.K. Gleerup.
Gatrell, A.C. (1983a) *Distance and Space: A Geographical Perspective*, London, Oxford University Press.
—— (1983b) 'Saving space: a perspective for geographical enquiry', *Area*, 15 (3), 251–6.
—— and Cole, K.J. (1983) 'Quantitative geographical explorations of small-area Census data,' CAMPUS Census Project, Research Note 2, Department of Geography, University of Salford.
Golledge, R.G. and Rayner, J.N. (eds) (1983) *Proximity and Preference: Problems in the Multidimensional Analysis of Large Data Sets*, Minneapolis, Minnesota, University of Minnesota Press.
—— and Spector, A.N. (1978) 'Comprehending the urban environment: theory and practice', *Geographical Analysis*, 10, 403–26.
Guy, C.M. (1983) 'The assessment of access to local shopping opportunities: a comparison of accessibility measures', *Environment and Planning B*, 10, 219–38.
Hägerstrand, T. (1967) *Innovation Diffusion as a Spatial Process*, trans. A. Pred, Chicago, Ill., University of Chicago Press.
Haggett, P. (1965) *Locational Analysis in Human Geography*, London, Edward Arnold.
—— Cliff, A.D. and Frey, A.E. (1977) *Locational Analysis in Human Geography*, London, Edward Arnold.

Humphreys, D.C., Mason, C.M. and Pinch, S.P. (1983) 'The externality fields of football grounds: a case study of The Dell, Southampton', *Geoforum*, 14, 401–11.
Jones, K. and Kirby, A. (1980) 'The use of chi-square maps in the analysis of census data', *Geoforum*, 11, 409–17.
Kruskal, J.B. and Wish, M. (1978) *Multidimensional Scaling*, London, Sage Books.
Lynch, K. (1960) *The Image of the City*, Cambridge, Mass., MIT Press.
Morrill, R.L. (1970) *The Spatial Organization of Society*, Belmont, California, Duxbury Press.
Nipper, J. and Streit, U. (1982) 'A comparative study of some stochastic methods and autoprojective models for spatial processes', *Environment and Planning A*, 14, 1211–31.
Openshaw, S. (1984), *The Modifiable Areal Unit Problem*, Norwich, CATMOG No. 38, Geo Abstracts.
—— and Taylor, P.J. (1979) 'A million or so correlation coefficients: three experiments on the modifiable areal unit problem', in N. Wrigley (ed.), *Statistical Applications in the Spatial Sciences*, London, Pion.
Page, R. (1979) 'Rural and urban spatial cognition', BA dissertation, Department of Geography, University of Salford.
Richardson, G.D. (1981) 'The appropriateness of using various Minkowskian metrics for representing cognitive configurations', *Environment and Planning A*, 13, 375–85.
Rushton, G., Goodchild, M.F. and Ostresh, L.M. (1973) *Computer Programs for Location-Allocation Problems*, Department of Geography Monograph 6, Iowa City, University of Iowa.
Springer, T.M. (1980) 'The optimal location of fire stations in Manchester and Salford', BA dissertation, Department of Geography, University of Salford.
Streit, U. (1981) 'Analysing spatial data by stochastic methods', in G. Bahrenberg and U. Streit, *German Quantitative Geography*, Paderborn, Ferdinand Schöningh, 35–44.
Unwin, D. (1981) *Introductory Spatial Analysis*, London, Methuen.
Wrigley, N. (1977) 'Probability surface mapping: a new approach to trend surface mapping', *Transactions, Institute of British Geographers*, NS 2 (2), 129–40.

III
GEOGRAPHY FOR SOCIETY

11
QUANTIFICATION AND RELEVANCE[1]
R.J. Bennett

Geography, like many of the social sciences, went through a form of quantitative revolution in the late 1950s and 1960s. In retrospect this can be seen as part of the post-war surge in interest in technological innovation, technical approaches to management, and planning. In the 1970s, rather later than some other disciplines, geography experienced an attempt at a 'radical' revolution. The intention in this case was to displace supposed value-free approaches to scientific method by ones which explicitly acknowledged a value system based upon, in most cases, a theory of labour value. Much of the early vigour of this radical group of geographers was concentrated on attacking quantitative geography and what was termed 'spatial science'. The 1980s, however, have seen the emergence of a more mature approach by both quantitative workers and their critics. The thrust of applied research has played a critical part in the formation of this new base. It is thus highly pertinent to take up in this chapter, if only briefly, the interrelated questions of quantification, relevance and utility. In doing this I wish to develop three main points, and from these to derive a fourth:

1 The necessity for a 'paradigm shift' in geography as a whole away from the view that the core of the discipline is framed by *spatial geometry*.
2 The necessity of reappraising the nature and appropriateness of the methods of statistical inference within quantitative geography, especially those derived from the statistical theories of Neyman and Pearson.

3   The need of geography to respond to the increasing stimuli provided by 'applied', social, political and policy issues which are tending to shift the emphasis of the subject towards frame disciplines and away from a central geographical core.
4   The need to put in place an appropriate response to the critique of quantitative geography as positivism. This is a somewhat separate issue from the previous three, but derives from each of them.

I hope that not too much injustice will be done to each theme by a discussion that, in the limited space available, is necessarily brief.

PARADIGM SHIFT AWAY FROM SPATIAL GEOMETRY

William Bunge, in his 1962 *Theoretical Geography*, introduced the concept of structuring geographical data in terms of their elementary geometrical properties of point, line, flow, zone and surface. Much of Bunge's thinking was inspired by Fred K. Schaefer who, in a seminal paper in 1953 and in writings reported in Bunge (1962, 1966), propounded the thesis that *spatial patterns were morphological laws*. The proposal was that generalities of location link unique relative positions into geometrical and topological regularities which apply across space in a repetitive and predictable form. This was part of a vehement attack on what he saw as a stultifying tradition emphasizing the uniqueness of location and undermining the capacity to propound generalizations. Schaefer's influence, however, together with that of many other writers of this period, stimulated an approach to spatial data primarily emphasizing the inductive search for pattern which was matched against prior theory derived from geometric and topological properties of space. Both Schaefer and Bunge exerted a considerable influence on Peter Haggett's 1965 *Locational Analysis in Human Geography*, which came to dominate much subsequent thinking, especially in the UK. The objectives of this lineage of work are the detection and analysis of spatial pattern, and it is this lineage which is most characteristically termed 'spatial science'. Spatial science is clearly defined by Haggett *et al.* (1977, 1) as (1) describing aggregate arrangements in space, (2) methods for detecting these arrangements and (3) using this information in applications.

This approach to spatial problems using the structures of point, line, flow, zone and surface has immensely stimulated geographical research, but it has tended to divert attention from analysis of spatial process, and to emphasize inductivism.

As a result of dissatisfaction with this emphasis on spatial pattern and inductivism, many researchers have sought alternative frameworks to attack geographical problems. At the extremes these alternatives have been based on an outright rejection of spatial science. For example, in human geography the spatial view has been characterized as a form of fetishism which obscures the more fundamental social questions. This view is vehemently pressed by some Marxist sociologists. For example Castells (1976) claims that there is no specifically spatial theory that is not part of a more general social theory; in other words human geography has no valid methodological base within the social sciences. For physical geography the rejection of spatial science comes from a different direction, but is no less extreme. There, the methodology of other subjects has increasingly overridden any rather weak spatial view the geographer might adopt. In particular, the criteria of intellectual status have often dictated publication of research findings in journals outside geography, and this has introduced non-geographical criteria into research in physical geography. Whilst useful in many ways, this influence has eroded the geographical core of the discipline.

Apart from methodological critiques, this separate development of human and physical geography erodes the unity of the subject. Where physical and human issues were frequently linked as part of population–environment concerns, now many human geographers grasp the methods of social science which, in the most extreme Marxist and critical theorist cases, reject any substantial role for the influence of the physical environment; and the physical geographers largely seek theoretical development divorced from social and human needs.

It is now accepted by most geographers that the spatial geometric base of spatial science has little major utility, except in a few special situations. However, what must remain very relevant to spatial science are spatial relations, spatial processes, the significance of place, and place differences. To take as an example the definition of a spatial process, this is now usually interpreted not in terms of geometric or topological situations, but instead as one realization of an underlying generating process. Part of the modern position is stated well by Haining (1981). He distinguishes two levels of conception. At the first level is the *mathematical theory of spatial process*, which is expressed in terms of a structure of variables, relationships between variables and the environment of variables (or parameters). The value of a variable then defines the system, and a *process* is defined by the rules governing the changes over time in the system as a chain of events. *Spatial pattern* is the evidence of the second level, the map pattern of a single realization of the underlying spatial process. This is the surface of relations constituting the data which can be used for empirical analysis. Analysis is thus founded on analysing the generating process, with the map pattern constituting the data used to assess realizations, not vice versa. A wider view of spatial process developed by Sack (1974, 1980) is discussed more fully in my conclusion.

REAPPRAISAL OF STATISTICAL INFERENCE

Associated with the approaches of spatial science, but not essential to it, was an overburdening emphasis on statistical inference in much geographical research of the 1960s and 1970s. In practice many quantitative researchers never used methods of inference, but a number of widely quoted papers and all the text-books on quantitative geography emphasized this approach, so that *inference* became the dominantly voiced methodology. Certainly it dominated to the extent that Harvey, in his *Explanation in Geography* (1969), represented it as the paradigm within which geography had come to work, and Gould (1970) attacked it by asking whether *statistix inferens* was the name of a geographical 'wild goose'. Wrapped up in the difficulties of spatial science are many of the problems which result from an unthinking use of statistical inference; and the emphasis upon inference again allowed a more ready conjunction of scientific method and positivist methodology, as discussed later. It is now clear that emphasizing inference *per se* produced numerous difficulties.

Within all scientific method there is a complex interplay between induction and deduction as the two main approaches by which theories are developed about phenomena. This applies to most physical and social science. In general the two approaches must be seen as inseparable so that, although we may choose to emphasize a more inductive style of argument in one case and a more deductive style in another, it is inconceivable that the development of theory is seen as solely an inductive or deductive exercise. Instead the best of geographical analysis draws deeply upon both the analyst's inductive experience of empirical reality and upon his prior deductive theory. It is to be hoped that he will be sufficiently emancipated from each to respond both to reality and to the philosophical positions of his theory. In its simplest terms the division arising from an emphasis on either the inductive or the deductive approach to statistics has given rise to two major schools of statistical theory. Both lead to similar estimators, but there is an important line of division in the premises of their approaches.

The first of these approaches is the 'classical' theory of inferential statistics and hypothesis-testing as developed by Neyman and Pearson in the 1920s. This often appears to be primarily inductive. It has a 'frequency' interpretation in which probabilities can be asociated only with events arising from repeatable experiments. Hence probability is interpreted as relative frequency over a large number of repetitions of an experiment. The theory as developed by Neyman and Pearson has the properties of a decision theory. A prior cut-off is specified by statistical significance level, which defines a critical region. Then, under a specified null hypothesis and test statistic, a calculated result for the test statistic yields a result which permits a decision on whether a sample statistic accords or does not accord with the properties of a parent population. For such an approach, various, usually rather rigorous, assumptions are required. Particularly important are (1) that the sample is random, (2) that the decision experience is repeatable and (3) that the sample is not the entire population.

The second main approach in statistical theory is that utilizing Bayesian methods. In this approach the intention is not to reach a decision about whether to accept or reject a null hypothesis, but instead to reach a conclusion as to the degrees of belief which can be placed in a given statement. The outcome of this approach is a posterior probability density function (pdf) which expresses the degree of belief. This is based on the prior pdf, the unconditional pdf and the conditional pdf or likelihood function. The *prior pdf* is the best estimate we have at the start of the experiment. It incorporates *a priori* information and theory and places no restrictions on the form of information that can be employed. It thus overcomes many of the restrictive, statistical assumptions of the Neyman and Pearson theory. The *unconditional pdf* is the observed distribution of the data for all possible probabilities of given parameters that might affect the situation under study. The *conditional pdf*, also termed the *likelihood function*, is, perhaps, the most important element, and it expresses the probability of given experimental outcome in terms of the unknown parameters. Hence it gives us the entire evidence of our experiment in an empirical situation.

The advantage of the Bayesian theory is that it is a unified approach. It concentrates attention on the practical process of research design and refinement of hypothesis, which it regards as a sequential process moving not to a final decision (as in the Neyman and Pearson theory) but to a level of support for a particular set of ideas. It is thus symmetrical to the view of geography as an inductive-deductive process of refining theory. The method has wide application since the approach does not rely on rigid statistical assumptions, on large sample properties, or on precise repeatability of experiments. Hence, it has a greater appropriateness to most geographical problems than the Neyman and Pearson theory. Yet, despite its advantages, Bayesian theory has been developed to a much lesser extent in geography and planning than the theory of statistical inference. (A recent exception is the approach of Wilson and Bennnett 1985.)

The Fisher approach to statistical analysis also leads to a level of support for a theory, rather than to a Neyman and Pearson style of final decision. In addition, there are methods of rather looser analysis, frequently referred to as exploratory data analysis. These do not seek significance tests at all but, instead, by simple methods of ratios and percentages seek to reveal the properties of the data characterizing a particular situation.

As a result of a greater awareness of this range of alternative approaches, a considerable internal debate within the corpus of quantitative geographers has led to the rejection of most of the earlier text-book presentations of quantitative methods, and also the terminology and methods of statistical inference as the dominant mode of quantitative analysis. In this reappraisal three features have stood out:

1. The inappropriateness of most classical statistics for situations in which: first, we have considerable prior knowledge, self-knowledge, or phenomenological understanding; second, experiments are normally *not* repeatable; and, third, the sample is often the entire population.
2. Many geographical analyses have been far too sophisticated for the data or situation to which they relate.
3. There has been too little concern with the specifically spatial problems of handling data – that is geography is frequently unique in the nature of the problems it brings into play in quantitative analyses.

The last of these points, because it is so frequently overlooked, perhaps merits a brief discussion. It draws us back to aspects of the core of spatial science, but now with emphasis on how data which are spatial in context can be used in any quantitative analysis. The problem, then, is that geographical data are mediated by the structure of data collection. As a result, conclusions drawn are usually specific and unique to the region or problem in hand. This affects point data for the coefficients of gravity and spatial interaction models. Similarly with area data, the statistical measures (such as the averages) of zonal characteristics, and any statistical tests of association between zonal data, will be specific to the area analysed. Hence there is, in any spatial map pattern of an observed data set, a convolution of two effects: (1) the underlying spatial process which generates a particular spatial realization; and (2) the spatial structure of the data as collected. In the case of interactions between point data this is sometimes referred to as the *map-pattern* or *Curry effect* (Curry 1972). In the case of zonal data the difficulty raised is referred to as the *modifiable areal unit problem* (Openshaw and Taylor 1981). This problem arises when only aggregated data are available, for example from census data tracts. As a result, the objects of study cannot be assessed directly. The problem is that any set of initial entities can be aggregated in a very large number of ways to form regional or zonal data. The areal unit problem can be disaggregated into two separate effects, *scale* and *aggregation*, and it is their combined effect which gives rise to the areal unit problem. For example, different *scales* of zoned data may have the same number of zones, but different numbers of individuals (or areas) in each zone. Alternatively, *aggregations* may differ: the same number of individuals are contained within a uniform number of zones, but the way in which zonal boundaries are drawn differs. The *modifiable areal unit* problem results when the number of individuals in zones and the number of zones are both variable. This is characteristic of most censuses and other geographical data sources. The particular problem arising from these difficulties is that it is seldom possible to infer from zonal to individual data.

In conclusion, then, quantitative methods should not be seen as only or even essentially those of statistical inference. Alternative methods are available, and these are the ones which are most appropriate for the most common geographical situations of non-repeatable experiments, non-random samples which constitute the entire population, and where difficulties such as spatial autocorrelation and the modifiable areal unit problem frequently occur.

THE ROLE OF APPLIED RESEARCH

Underpinning the shift away from both spatial science and statistical inference as central methodologies for geography has been an increasing emphasis on its applied or 'frame' disciplines. By such a shift, geography naturally becomes meshed with the methodologies of other disciplines. For example, applied research or policy related to the location of industry inevitably requires the linkage of research on location analysis with research on industrial decision-making and governmental effects. This development is significant in three main ways. First, it has been stimulated by the increasing emphasis within geography as a whole upon 'relevance' and questions of social distribution. Second, it has allowed the reassertion of subjectivity within the discipline. The knowledge and experience of the analyst is crucial for applied research even though it is of course not value-free. The critical and logical attack on a problem is combined by the analyst with a subjective assessment of facts, judgements about the significance of inaccuracies, errors and the means by which data have been obtained, and in human geography phenomenological understanding of the social situation. The more geography becomes enmeshed in its applied areas, the more important this subjective analysis becomes, deriving from and used in conjunction with the techniques and methods of quantitative analysis.

The third consequence is potentially deleterious. This is the tendency to fragmentation of the subject as a result of its interlinkage with the methodologies of other disciplines. Thus, for example, applied research on the economics of location is in danger of being subsumed within economics; research on the tax burdens or welfare benefits of governmental activity might be subsumed within economics or politics; research on climatic variability might be swallowed by atmospheric physics; research on slope and soil mantle properties might be absorbed by engineering soil mechanics; and so on. These potential dangers are the greater because other disciplines have been more effective at academic imperialism than geography, and the overall trend of academic values has tended towards sub-specialisms and away from the broader, all-embracing approach which was essential to the traditional nature of geography and must still be retained as the essence of geographic method. Thus, although it is essential for geography to become involved in applied research, this is not without its dangers, in the present intellectual climate. I return to this dilemma in the conclusion, where I aruged that it can be resolved only by reasserting the geographical core.

THE CRITIQUE OF POSITIVISM

This critique has been one of the major themes in the literature of radical geography, which has seen spatial science and inductivism as central targets for attack, but has largely ignored applied research, both at a critical level and in any major attempt to contribute at an applied level. Major examples of this critique are Harvey (1973), Sayer (1976), Gregory (1978, 1981), and contributions to Gregory and Urry (1985). A largely unstated assumption of these writers is that quantitative geography is positivist of necessity. As such it is alienating in distracting researchers from the central question of social distribution. First, it creates a false sense of objectivity by attempting to remove the observer from the observed, which in turn facilitates the control and manipulation of society. Second, its use of computers, machines and techniques subsumes people under a panoply of mathematics and machinery, reducing them to an atomistic level by eliminating considerations of soul and humanistic concerns. Third, it is descriptive of existing behaviour and hence supports the status quo in society, especially the social distribution of well-being. Fourth, it allows no consideration of values and hence of the norms by which society *should* be organized. Fifth, it attempts to construct models and theories of universal generality by a mode of inductive logic moving from the particular to the general.

Some authors have gone even further and challenged the methodological base of geography as a whole. For example, Peet and Slater (1980, 543–4) claim that, 'Geography has no theoretical object that is specific to it, and its "theory" is no more than a heterogeneous amalgam of spatial models developed by a variety of bourgeois scholars.' Eliot Hurst (1980) argues for a *de-definition* of geography: 'Geography is revealed for the most part as merely descriptive and scientifically bankrupt ... [and] is nothing more than a theoretical ideology based on technical practice alone' (p. 7). Instead, a truly multi-disciplinary approach (dialectical and historical materialism) is required which can uncover 'the true relationships between economy and society in the totality of social existence' (p. 12). Geography is a 'fetishized domain', and even 'the co-opted Marxism of "radical" geography and "Marxist Geography" ... is superficial and doctrinaire, incapable of transcending the orthodox problematic's boundaries' (p. 16). Thus the rejection of positivism also entails the rejection of space, and hence of geography as a whole.

It is my contention here that this critique of quantification in geography as positivism is almost entirely ill-founded. Whilst it has had some useful benefits in redirecting attention to social questions and to applied areas, it is seriously flawed. The flaws derive from two sources: first, an uncritical acceptance of the argument that statistical inference, its 'spatial science' relationship, and hence positivism, form the foundation of quantitative geography; and, second, ironically, an uncritical acceptance of the methodological arguments of other disciplines. These two flaws lead to three important conclusions for applied geography and geography as a whole.

The first major conclusion must be that the representation of geography as positivism is not at all adequate. Despite the acceptance of this equivalence by many authors, there is not a close or one-to-one correspondence between what quantitative geography should be and positivism. I have developed this argument at length elsewhere (Bennett 1981) and shall only summarize it here. One major aspect of the confusion seems to arise from the representation of quantitative geography purveyed in Harvey's (1969) *Explanation in Geography*. Although a wide-ranging and an internally contradictory book, a clear message derives from its text: namely, first, that geography *is* primarily inductive; second, that geography *is* searching for universal laws; and, third, that geography *is* an objective science. Moreover, in geography where phenomena do not seem to fit universal laws, they can be treated *as if they do*. These aspects of Harvey's 1969 presentation all have some aspect of truth, but give only a partial representation of geography. Moreover, it is a representation of the quantitative literature based on statistical methods of inference, rather than on the wider and more recently applied methodologies of Bayesian, Fisherian, or exploratory analysis. Unfortunately, it is a caricature which has been widely accepted. For example, Johnston (1979, 61) quotes Harvey's (1969, 34) two routes to scientific explanation without criticism and later suggests that 'the approach of positivist spatial science ... [is] codified in Harvey's *Explanation in Geography*' (Johnston 1979, 404). Eliot Hurst (1980, 3) refers to Harvey's book as a 'quasi-philosophical rationalisation' of geography leading 'to a distinctive practice within geography which searched for some surrogate landscape of mathematical spatial conformity'. Johnston (1979, 1981, 1983) has accepted that the Anglo-American tradition *was* positivist; even Haggett *et al.* (1977, x) accept this caricature and apologize that their book is 'narrow and positivist in its philosophy'.

The dangers implicit in Harvey's representation are well-evidenced by Gregory (1978). Although now modifying his opinions (Gregory 1981) he looks to the nineteenth-century view of positivism as defined by Comte. In particular, he makes parallels between geographical writing and Comte's concepts. He is certainly able to find some confirmation of this approach. However, the comparison is inexact, based on specifying a critical entity (such as 'systems theory', 'positivism', or 'neoclassical theory'), then isolating significant characteristics of it (such as those defined by Comte) and asserting that the source within geography is represented solely in terms of these characteristics. This is an inductive fallacy which amounts to constructing a form of 'straw man'. As a result, Bartels (1982) has questioned the usefulness of this literature, and Bennett and Wrigley (1981) have termed much of the critique of positivism 'at best a misrepresentative irrelevance, and at worst a fatuous distraction'; they argue that the critique has been directed at an abstract and misrepresented view of much of quantitative geography, one which attributes methods, views and conclusions to quantitative geographers which most have never held, or, if they did ever hold them, have since abandoned, or, if they still hold them in some form, do so only in part alongside wider views. For teaching purposes and for convenience of reference it often seems useful to seek a caricature of items of academic work; for example, 'positivist', 'functionalist', or 'realist'. This is a form of shorthand. But for useful academic debate such caricatures do a disservice if they do not in *all* important respects give a fair and holistic representation of the academic work in question.

A second danger which derives from the first is that the critique of geography as positivism relates almost exclusively to human geography, and hence excludes from discussion most of the core of geography as it is traditionally viewed. For example, relations between society and the environment, and the study of area in all its manifestations as place, which were traditionally central concerns of the core definition of the subject, come to be defined as meaningless pursuits. Harvey (1974), for example, sees the effects of the physical environment as merely a product of the social system; that is, scarcity is a result of social control; there is no real resource squeeze, only that deriving from class power. This ignoring of physical questions is sometimes justified by the view that the concept of 'human agency' can be applied only to 'human matters'; if this is so then this justification accepts social theory as at best a partial theory and hence of limited relevance to geography which has physical–human relations at its core. This concentration on social questions at the expense of all others is perhaps the most important danger of all for geography, since (a) it is partial, (b) it is based on a social theory of value only (labour value), and (c) it ignores the physical knowledge of environmental systems which is essential for resolving resource questions.

The third danger, which derives from the previous two, is the uncritical acceptance of geography as social science. For example, some authors have concluded that geography has no place in the study of social systems (see e.g. Sayer 1982). As a result, there is a strong tendency to employ theory borrowed from social science as a whole, which has been developed without thought for how the geographical objects or interests of enquiry can be accommodated: of space or society and the environment. For example, Harvey's (1973) *Social Justice and the City* claims that inner-city dwellers are exploited by suburbanites through the tax system; suburb-dwellers are able to locate so as to avoid the redistributive tax burden of supporting the poor who are trapped in the inner city. This is an interesting hypothesis, one that demands analysis, but most empirical results to date suggest that it is far from confirmed (see e.g. Bennett 1980). This demonstrates the dangers of searching for and accepting theoretical explanations without recourse to empirical analysis; that is, accepting an assumption because it accords with theory. In the example of inner-city exploitation, the problem can be resolved only at an empirical level and by relating locational decisions to social decisions, not by explaining locational decisions as solely social (or class) decisions.

CONCLUSION: THE REASSERTION OF THE GEOGRAPHICAL CORE WITHIN APPLIED QUANTITATIVE GEOGRAPHY

The preceding discussion has brought us to four important conclusions. First, spatial geometry and spatial science were an inadequate base for geographical enquiry. Geography cannot be distinguished, as Haggett (1965) asserted, solely by its spatial approach. Although space is clearly a major aspect of geographical enquiry, space as such is merely a geometry or pattern and yields no process understanding. Second, statistical inference as developed in the 1960s within geography, and still largely dominant among quantitative workers, is almost entirely inadequate for geographical problems. This is due to the non-repeatability of the experiments and the data; and to the facts that the sample frequently constitutes the population; that much prior knowledge is available and should be used; and that the spatial character of geographical data undermines most of the assumptions of inferential statistics (especially because of autocorrelation and the modifiable areal unit problem). Third, the recent emphasis on applied and policy-related research has brought specialist sub-areas of the discipline to the fore and opens the discipline to the danger of being embraced within the methodologies of frame disciplines outside the geographic core. Fourth, the critique of geography as positivism, though it has been useful in highlighting some of these weaknesses and in asserting the importance of questions of social distribution, has further weakened geography as a whole: it has set up a falsely based and extreme methodological debate, induced divisiveness by disregarding physical geography and its relation with human geography, and accepted uncritically the methodology of social science, which has been developed without regard for geography's concerns.

These conclusions point to the need to reassert the *geographical core* rather than other disciplines. This core, as in traditional approaches to the subject, combines physical and human geography; that is, it defines the role of geography as an evironmental *and* social study, not as mere social science. Within that core, however, the emphasis must develop three attributes. First, there must be a process meaning to the space under study, as expressed, for example, in Sack's (1980, 197) view that geographers must essentially 'consider the full range of meanings of space and the nature of their synthesis as part of geography's general task of understanding the earth's surface as the home of man'. This task is to be tackled, Sack argues, by integrating the effect of society on modes of thought with the conception of space deriving from any mode of thought. This involves fusing what Sack terms the objective and subjective as well as space and substance; that is, the fusing of the modes of thought characteristic of the sciences with those characteristic of the arts; the one treating symbols and attaching no meaning to them, and the other treating symbols as having not only meaning but a magic and myth which are as important as the objective as foci of concern. It is only by such fusion that geography can achieve a complete development of environmental and social study. The second and related attribute required is the reassertion of the social meaning of space and area; the reassertion of *place* and the idiographically unique aspects of area or region, but within a framework of generalizations and theory. Third, this must be achieved by the combining of empirical analysis and prior theory, not merely by the assertion of theory, and not by the mindless collection of data.

The conclusion is, therefore, that geography cannot evolve through a pluralism of positivism and radical methodologies, as Johnston (1981) has asserted. Instead, geography must return to its roots and revive its concern with the interrelationship of environmental and social concern within place, area, or context. Within this concern the role of quantitative geography is to assist the validation of theory, and not to promote a methodological core in itself (as positivism or spatial science). Quantitative geography is hence not the prerogative of any one method or philosophy, such as positivism, but a tool developed for use within the value position or the research question in hand. In itself it is neither a ready-made philosophy nor a value-free method of research. Instead, it is a servant to the wider geographical questions derived from reasserting the distinctive geographical core and its application to problems in place.

NOTE

1   An earlier version of this chapter was delivered at the Institute of British Geographers Annual Conference in Edinburgh in January 1983.

REFERENCES

Bartels, D. (1982) 'Geography: paradigmatic change or functional recovery? A view from West Germany', in P. Gould and G. Olsson (eds), *A Search for Common Ground* London, Pion.
Bennett, R.J. (1980) *The Geography of Public Finance: Welfare under Fiscal Federalism and Local Government Finance*, London, Methuen.
—— (ed.) (1981) *European Progress in Spatial Analysis*, London, Pion.
—— and Wrigley, N. (1981) 'Retrospect and prospect on quantitative geography', in N. Wrigley and R.J. Bennett (eds), *Quantitative Geography: A British View*, London, Routledge & Kegan Paul.
Bunge, W. (1962) *Theoretical Geography*, Lund University Studies in Geography, Lund, Sweden.
—— (1966) *Theoretical Geography*, 2nd edn, Lund University Studies in Geography, Lund, Sweden.
Castells, M. (1976) *The Urban Question*, London, Edward Arnold.
Curry, L. (1972) 'A spatial analysis of gravity flows', *Regional Studies*, 6, 131–47.

Eliot Hurst, M.E. (1980) 'Geography, social science and society: towards a de-definition', *Australian Geographical Studies*, 18, 3–21.

Gould, P. (1970) 'Is "statistix inferens" the geographical name for a wild goose?', *Economic Geography*, 46, 439–48.

Gregory, D. (1978) *Ideology, Science and Human Geography*, London, Hutchinson.

—— (1981) 'Human agency and human geography', *Transactions, Institute of British Geographers*, NS 6, 1–18.

—— and Urry, J. (eds) (1985) *Social Relations and Spatial Structure*, London, Macmillan.

Haggett, P. (1965) *Locational Analysis in Human Geography*, London, Edward Arnold.

——, Cliff, A.D. and Frey, A. (1977) *Locational Analysis in Human Geography*, 2nd edn, London, Edward Arnold.

Haining, R.P. (1981) 'Analysing univariate maps', *Progress in Human Geography*, 5, 58–78.

Harvey, D. (1969) *Explanation in Geography*, London, Edward Arnold.

—— (1973) *Social Justice and the City*, London, Edward Arnold.

—— (1974) 'Population, resources, and the ideology of science', *Economic Geography*, 50, 256–77.

Johnston, R.J. (1979) *Geography and Geographers: Anglo-American Human Geography since 1945*, London, Edward Arnold.

—— (1981) 'The development of Anglo-American quantitative geography', in R.J. Bennett (ed.), *European Progress in Spatial Analysis*, London, Pion.

—— (1983) *Geography and Geographers: Anglo-American Human Geography since 1945*, 2nd edn, London, Edward Arnold.

Openshaw, S. and Taylor, P. (1981) 'The modifiable areal unit problem', in N. Wrigley and R.J. Bennett (eds), *Quantitative Geography: A British View*, London, Routledge & Kegan Paul.

Peet, R. and Slater, D. (1980) 'Reply to the Soviet Review of *Radical Geography*', *Soviet Geography*, 21, 541–5.

Sack, R.D. (1974) 'The spatial separatist theme in geography', *Economic Geography*, 50, 1–19.

—— (1980) *Conception of Space in Social Thought*, London, Macmillan.

Sayer, A. (1976) 'A critique of urban modelling: from regional science to urban and regional political economy', *Progress in Planning*, 6, 187–254.

—— (1982) 'Explanation in economic geography: abstraction versus generalization', *Progress in Human Geography*, 6, 68–88.

Schaefer, F.K. (1953) 'Exceptionalism in geography: a methodological examination', *Annals, Association of American Geographers*, 43, 226–49.

Wilson, A.G. and Bennett, R.J. (1985) *Mathematical Methods in Human Geography and Planning*, London, John Wiley.

12
GEOMORPHOLOGY IN THE SERVICE OF SOCIETY
Denys Brunsden

> The geological environment is complex with so many facets that control its behaviour that the only way to achieve a full understanding is for there to be an interdisciplinary approach in which engineers and geologists work much more closely together. The meeting ground can, I believe, be found by both the professions concentrating on understanding the geomorphology of construction sites. (D.J. Henkel, Rankine Lecture to the British Geotechnical Society 1982)

The present political and socio-economic world context for geomorphology is the disparity between the wealth of nations, a rapid increase in the number of people living at subsistence levels, and a determination that the standard of living of the people of the developing countries should be raised (Tricart 1956; Matley 1966; Ehrlich 1977; Verstappen 1983).

This concern coincides with a revived desire and capability of geomorphologists to apply their techniques, methods of analysis and conceptual understanding of the behaviour of earth surface materials, processes and landforms to these problems. The aims are to assist in the efficient discovery, assessment and wise management of the earth's finite resources, to prevent environmental deterioration and to avoid or prevent natural hazards.

Verstappen (1983) summarized the diverse applications of geomorphology under the following headings:

1 Applications in the field of the earth sciences, including topographic and thematic mapping of natural resources.
2 Applications in the field of environmental studies surveying, in particular, natural hazards, landslides, avalanches, earthquakes, volcanism, land subsidence, flooding and drought.
3 Applications in the field of rural development and planning, emphasizing land utilization, erosion control, conservation and river basin development.
4 Applications in the field of urbanization for urban extension, site selection, or mining.
5 Applications in engineering including assessments for communication networks, river and coastal engineering.

The work involved in these applications (table 12.1) is always factual, precise and functional, requiring detailed descriptions of the spatial extent, forms and interrelations of the geomorphological features involved, and the quantification or numerical analysis of forms, materials, processes and their effects. For this to be effective it is usual and essential that the geomorphologist is called in early in an investigation, is given a clear brief, has continuing contacts with the client and works as part of a team, including engineers, planners, geologists, soil scientists, agriculturalists, hydrologists or conservationists. All of these have their own contribution to make but the synthesis achieved by co-operation opens new vistas and imaginative solutions.

MAN AS A GEOMORPHOLOGICAL AGENT

The recognition that man is a powerful agent of landscape change was recorded by Marsh (1864) who made early, perceptive comments on the influence of forest clearance on soil erosion, sedimentation and flooding and the importance of slope steepness and rainfall intensity as controlling parameters. It was not until the early twentieth century, however, that detailed attention was paid to the subject and even then the main theme was on the effects of man as a geomorphological agent, often with an implication that his work was dangerous or wrong, rather than on *practical* attempts to improve available methods of modifying nature. The concept of man as a disturbing force was outlined in several formative papers. Woikof (1901) described the 'dissociation between man and the soil' following the destruction of natural vegetation. Fischer (1915) estimated the volume of mineral resources utilized by man and Sherlock (1922) gave a full summary of the effects of forestry, grazing, agriculture and industry including estimates of the rates of erosion and

Table 12.1 Examples of geomorphological contributions to practical problems

Subject	Case study examples
Geomorphological surveys	Doornkamp et al. 1980, Chartres 1982
Resource surveys	Christian et al. 1957, Mitchell 1973
Military applications	Stewart 1968, Perrin and Mitchell 1969
Trafficability	Dowling and Beavan 1969
Aggregate surveys	McLellan 1967, Brunsden et al. 1979
Foundation assessment	Holland and Stevenson 1963
Minerals surveys	Cole et al. 1973, UNESCO 1968
Agriculture	Prokopovich 1969
Urban development	Leggett 1973, Cleveland 1971
Land reclamation	Beaver 1961, Wallwork 1974
Building weathering	Fookes and Collis 1975, 1976, Cooke et al. 1982
Soil development	Ruhe 1967, Young 1973a, b
Highway location	Brunsden et al. 1975a
Dam surveys	Brunsden et al. 1978
Water resources	Beaumont 1974, Taylor 1981
Soil erosion	Hudson 1971, Thornes 1976
Water storage	Emmett 1974
Debris flow hazard	Schick 1974a, Bull 1977
Effect of urbanization	Hollis 1974, Douglas 1976
Flood alleviation	Penning-Rowsell and Chatterton 1977
Water quality	Edwards and Thornes 1973
Meander development	Ackers and Charlton 1970
Geochemical status	Nortcliff et al. 1979
Extreme events	Thornes 1977
Wind erosion	Rapp 1974, Cooke and Warren 1973
Loess problems	Fookes and Best 1969
Coast erosion	Berkman 1975, Mitchell 1968
Protection	Newman 1974, Price and Tomlinson 1969
Coastal engineering	Willis and Price 1975
Dredging	Joliffe 1974
Landslides	Hutchinson 1969, Hails 1977, Skempton and Hutchinson 1969
Hydrocompaction	Lofgren 1969
Subsidence	Coates 1971, 1983, Wallwork 1960
Piping	Parker and Jenne 1967, Fletcher et al. 1954
Cut slope design	Skempton and Chandler 1974, Skempton and DeLory 1957
Dessication	Yaalon and Kalmar 1972
Water repellancy	Qashu and Evans 1969, Krammes and Osborn (n.d.)
Rock slope design	Hoek and Bray 1974, Franklin and Denton 1973
Periglacial problems	Ferrians et al. 1969, Melinkov and Tolstikhin 1974

sedimentation. These aspects, of a subject which came to be called anthropic or anthropogenetic geomorphology, have continued to receive sporadic attention up to the present day (Aufrére 1929; Fels 1935; Mensching 1951; Tricart 1953; Thomas 1956; Golomb and Eder 1964; Smith and Bridges 1965; Jennings 1965; Cailleux and Hamelin 1969; Brown 1970; Collier 1972; Strahler and Strahler 1973; Goudie 1981; Dov Nir 1983), (figure 12.1). They are, however, comments on man's impact on the landscape and not summaries of geomorphology in the service of man.

GEOMORPHOLOGY IN PRACTICE

Early practical geomorphology, in which geomorphological knowledge was used to design, improve or control projects, was usually carried out by agriculturalists, foresters and engineers who were involved with coastal protection, land management or river basin control. For example, in 1906 the Royal Commission on Coast Erosion and Afforestation was established in Britain to report (HMSO 1911) on coastal erosion and sedimentation and to recommend 'what measures are desirable for the prevention of such damage'. Similar official surveys were carried out by agencies such as the US Soil Conservation Service, the New Zealand Soils and River Control Council and the Land Research Division of the CSIRO Australia, who developed much of our early knowledge on earth surface processes. This work was continued by surveyors on soil erosion, land resource or watershed management projects. (Glenn 1911; Bryan 1925; Bennett 1928, 1939, 1955; Meginnis 1935; Jacks and Whyte 1939; Thornbury 1954; Christian et al. 1957; Wischmeier and Smith 1958; Hudson and Jackson 1959; Mabbutt and Stewart 1963; Ryabehikov 1964; Christian and Stewart 1968; Stewart 1968; Young 1968; Prokopovitch 1969.)

The work was enhanced by the use of process monitoring studies, air photograph analysis (Verstappen 1963, 1977), remote sensing, modern laboratory techniques and sophisticated land survey programmes (Tricart 1962, 1968). Today we tend to take these ideas as standard practice and there are many modern, comprehensive reviews of the work achieved in the last thirty years, a period which has witnessed an encouraging explosion of geomorphological applications. (Coates 1971, 1981; Mitchell 1973; Mitchell 1979; Cooke and Doornkamp 1974; Hails 1977; Brunsden et al. 1978; Thornes 1979; Jones 1980; Craig and Craft 1982; Gregory 1982; Jones 1983; Verstappen 1983; Doornkamp 1983.)

Despite this, however, applied geomorphology is by no means fully established, accepted or utilized and the geomorphological literature is replete with strictures to geomorphologists to 'practise what they preach' and appeals to clients to 'understand and use' the subject. (Thornbury 1954; Dylik 1957; Klimaszewski 1961; Tricart 1962; Dixey 1962; Dury 1972, 1978; Brunsden et al. 1978.) Even in 1982, Henkel, a practising geotechnical engineer, felt it necessary to state:

> Over the years I have been involved in a wide variety of construction projects and those which have proved to be the most demanding and stimulating and have contributed most to my education have always been associated with the need to bring together geology, geomorphology and geotechnical engineering. . . .
> My message is that, in order to define the fundamental assumptions, there is no substitute for painstaking study in the field of geology and geomorphology. We still need something of the Victorian virtue known as 'eye for the ground'. (Henkel 1982, 177, 193)

Hutchinson (1982) commented on this paper that:

> geomorphology has been seriously neglected and there is now a pressing need to give this discipline its proper place in the geotechnical spectrum. . . . Dr Henkel suggests that the discipline of geomorphology can act as a long needed catalyst to bring about a more effective combination between geology and geotechnics. I believe that these views deserve our serious consideration. (Hutchinson 1982, 194)

Clearly, geomorphologists have been fighting a long, uphill battle for acceptance and despite a recent rapid growth of expertise and experience it is obvious that many potential clients are still not aware of, or in a position to use, the available knowledge.

PURE OR APPLIED?

The rapid growth has had another serious effect, namely the emergence of internal criticism from authors who, rightly and valuably, query whether there are dangers to the health of the subject by a too rapid, too uncritical or too embracing 'quest for utility'.

The most mature statement was made by Chorley (1978) whose comments stimulated the direction of this chapter and should be widely quoted.

a

c

b

d

Figure 12.1 West Bay, Dorset, looking west. a View taken soon after the open lattice-work of the harbour jetties was filled in. b By the turn of the century significant coastal retreat had occurred. c In the 1920s–1940s coastal erosion required sand drift prevention measures. d In the 1970s storms were able to overwhelm coast defences, and the 1980s are characterized by massive expenditure on promenade walks and other civil engineering works.

Figure 12.1 (*cont.*) West Bay, Dorset, looking east. The photographs show the retreat of the shore until the late 1970s and the development of expensive promenade walls and groyne systems, all of which are now replaced or redundant.

> First, it is clear that, for reasons of syllabus, popularity (professional, student and lay), finance, 'relevance' and the like, geomorphology is becoming increasingly committed, if not wholly then in significant part, to a role involving its relationship to human well-being and aspirations. (Chorley 1978, 10)

Chorley pointed to a 'gulf between the theoretical bases dictated by the scientific methods of logical positivism and those prompted by utilitarian and social doctrinaire theories', a gulf 'which may truly divide radically different environmental aims' (p. 11). He developed this argument in an elegant way by suggesting that the growth of theory in any subject is related to the existing intellectual climate and that the type of work we do, together with its 'theoretical underpinnings, is vulnerable to the pressures of convention'. Chorley went on to claim that:

the quest for utility not only conditions the objects of our work, but also the manner in which theory may emerge. ... The wheel of theory may be coming full circle from one teleology to another, from an old religious, to a new social orthodoxy. (Chorley 1978, 11)

These statements raise important issues for the development of geomorphology which deserve very serious consideration. Two questions seem to be important. Is it true that we are being subjected to the pressures of convention? Is this necessarily bad for the subject?

THE PRESSURES OF CONVENTION

Geomorphological research in the UK and in many other countries is funded by government (through the Natural Environment Research Council in the UK), by private university sources, by grants from charitable trusts, by international agencies, and by commerce and industry. In the UK following the Rothschild Report (HMSO 1971), the British government followed other nations in adopting, in part, a 'customer-contract' method of funding. The government became a 'client' who commissioned research from its institutions, universities and private industry.

Inevitably there has been a pragmatic view towards contract research, and a tendency towards 'relevant research' has also emerged for unsolicited research applications. At least some applicants have perceived that if they emphasized a practical use, a short-term relevance or a training element in their research proposals they stood a greater chance of scoring points in a funding competition.

Whilst it is true that many substantial funds are earmarked for pure research in universities and for research student training awards, in Britain considerable funds were awarded to the government's own institutions such as the Hydraulic Research Station, Institute of Oceanography or British Geological Survey (Institute of Geological Science) which are charged with a public duty to give advice on practical and environmental problems.

Today, under the financial constraints of reduced public expenditure the situation is again changing as university funds are reduced, early retirements encouraged, research posts frozen, departments 'rationalized' and effective maintenance funds lowered. In addition, many organizations and research funds are being privatized to industry and away from the institutes and universities, all of whom must compete for funds. In this climate the research councils have done well to preserve the level of pure research funding that they can make available to universities and similar institutions.

This financial constraint means that research workers in UK universities must seek alternative sources, which generally means industry and contract research. Many organizations are often, against their will and without proper debate on the academic desirability of their decisions, changing character to attract contracts, meet deadlines, to provide training courses for foreign students (i.e. lucrative contracts attracting high-fee students) and ironically paying less attention to the training of UK students to meet the future needs of a 'technology-led recovery'. Geomorphologists together with other scientists are here undoubtedly affected by 'utilitarian and social doctrinaire theories' and the existing intellectual climate.

The privatization of research is heavily directed at consumer-relevant studies. For example, in the UK the Department of the Environment now regularly issues invitations to tender for research contracts as part of its planning research programme. In the early stages any organization (commercial, institute, university or consortia) can express an interest in tendering. Later, selected bodies are invited to make a bid for the contract in which they must meet firm departmental guidelines, completion dates, costs and products and must demonstrate the suitability of their staff, specifications, equipment, resources and experience. Virtually all these contracts are for development or planning purposes and they are only gained in competition with very professional organizations who must make a profit to survive. Yet many of the projects involve a high level of original research which could be, and in some cases can only be, carried out by university research staff. Not surprisingly academic scientists, including geomorphologists, find themselves increasingly sought after to supply relevant expertise to commercial organizations.

It seems clear, then, that geomorphology along with other disciplines is increasingly affected and, in part, necessarily committed to a pragmatic role. The question next arises, 'Is this a bad thing?'

DANGERS OF A NEW SOCIAL ORTHODOXY

In the British Geomorphological Research Group's conference on the present problems and future prospects of Geomorphology (Embleton et al. 1978) two other authors briefly supported Chorley's concern for the subject: 'it would be a great pity if the flowering of theory was nipped in the bud by a complete shift in emphasis to "relevant" or "applied" research' (Thornes 1978, 19), and

> as applied research is aimed at solving specific problems, usually those with pressing economic significance, it is easily and often emphasized at the expense of basic research. Such emphasis will

almost certainly lead to fewer fundamental discoveries, discoveries that actually make applied research meaningful. (Walker 1978, p. 203)

Both authors agreed that the greatest threat, if research became closely tied to the state of the economy, was to the development of theory, and 'pure' discoveries. To examine this threat it is necessary to examine whether applied geomorphology has inhibited the growth of theory, whether it has contributed to new theory and whether it reflects and supports recent developments in the theoretical bases of the subject.

During the 1960s many branches of geomorphology developed within the useful paradigm that geomorphological concepts should be based on the idea of equilibrium behaviour. Solutions were often sought within a systems framework supported by an understanding of mechanics and process. It was generally accepted that, in frequency and magnitude terms, the moderate event achieved most work and even in applied geomorphology such ideas as regime theory or limiting equilibria were used to determine design criteria.

It was rapidly realized, however, that many systems were in disequilibrium and that unsteady behaviour in systems deserved wider attention (Brunsden and Thornes 1979). Today the dominant conceptual ideas of the subject stress such in-words as 'thresholds' (Schumm 1979), 'episodic change' (Schumm 1976), 'transient behaviour' (Brunsden and Thornes 1979), 'neocatastrophism' (Dury 1975), 'neotectonics' (Fairbridge 1981), 'relaxation time' (Thornes and Brunsden 1977), 'recovery times' (Brunsden and Kesel 1973; Allen 1974), 'dominance and formative events' (Wolman and Gerson 1978; Brunsden and Jones 1979).

Throughout science attention (Berggren and Van Couvering 1984) is being given to the new concepts of 'catastrophe' (Thom 1975), 'punctuated equilibria' (Gould 1974, Eldredge and Gould 1972) and extinctions (Ager 1973). It is relevant to note that geomorphological applications are often concerned with disequilibrium, disasters, the thresholds of change and with the recovery of systems to a more stable state. We can, at least, argue that applied work is a product of, or in sympathy with, the theory of the day. In many cases it is possible to argue that much of the stimulus for this theoretical development has derived from a need to find pragmatic solutions to unsteady or 'formative event' situations. Certainly an applied approach has not inhibited theoretical growth since we are currently experiencing the most rapid change in our conceptual framework since the introduction of the general systems theory in 1954.

A further important point is that applications of existing theory and technology are the best form of hypothesis testing since a failure to solve a practical problem in the natural landscape laboratory will quickly reveal deficiencies in the theory and stimulate new discoveries which model more exactly the real situation.

Applied problems have been responsible also for the development of much new theory. The best examples are perhaps found in fluvial geomorphology and hydrology where most of our understanding of regime theory or hydraulics was developed by canal and river engineers. Similarly, concepts of limiting equilibrium, loading, passive behaviour of earth materials and slope stability all developed for practical reasons. Similar examples can be cited for shoreline management and beach transport (Beach Erosion Board 1933; Williams 1960; Coastal Engineering Research Center 1966); river regulation, head cutting and meandering (Daniels 1960; Ruhe 1971); agriculture and infiltration or overland flow processes (Bennett 1939; Musgrave 1947; Zingg 1949; Phillip 1957; FAO 1965); road construction and permafrost behaviour (Muller 1947; Demek 1969; Becker 1972); building design and sand deposition (Gandemer 1977); mining and siltation (Gilbert 1914, 1917); mining, tunnelling and subsidence theory (Jicinsky 1884; Hoek and Brown 1980) or excavation and rock control theory (Hoek and Bray 1974).

Similarly, applied studies can be shown to have *strengthened* theoretical development in the improvement of inventory (Carrera and Merenda 1976), mapping (Mahr and Malgot 1978), calibration (Kienholz 1978), evaluation (Stevenson 1977; Stevenson and Sloane 1980), susceptibility (Brabb et al. 1972) and probability distributions (Ward et al. 1980) for landslide hazard surveys which have made significant attempts to extend the conventional approaches to landslide studies based on theories of limiting equilibrium (figure 12.2).

Another example is the work of the Bahrain Surface Materials Resources Survey (Doornkamp et al. 1980), which developed a theoretical basis for the development of saline ground and created a salt hazard mapping system subsequently used in Suez and Dubai for planning purposes (Cooke et al. 1982). Such examples enhance the view that applied projects have much to offer theory in general.

THE GEOMORPHOLOGIST AS TECHNICIAN

New theoretical developments are not always necessary for applied geomorphology to provide a useful input to the future of the subject. In many situations the work will require routine methods, standard surveys and inventory or taxonomic procedures (figure 12.3). For some scientists this will be unstimulating, repetitive or even intellectually barren. Such an attitude neglects three important points. First, the work will almost certainly take place in an unusual, perhaps exciting, environment which

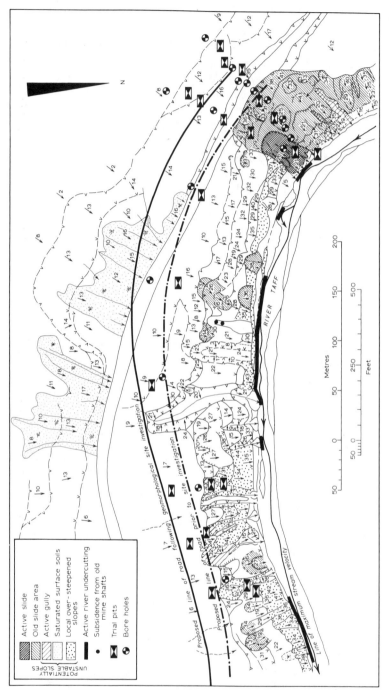

Figure 12.2 Three examples of landslide hazard maps prepared for practical purposes: a Prince Llewellyn in the Taff Vale Trunk Road showing how a geomorphological survey can be used to design a site investigation (Brunsden *et al.* 1975).

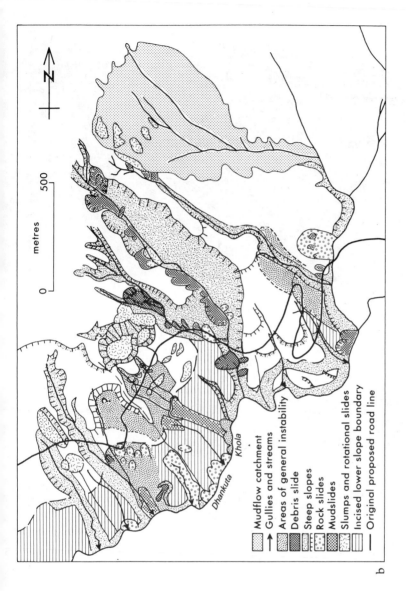

Figure 12.2 (cont.)
b Dharan–Dhankata Highway, Nepal, showing route evaluation techniques (Brunsden et al. 1975).

Note
(a) and (b) were prepared by Messrs Rendel Palmer & Tritton Consulting Engineers who are pioneers in the use of these techniques. (Map (a) is reproducted by kind permission of Rendel Palmer & Tritton and the Welsh Office; map (b) is reproduced by kind permission of Rendel Palmer & Tritton.)

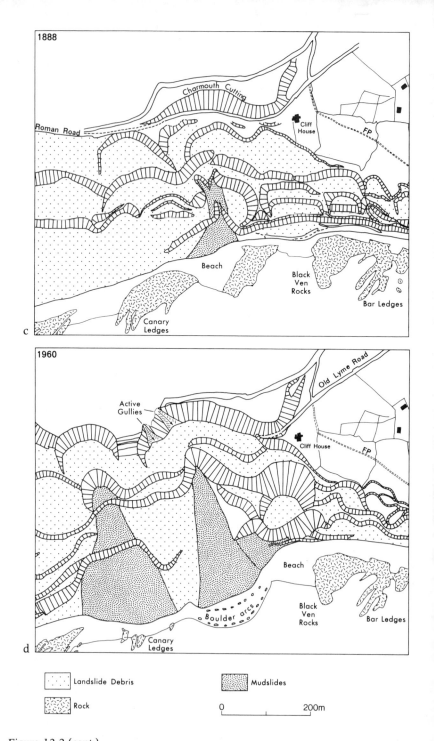

Figure 12.2 (cont.)
c and d Black Ven, Dorset, 1888 and 1960, showing the use of historical records.

GEOMORPHOLOGY IN THE SERVICE OF SOCIETY 241

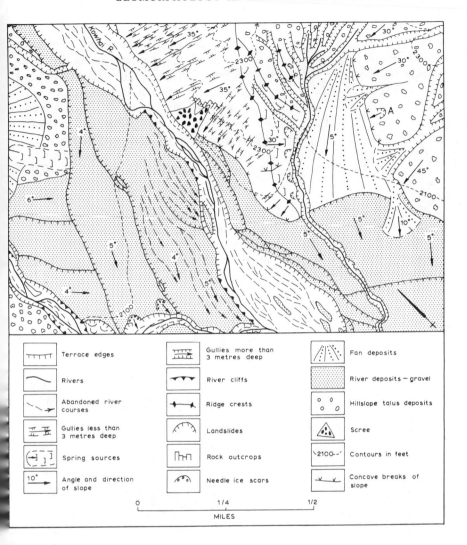

Figure 12.3 Typical, standard, geomorphological inventory map for an area of soil erosion in Canterbury, New Zealand

would not otherwise have been documented for science. Second, it will be carried out with the full support of a powerful logistic and financial framework. The funding, machinery and technology available usually far surpass that available for pure research. Third, it is essential that the work is carried out to high professional standards, adheres to recognized codes of practice and is technically competent, since a client

242 THE FUTURE OF GEOGRAPHY

expects a degree of skill which is expected of a competent practitioner in the profession (Sherrell 1976). Since the test of competence is 'contemporary practice as judged by the standards of experts' it follows that applied geomorphologists must endeavour to maintain very high standards. This can only be good for the subject; it emphasizes objectivity, improves technical standards and precision since the cost of giving poor advice is high, and must, in an unseen way, improve our capacity to act as pure scientists. A more rigorous methodology and practice at making decisions, on which huge financial investment or risk to property and life may depend, reflects on pure research and improves the reputation of the subject.

Figure 12.4 Landsat image of 14 February 1973, Band 7, for part of Somalia (left) and an example of how a geomorphological structure-landform survey prepared for hydrocarbon exploration can yield information of great value to 'pure' studies of landform evolution (right). Particularly note the relationships between structure and drainage and the way the central lava flow preserves an earlier course of the River Giuba.

The need for applied technical expertise also has abundant spin-off for the subject as a whole. For example, the use of remote sensing technology to evaluate, using geomorphological methods, the structural framework of an area for hydrocarbon exploration is also valuable for 'pure' studies of landscape evolution (figure 12.4). The development of side-scan sonar surveys for sea-bed mapping and the location of offshore structures, pipelines and cables has yielded innovative studies of the nature of submarine landsliding and delta development (Henkel 1970; Prior and Suhayda 1979). The mapping of hazards on alluvial fans (Schick 1971, 1974a, b) and the assessments of the hydraulic geometry and discharge of

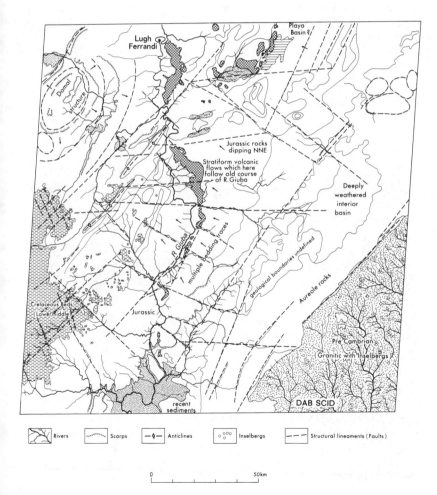

wadi systems on irrigation or highway location (Griffiths 1978) have greatly improved our knowledge of the nature of formative events in arid areas (figure 12.5).

Field measurement techniques such as mapping expertise (Brunsden et al. 1975), hazard surveys (Penning-Rowsell and Chatterton 1977), sediment yield modelling (Pickup 1980; Russell 1981; figure 12.6), hardware modelling (Ackers and Charlton 1970; Goudie 1974, 1977), simulation studies and water quality data analysis (Edwards and Thornes 1973) and forecasting (La Chapelle 1977) are further examples where applied or applicable research has advanced the subject.

Figure 12.5 Effects of flood on Wadi Abha, Saudi Arabia. A catastrophic flood in March 1982 destroyed fifteen bridges and many kilometres of highway. Study of such phenomena in the course of an applied geomorphological study yields new information on the nature of formative events in wadi systems and encourages research on effectiveness, work, persistence and recovery times as theoretical concepts.

INTERDISCIPLINARY RESEARCH AND RELATIONS

One of the driving forces for the development of applied studies has been the need for geomorphology to re-examine its position at the interface of earth, environment, social and economic sciences. Since Stamp (1966) first described geography's 'civil war' and Chorley (1971) remarked on his 'ever deepening dilemma', many authors have debated the changing roles and relations of physical and human geography, the nature of the divisions within geomorphology and the relative position of geomorphology with respect to other disciplines. (Chorley 1971; Cooke 1971; Hare 1973; Gould 1973; Chorley 1973; Hewitt and Hare 1973; Mikesell 1974; Falconer et al. 1974; Brown 1975; Atkinson 1975; Hard 1976; Veen 1976; Jones 1976; Smith and Ogden 1977; Ward 1978; Lawton 1978; Price 1978; Worsley 1979; Clayton 1980; Brown 1980; Jones 1983; Johnston 1983.)

The need to be more relevant to geography, one of its parent disciplines, has prompted many geomorphologists to work in areas where they can contribute to the interests of human geographers, planners and environmentalists. It is obvious that geomorphological applications are in harmony with the growing awareness by human geographers and with public and political interest in environmental management problems. Geomorphologists are now required to have an understanding of the socio-economic and technical planning of their surveys so that they are fully compatible with the needs of the social science disciplines and especially of those of urban planners (McGill 1964; US Department of Housing and Urban Development 1969; Betz 1975; Akili and Fletcher 1978).

As Jones (1983) has so beautifully argued, there is a clear area, the man–environment arena, where the geosciences and the socio-economic political sciences overlap and there is an urgent need for research on what he defined as:

a *Natural hazard impact* – to change planning and management practices and influence policy formulation.
b *Environmental auditing* – to assess changes in natural environment systems and evaluate the need for new practices.
c *Resource assessment* – to provide a basis for development.
d *Impact assessment* – to predict future changes (includes environmental impact statements and impact avoidance assessments).
e *Ex post audits* or *ex post assessments* – to review in hindsight the accuracy of prediction and the success of projects and policies.

Such work, all of an applied nature, clearly states one possible future for geomorphology and indeed the earth sciences as a whole in which it strives to reestablish its links with the 'human' sciences.

A second element of the applied 'forcing function' is the awareness that there is much to be gained from a close association with the other geosciences – geology, geophysics, sedimentology, ecology, soil science, Quaternary geology – and with the various branches of civil engineering, soil mechanics, rock mechanics and hydrology.

One of the joys of modern geomorphology is the willingness of scientists in other disciplines to assist its efforts to achieve a greater understanding of natural environment systems. The appointment of Professor J.N. Hutchinson to the first chair of Engineering Geomorphology in the Department of Civil Engineering at Imperial College and the delivery of the prestigious Rankine Lecture by D. Henkel (1982) on 'Geology, geomorphology and geotechnics' suggest that applied geomorphology may be 'coming of age'. Geographical geomorphologists are doing much to build bridges between disciplines by ensuring that their work is compatible to the aspirations of other scientists. There are real benefits to be gained from an association and incorporation of the methods and concepts, as well as the numerical rigour of disciplines such as soil mechanics.

It is also worth recording that the tender documents for many international development and environmental projects issued by such agencies as the DoE, UNDP and World Bank, now include clauses requiring an understanding of the geomorphology of a development site. The documents further require that the knowledge of landforms, materials and earth surface processes be utilized in the design of structures, remedial works, planning frameworks, land zoning plans, building ordinances and codes of practice. Geomorphological knowledge is now being utilized as an invaluable input to the planning, development and construction process with a consequent improvement to the success of the project and the wise use of the earth as the home of man. This is a most powerful argument for developing interdisciplinary

Figure 12.6 A sediment yield study for the new Kotmale Dam catchment in the Mahaweli Ganga Development Plan. The sediment yields were estimated using equations developed for theoretical geomorphological problems by Fournier & Kirkby (see Russell 1981). Detailed sediment yields within the catchment were estimated using a scoring system based on slope angle (A) and land use (B) to yield estimates of high scoring areas requiring treatment (C). This work aids our understanding of the sensitive areas of landscape change.

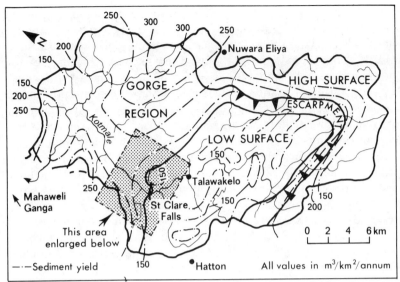

The Kotmale drainage basin with the probable sediment yield pattern

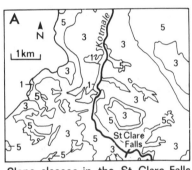

Slope classes in the St Clare Falls area

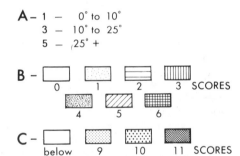

A — 1 — 0° to 10°
3 — 10° to 25°
5 — 25° +

B — 0 1 2 3 SCORES
4 5 6

C — below 9 9 10 11 SCORES

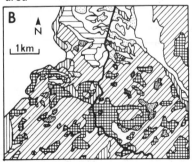

Land use classes

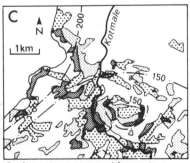

Sediment scores with contours showing mean yield values

applied research since it suggests that geomorphology is adapting to the needs of society and growing in status and responsibility. Modern geomorphologists take this role very seriously and point out (Brunsden et al. 1978; Jones 1983) that although there is not a big market for geomorphologists to be employed as geomorphologists, many will obtain employment outside the subject. They will be in a powerful position to play a positive and beneficial role as decision-makers in commerce, industry, government and administration by ensuring that the needs and dangers of the physical environment receive due consideration.

It is important that those who make decisions which affect the environment should be aware of the consequences of their actions. Irrespective of whether or not geomorphologists contribute their knowledge to the development process, it should be recognized that decisions *will* be made and it is a valid view to suggest that the geomorphological input should be made by those best able to provide it – the geomorphologists – and not by geologists, engineers, sedimentologists or ecologists, however well meaning or competent in their own fields they may be. When viewed in this light applied geomorphology is seen as an educational ideal and a national service.

REFERENCES

Ackers, P. and Charlton, F.G. (1970) 'Meandering geometry arising from varying flows', *Journal of Hydrology*, 11, 230–52.
Ager, D.V. (1973) *The Nature of the Stratigraphic Record*, London, Macmillan.
Akili, W. and Fletcher E.H. (1978) 'Ground conditions for housing foundations in the Dahran region, Eastern Province, Saudi Arabia', *Proceedings, International Association of Housing Science Conference*, 2, 532–46.
Allen, J.R.L. (1974) 'Reaction, relaxation and lag in natural sedimentary systems: general principles, examples and lessons', *Earth Science Review*, 10, 263–342.
Atkinson, B.W. (1975) In discussion of E.H. Brown, 'The content and relationships of physical geography', *Geographical Journal*, 141, 35–48.
Aufrère, L. (1929) 'Les rideaux – étude topographique', *Annales de Géographie*, 529–60.
Beach Erosion Board (1933) *Interim Report*, Washington, DC, Government Printing Office.
Beaumont, P. (1974) 'Water resource development in Iran', *Geographical Journal*, 140, 418–31.
Beaver, S.H. (1961) 'The reclamation of industrial waste land for agriculture and other purposes', *Problems of Applied Geography*, Warsaw, PWN.
Becker, J.C. (1972) 'Alaska builds highway over muskeg and permafrost', *Civil Engineering*, American Society of Civil Engineers, 42, 75–7.
Bennett, H.H. (1928) 'Soil erosion – a national menace', *US Department of Agriculture Circular 33*, reprinted in *Benchmark Papers in Geology*, 8, Strouds-

burg, Penn., Dowden, Hutchinson & Ross, 57–83.
—— (1939) *Soil Conservation*, New York, McGraw-Hill.
—— (1955) *Elements of Soil Conservation*, New York, McGraw-Hill.
Berggren, W.A. and Van Couvering, J.A. (eds) (1984) *Catastrophes and Earth History. The New Uniformitarianism*, Princeton, NJ, Princeton University Press.
Berkman, L. (1975) 'Erosion puts southland beaches on endangered list', *Los Angeles Times*, 10 August 1975.
Betz, F. Jr (1975) *Environmental Geology*, Stroudsburg, Penn., Dowden, Hutchinson & Ross.
Brabb, E.E., Pampeyan, E.H. and Bonilla, M.G. (1972) 'Landslide susceptibility in San Mateo County, California', *US Geological Survey Miscellaneous Field Studies*, Map MF344.
Brown, E.H. (1970) 'Man shapes the earth', *Geographical Journal*, 136, 74–85.
—— (1975) 'The content and relationships of physical geography', *Geographical Journal*, 141, 35–48.
—— (ed.) (1980) *Geography: Yesterday and Tomorrow*, London, Oxford University, Press.
Brunsden, D. and Jones, D.K.C. (1979) 'Relative time scales and formative events in coastal landslide systems', *Zeitschrift für Geomorphologie*, IGU-AAG Coastal Symposium, New Orleans Suppl., 34, 1–19.
—— and Kesel, R.H. (1973) 'The evolution of a Mississippi river bluff in historic time', *Journal of Geology*, 81, 576–97.
—— and Thornes, J.B. (1979) 'Landscape sensitivity and change', *Transactions, Institute of British Geographers*, NS 4 (4), 463–84.
——, Doornkamp, J.C., Fookes, P.G., Jones, D.K.C. and Kelly, J.M. (1975) 'Large scale geomorphological mapping and highway engineering design', *Quarterly Journal of Engineering Geology*, 8, 227–53.
——, Doornkamp, J.C., and Jones, D.K.C. (1978) 'Applied geomorphology: a British view', in Embleton, C, *et al.* (eds), *Geomorphology, Present Problems and Future Prospects*, London, Oxford University Press, 251–62.
——, Doornkamp, J.C. and Jones, D.K.C. (1979) 'The Bahrain Surface Materials Resources Survey and its application to planning', *Geographical Journal*, 145, 1–35.
Bryan, K. (1925) 'The Papago Country, Arizona', *US Geological Survey Water Supply Paper*, 499, 1–436.
Bull, W.B. (1977) 'The alluvial-fan environment', *Progress in Physical Geography*, 1, 222–70.
Cailleux, A. and Hamelin, L.E. (1969) 'Poste de Baleine (Nouveau-Quebec): Example de géomorphologie complexe', *Rd. G. D.*, 3, 129–50.
Carrera, A. and Merenda, L. (1976) 'Landslide inventory in northern Calabria, southern Italy', *Geological Society of America Bulletin*, 87, 1153–62.
Chartres, C.J. (1982) 'The role of geomorphology in land evaluation for tropical agriculture', *Zeitschrift für Geomorphologie*, NF Suppl., 44, 21–32.
Chorley, R.J. (1971) 'The role and relations of physical geography', *Progress in Geography*, 3, 87–109.
—— (1973) 'Geography as human ecology', in Chorley, R.J. (ed.), *Directions in Geography*, London, Methuen, 115–70.

—— (1978) 'Bases for theory in Geomorphology', in Embleton, C. et al. (eds), *Geomorphology: Present Problems and Future Prospects*, London, Oxford University Press, 251–62.
Christian, C.S. (1957) 'The concept of land units and land systems', *Proceedings, 9th Pacific Science Congress*, 20, 74–81.
——, Jennings, J.N. and Twidale, C.R. (1957) 'Geomorphology', in Dickson, B.T. (ed.), *Guide Book to Research Data for Arid Zone Development*, Paris, UNESCO, 51–65.
—— and Stewart, G.A. (1968) 'Methodology of integrated surveys', *Aerial Surveys and Integrated Studies, Proceedings, Toulouse Conference*, Paris, UNESCO, 233–80.
Clayton, K.M. (1980) 'Geomorphology', in Brown, E.H. (ed.), *Geography: Yesterday and Tomorrow*, London, Oxford University Press, 167–80.
Cleveland, C.B. (1971) *Regional Landslide Prediction, California*, Resources Agency, Sacramento.
Coastal Engineering Research Center (1966) *Shore Protection, Planning and Design*, Technical Report, 4, 3rd edn, Washington.
Coates, D.R. (1971) *Environmental Geomorphology*, New York, Binghamton.
—— (1981) *Environmental Geology*, New York, John Wiley.
—— (1983) 'Large scale land subsidence', in Gardner, R.A.M. and Scoging, H. (eds), *Mega Geomorphology*, Oxford, Clarendon Press.
Cole, M.M., Owen Jones, E.S. and Custance, N.D.E. (1973) 'Remote sensing in mineral exploration', in Barrett, F. and Curtis, L.F. (eds), *Environmental Remote Sensing: Applications and Achievement*, London, Methuen.
Collier, C.R. (ed.) (1972) Committee on Geological Sciences, *The Earth and Human Affairs*, National Academy of Sciences USA, Division of Earth Science, San Francisco.
Cooke, R.U. (1971) 'Systems and physical geography', *Area*, 3, 212–16.
——, Brunsden, D., Doornkamp, J.C. and Jones, D.K.C. (1982) *Urban Geomorphology in Drylands*, Oxford, United Nations University and Oxford University Press.
—— and Doornkamp, J.C. (1974) *Geomorphology in Environmental Management*, Oxford, Clarendon Press.
—— and Warren, A. (1973) *Geomorphology in Deserts*, London, Batsford.
Craig, R.G. and Craft, J.L. (eds) (1982) *Applied Geomorphology*, London, Allen & Unwin.
Daniels, R.B. (1960) 'Entrenchment of the Willow River drainage ditch, Harrison County, Iowa', *American Journal of Science*, 258, 161–76.
Demek, J. (1969) 'Beschleunigung der geomorphologischen Prozesse durch die Wirkung des Menschen', *Geologische Rundschau*, 58, 111–21.
Dixey, F. (1962) 'Applied geomorphology', *South African Geographical Journal*, 44, 3–24.
Doornkamp, J.C. (1983) *Applied Geography*, Nottingham Monographs in Applied Geography No. 1, Nottingham.
——, Brunsden, D. and Jones, D.K.C. (eds) (1980) *Geology, Geomorphology and Pedology of Bahrain*, Norwich, Geo Books.
Douglas, I. (1976) 'Urban hydrology', *Geographical Journal*, 142, 65–72.
Dov Nir (1983) *Man, a Geomorphological Agent*, Dordrecht, Reidel.

Dowling, J.W.F. and Beavan, P.J. (1969) 'Terrain evaluation for road engineers in developing countries', *Journal of the Institute of Highway Engineers*, 16 (6), 5–22.
Dury, G.H. (1972) 'Some current trends in geomorphology', *Earth Science Review*, 8, 45–72.
—— (1975) 'Neocatastrophism?', *Anias Academia Brasiliensis Ciences*, 47 (suppl.), 135–51.
—— (1978) 'The future of geomorphology', in Embleton, C., Brunsden, D. and Jones, D.K.C. (eds), *Geomorphology: Present Problems and Future Prospects*, London, Oxford University Press, 263–74.
Dylik, J. (1957) 'Dynamic geomorphology, its nature and method', *Bulletin de la Société des Sciences et des Lettres de Lodz*, Classe III, VIII (12), 1–42.
Edwards, A.M.C. and Thornes, J.B. (1973) 'Annual cycle in river water quality: a time series approach', *Water Resources Research*, 9, 1286–95.
Eldredge, N. and Gould, S.J. (1972) 'Punctuated equilibria', in Schopf, M. (ed.), *Models in Palaeobiology*, San Francisco, Freeman Cooper.
Embleton, C., Brunsden, D., and Jones, D.K.C. (eds) (1978) *Geomorphology, Present Problems and Future Prospects*, London, Oxford University Press.
Emmett, W.W. (1974) 'Channel aggradation in western United States as indicated by observations at Vigil Network sites', *Zeitschrift für Geomorphologie*, Suppl. 21, 56–62.
Erlich, P.R. et al. (1977) *Ecoscience: Population, Resources, Environment*, San Francisco, Freeman.
Fairbridge, R.W. (1981) 'The concept of Neotectonics, an introduction', *Zeitschrift für Geomorphologie*, NF suppl. 40, 7–12.
Falconer, A., Fahey, B.D. and Thompson, R.D. (1974) *Physical Geography: The Canadian Context*, Toronto, McGraw-Hill Ryerson.
FAO (1965) 'Soil erosion by wetter-zone measures for its control on cultivated lands', FAO, *Agricultural Development Paper*, 81.
Fels, E. (1935) *Der Mensch al Gestalter der Erde*, Leipzig.
Ferrians, O.J., Kachadoorian, R. and Greene, G.W. (1969) 'Permafrost and related engineering problems in Alaska', *US Geological Survey Professional Paper*, 678, 37.
Fischer, E. (1915) 'Der Mensch als geologischer Faktor', *Zeitschrifte Deutsche Geologische Gesellschafte*, 67, 106–48.
Fletcher, J.E., Harris, K., Peterson, H.B. and Chandler, V.N. (1954) 'Piping', *Transactions, American Geophysical Union*, 35, 258–63.
Fookes, P.G. and Best, R. (1969) 'Consolidation characteristics of some Late Pleistocene periglacial metastable soils of East Kent', *Quarterly Journal of the Geological Society of London*, 2, 103–28.
—— and Collis, L. (1975a) 'Problems and the Middle East', *Concrete*, 9 (7), 12–17.
—— and Collis, L. (1975b) 'Aggregates and the Middle East', *Concrete*, 9 (11), 14–19.
—— and Collis, L. (1976) 'Cracking and the Middle East', *Concrete*, 10 (2), 14–19.
Franklin, J.A. and Denton, P.E. (1973) 'The monitoring of rock slopes', *Quarterly Journal of Engineering Geology*, 6, 259–86.
Gandemer, J. (1977) 'Wind environment around buildings: aerodynamic con-

cepts', *Proceedings of the 4th International Conference on Wind Effects on Buildings and Structures*, 423–32.
Gilbert, G.K. (1914) 'The transportation of debris by running water', *US Geological Survey Professional Paper*, 86, 1–263.
—— (1917) 'Hydraulic mining debris in the Sierra Nevada', *US Geological Survey Professional Paper*, 105, 1–154.
Glenn, L.Ch. (1911) 'Denudation and erosion in the southern Appalachian region and the Monongahela Basin', *US Geological Survey Professional Paper*, 72, 10–30.
Golomb, B. and Eder, H.M. (1964) 'Landforms made by man', *Landscape*, 2, 4–7.
Goudie, A.S. (1974) 'Further experimental investigation of rock weathering by salt and other mechanical processes', *Zeitschrift für Geomorphologie*, NE Suppl., 21, 1–12.
—— (1977) 'Sodium sulphate weathering and the disintegration of Mohenjo-Daro, Pakistan', *Earth Surface Processes*, 2, 75–86.
—— (1981) *The Human Impact: Man's Role in Environmental Change*, Oxford, Basil Blackwell.
Gould, P.R. (1973) 'The open geographic curriculum', in Chorley, R.J. (ed.), *Directions in Geography*, London, Methuen, 253–84.
Gould, S.J. (1965) 'Is Uniformitarianism necessary?', *American Journal of Science*, 263, 223.
—— (1974) 'The evolutionary significance of "bizarre" structures: antler size and skull size in the "Irish Elk", *Megaloceros giganteus*', *Evolution*, 28, 191–220.
Gregory, K.J. (1982) 'Fluvial geomorphology, less uncertainty and more practical application', *Progress in Physical Geography*, 6, 427–38.
Griffiths, J.S. (1978) 'Flood assessment in ungauged semi-arid catchments as a branch of applied geomorphology', *King's College London Geography Department Occasional Paper* 8.
Hails, J. (1977) *Applied Geomorphology*, Amsterdam, Elsevier.
Hard, G. (1976) 'Physical geography – its function and future', *Tijdschrift voor Economische en Sociale Geografie*, 67, 358–68.
Hare, F.K. (1973) 'Energy-based climatology and its frontier with ecology', in Chorley, R.J. (ed.) *Directions in Geography*, London, Methuen, chapter 8.
Henkel, D.J. (1970) 'The role of waves in causing submarine landslides', *Geotechnique*, 20 (1), 75–80.
—— (1982) 'Geology, geomorphology and geotechnics', *Geotechnique*, 32 (3), 175–94.
Hewitt, K. and Hare, F.K. (1973) *Man and Environment: Conceptual Frameworks*, Committee on College Geography Resource Paper 20.
HMSO (1911) *Royal Commission on Coast Erosion and Afforestation*, Third (and final) Report of the Royal Commission appointed to inquire into and report on certain questions affecting coast erosion, the reclamation of tidal lands and afforestation in the UK, vol. 2, part 1, HMSO, London.
—— (1971) 'The organisation and management of government R and D', *Framework for Government Research and Development*, Cmnd 4814, London.
Hoek, E. and Bray, J.W. (1974) *Rock Slope Engineering*, London, Institute of Mining and Metallurgy.

—— and Brown, E.T. (1980) *Underground Excavations in Rock*, London, Institute of Mining and Metallurgy.

Holland, J.L. and Stevenson, J.C. (1963) 'Foundation problems in arid-climate silts', *Geological Society of America Special Paper*, 76, 276–7.

Hollis, G.E. (1974) 'The effect of urbanisation on floods in the Canon's Brook, Harlow, Essex', in Gregory, K.J. and Walling, D.E. (eds), *Fluvial Processes in Instrumented Catchments*, London, IBG, 123–39.

Hudson, N.W. (1971) *Soil Conservation*, London, Batsford.

Hudson, N.W. and Jackson, D.C (1959) *Erosion Research*, Henderson Research Station, Report on Progress 1958–59, Federation of Rhodesia and Nyasaland, Ministry of Agriculture.

Hutchinson, J.N. (1969) 'A reconsideration of the coastal landslides of Folkestone Warren, Kent', *Geotechnique*, 19, 6–38.

—— (1982) 'Vote of thanks', after Henkel, D.J., 'Geology, geomorphology and geotechnics', *Geotechnique*, 32 (3), 194.

Jacks, G.V. and Whyte, R.O. (1939) *The Rape of the Earth – a World Survey of Soil Erosion*, London, Faber.

Jennings, J.N. (1965) 'Man as a geological agent', *Australian Journal of Science*, 28, 150–6.

Jicinsky, W. (1884) Quoted in Sherlock, R.L. (1922).

Johnston, R.J. (1983) 'Resource analysis, resource management and the integration of physical and human geography', *Progress in Physical Geography*, 7, 127–46.

Joliffe, I.P. (1974) 'Beach–offshore dredgings: some environmental consequences', *Offshore Technology Conference Paper*, OTC, 2056.

Jones, D.K.C. (1980) 'British applied geomorphology: an appraisal', *Zeitschrift für Geomorphologie*, supplement 36, 48–73.

—— (1983) 'Environments of concern', *Transactions, Institute of British Geographers*, 8 (4), 429–57.

Jones, S.M. (1976) 'The challenge of change in geography teaching', *Geography*, 61, 195–205.

Kienholz, I. (1978) 'Maps of geomorphology and natural hazards of Grindewald, Switzerland, Scale 1:10 000', *Arctic and Alpine Research*, 10, 169–84.

Klimaszewski, M. (1961) 'Enquête sur les organismes de géomorphologie appliquée', *Revue de Géomorphologie Dynamique* 12, 43–5.

Krammes, J.S. and Osborn, J. (n.d.) 'Water repellant soils and wetting agents as factors influencing erosion', *US Pacific Southwest Forest and Range Experiment Station Notes*, 177–87.

La Chapelle, E.R. (1977) 'Snow avalanches, a review of current research and applications', *Journal of Glaciology*, 19, 313–24.

Lawton, R. (1978) 'Changes in university geography', *Geography*, 63, 1–15.

Leggett, R.F. (1973) *Cities and Geology*, New York, McGraw-Hill.

Lofgren, B.E. (1969) 'Land subsidence due to the application of water', in Varnes, D.J. and Kiersch, G. (eds), *Reviews in Engineering Geology, Geological Society of America*, Colorado, 2, 271–303.

Mabbutt, J.A. and Stewart, G.A. (1963) 'The application of geomorphology in resource surveys in Australia and New Guinea', *Revue de Géomorphologie Dynamique*, 14, 97–109.

McGill, J.T. (1964) 'The growing importance of urban geology', *US Geological Survey Circular*, 487.
McLellan, A.G. (1967) 'The distribution of sand and gravel resources in west central Scotland and some problems concerning their utilisation', Glasgow, University of Glasgow.
Mahr, T. and Malgot, J. (1978) 'Zoning maps for regional and urban development based on slope stability', *Proceedings, 3rd International Conference, Association of Engineering Geologists*, Section 1, 1, 124–37.
Marsh, G.P. (1864) *Man and Nature, or Physical Geography as Modified by Human Action*, ed., D. Lowenthal (1965), Cambridge, Mass., Belknap Press.
Matley, I.M. (1966) 'The Marxist approach to the geographical environment', *Annals, Association of American Geographers*, 56, 97–111.
Meginnis, H.G. (1935) 'Effect of cover on surface run-off and erosion in the loessial uplands of Mississippi', *US Department of Agriculture Circular*, 347.
Melinkov, P.I. and Tolstikhin, N.I. (1974) 'Obshchee merzlotovedenie', Nauka, Novosibirsk, 291.
Mensching, H. (1951) 'Akkumulation und Erosion niedersachsischer Flusse seit der Risszeit', *Erdkunde*, 5, 60–70.
Mikesell, M.W. (1974) 'Geography as the study of environment: an assessment of some old and new commitments', in Manners, I.R. and Mikesell, M.W. (eds), *Perspectives on Environment*, Association of American Geographers, 1–23.
Mitchell, B. (1979) *Geography and Resource Analysis*, London, Longman.
Mitchell, C.W. (1973) *Terrain Evaluation*, London, Longman.
Mitchell, J.K. (1968) 'A selected bibliography of coastal erosion, protection and related human activity in North America and the British Isles', *Natural Hazard Research Working Paper*, University of Toronto, Canada, 4, 66.
Muller, S.W. (1947) *Permafrost or Permanently Frozen Ground and Related Engineering Problems*, Ann Arbor, Edwards.
Musgrave, G.W. (1947) 'Quantitative evaluation of factors in water erosion – a first approximation', *Journal of Soil Water Conservation*, 2, 133–8.
Newman, D.E. (1974) 'A beach restored by artificial nourishment', *Proceedings, 14th Conference, Coastal Engineering*, Copenhagen.
Nortcliff, S., Thornes, J.B. and Waylen, M.J. (1979) 'Tropical forest systems: a hydrological approach', *Amazoniana*, 8 (1), 65–72.
Parker, G.G. and Jenne, E.A. (1967) 'Structural failure of Western US Highways caused by piping', US Geological Survey, Water Research Division, Washington, DC, 18 January 1967, for 67th Annual Meeting Highways Research Board.
Penning-Rowsell, E. and Chatterton, J.B. (1977) *The Benefits of Flood Alleviation*, Farnborough, Saxon House.
Perrin, R.M.S. and Mitchell, C.W. (1969, 1971) 'An appraisal of physiographic units for predicting site conditions in arid areas', *Military Experimental Establishment Report*, II, 2 vols.
Phillip, J.R. (1957) 'The theory of infiltration', *Soil Science*, 83, 345–57, 435–48; 84, 163–77, 257–64, 329–39; 85, 278–86, 333–6.
Pickup, G. (1980) 'Hydrologic and sediment modelling studies in the environ-

mental impact assessment of a major tropical dam project', *Earth Surface Processes*, 5, 61–75.
Price, R.J. (1978) 'The future of physical geography: disintegration or integration?', *Scottish Geographical Magazine*, 94, 24–30.
Price, W.A. and Tomlinson, K.W. (1969) 'The effect of groynes on stable beaches', *Proceedings, 11th Conference, Coastal Engineering*, 1, 518–25.
Prior, D.B. and J.N. Suhayda (1979) 'Application of infinite slope analysis to subaqueous sediment stability, Mississippi Delta', *Engineering Geology*, 14 (1), 1–10.
Prokopovitch, N.P. (1969) *Some Geologic Problems in Reclamation of Arid Lands*, US Department of the Interior, Bureau of Reclamation, Sacramento.
—— (1972) 'Land subsidence and population growth', *International Geological Congress*, 1972 (Montreal), Sect. 13, 44–54.
Qashu, H.K. and Evans, D.D. (1969) 'Practical implications of water repellancy of soils', *Progress in Agriculture*, Arizona 21, 3–5.
Rapp, A. (1974) 'A review of desertification in Africa – water, vegetation and man', Secretariat for International Ecology, Stockholm, SIES Report No.1.
Ruhe, R.V. (1967) 'Geomorphic surfaces and artificial deposits in southern New Mexico', State Bureau of Mines and Mineral Resources, New Mexico Institute of Mining Technology, Memorandum 18.
—— (1971) 'Stream regimen and man's manipulation', in Coates, D.R. (ed.), *Environmental Geomorphology*, New York, Binghamton, 9–23.
Russell, J.R. (1981) 'Sedimentation in the proposed Kotmale Reservoir, Sri Lanka', Problems of Soil Erosion and Sedimentation, Bangkok, Thailand, SE Asia Regional Symposium, 27–29 January.
Ryabehikov, A.M. (1964) 'On the interaction of the geographical sciences', *Soviet Geography*, 5 (10), 45–60.
Schick, A.P. (1971) 'A desert flood: physical characteristics, effects on man, geomorphic significance, human adaptation – a case study of the Southern Arava watershed', *Jerusalem Studies in Geography*, 2, 91–155.
—— (1974a) 'Alluvial fans and desert roads – a problem in applied geomorphology', Abhandlungen der Akademie Wissenschaften in Gottingen, Mathematisch-Physikalische, Klasse III, Folge No. 29, 418–25.
—— (1974b) 'Formation and obliteration of desert stream terraces – a conceptual analysis', *Zeitschrift für Geomorphologie*, Suppl., 21, 88–105.
Schumm, S.A. (1976) 'Episodic erosion, a modification of the geomorphic cycle', in Melhorn, W. and Flemal, R. (eds), *Theories in Landform Development*, New York, Binghamton, 68–85.
—— (1979) 'Geomorphic thresholds: the concept and its applications', *Transactions, Institute of British Geographers*, 4 (4), 485–515.
Sherlock, R.L. (1922) *Man's Influence on the Earth*, London, Witherby.
Sherrell, F.F. (1976) 'Professional liability and professional indemnity for consultants', *British Geologist*, 2, 35–6.
Skempton, A.W. and Chandler, R.J. (1974) 'The design of permanent cutting slopes in stiff fissured clays', *Geotechnique*, 24, 457–66.
—— and DeLory, F.A. (1957) 'Stability of natural slopes and embankment sections', *Proceedings, 7th International Conference, Soil Mechanics and Founda-*

tion Engineering, State-of-the-Art volume, 291–340.

—— and Hutchinson, J.N. (1969) 'Stability of natural slopes and embankment foundations. State-of-the-Art Report.' *7th International Conference of Soil Mechanics and Foundation Engineering*, Mexico, 291–335.

Smith, D.I. and Bridges, E.M. (1965) Report of a symposium on 'Rates of erosion and weathering in the British Isles', Bristol, 1965, *Transactions, Institute of British Geographers*, 36, xiii–xiv.

Smith, D.M. and Ogden, P.E. (1977) 'Reformation and revolution in human geography', in Lee, R. (ed.), *Change and Tradition: Geography's New Frontiers*, Department of Geography, Queen Mary College, London, 47–58.

Stamp, L.D. (1966) 'Ten years on', *Transactions, Institute of British Geographers*, 40, 11–20.

Stevenson, P.C. (1977) 'An empirical method for the evaluation of relative landslide risk', *Proceedings, Symposium of the International Association of Engineering Geology. Praha International Association of Engineering Geology, Bulletin* 16, 69–72.

—— and Sloane, P.J. (1980) 'The evolution of a risk-zoning system for landslide areas in Tasmania, Australia', *Proceedings, 3rd Australian and New Zealand Geomechanics Conference*, Wellington.

Stewart, G.A. (1968) *Land Evaluation*, Melbourne, Macmillan.

Strahler, A.N. and Strahler, A.H. (1973) *Environmental Geoscience: Interactions between Natural Systems and Man*, Santa Barbara, Freeman.

Taylor, L.E. (1981) 'Environmental aspects of water resources planning in England and Wales', *Journal of Hydrology*, 51, 231–43.

Thom, R. (1975) *Structural Stability and Morphogenesis*, New York, Benjamin-Addison Wesley.

Thomas, T.M. (1956) 'Wales: land of mines and quarries', *Geographical Review*, 46, 58–81.

Thornbury, W.D. (1954) *Principles of Geomorphology*, New York, John Wiley.

Thornes, J.B. (1976) 'Semi-arid erosional systems: case studies from Spain', London School of Economics, Geography Department, Paper 7.

—— (1977) 'Channel changes in ephemeral streams, observations, problems and models', in Gregory, K.J. (ed.), *River Channel Changes*, London, John Wiley, 317–35.

—— (1978) 'The character and problems of theory in contemporary geomorphology', in Embleton, C. *et al.* (eds), *Geomorphology, Present Problems and Future Prospects*, London, Oxford University Press, 14–24.

—— (1979) 'Research and application in British Geomorphology', *Geoforum*, 10, 253–9.

—— and Brunsden, D. (1977) *Geomorphology and Time*, London, Methuen.

Tricart, J. (1953) 'La géomorphologie et les hommes', *Rd. G. D.*, 4, 113–56.

—— (1956) 'La géomorphologie et la pensée marxiste', *La Pensée*, 69, 3–24.

—— (1962) *L'epiderme de la terre. Esquisse d'une géomorphologie appliquée*, Paris, Masson et Cie.

—— (1968) 'Aspects méthodologiques des études de ressources pour le développement', *O Tulippe Memorial Volume*, Gembloux, Duculot, 345–61.

UNESCO (1968) *Geology of Saline Deposits*, Paris, UNESCO.

US Department of Housing and Urban Development et al. (1969) *Environmental Planning and Geology*, Washington, DC, US Government Printing Office.
Veen, A.W.L. (1976) 'Geography between the devil and the deep blue sea' *Tijdschrift voor Economische en Sociale Geografie*, 67, 369–80.
Verstappen, H.Th. (1963) 'The role of aerial survey in applied geomorphology', *Revue de Géomorphologie Dynamique*, 10, 237–52.
—— (1968) 'Geomorphology and environment', inaugural address, Waltmann, Delft, 1–23.
—— (1977) *Remote Sensing in Geomorphology*, Amsterdam, Elsevier.
—— (1983) *Applied Geomorphology: Geomorphological Surveys for Environmental Development*, Amsterdam, Elsevier.
Walker, H.J. (1978) 'Research in coastal geomorphology: basic and applied', in Embleton, C. et al. (eds), *Geomorphology: Present Problems and Future Prospects*, London, Oxford University Press, 203–23.
Wallwork, K. (1960) 'Some problems of subsidence and land use in the mid-Cheshire industrial area', *Geographical Journal*, 126, 191–9.
—— (1974) *Derelict Land*, Newton Abbott, David & Charles.
Ward, R.C. (1978) 'Q_{max} or MAF? – some reflections on hydrology today', *Geography*, 63 (4), 301–13.
Ward, T.J., Li, R.-M. and Simons, D.B. (1980) 'Landslide potential delineation', in Soil Conservation Service, Short Course, Colorado State University, Fort Collins, Colorado, ch. 13, 1–53.
Watts, D. (1978) 'The new biogeography and its niche in physical geography', *Geography*, 63, 324–37.
Williams, W.W. (1960) *Coastal Change*, London, Routledge & Kegan Paul.
Willis, D.H. and Price, W.A. (1975) 'Trends in the application of research to solve coastal engineering problems', in Hails, J. and Carr, A. (eds), *Nearshore Sediment Dynamics and Sedimentation*, London, John Wiley, 110–22.
Wischmeier, W.H. and Smith, O.D. (1958) 'Rainfall energy and its relationship to soil loss', *Transactions, American Geophysical Union*, 39, 285–91.
Woikof, A.I. (1901) 'De l'influence de l'homme sur la terre', *Annales de Géographie*, 10, 97–114, 193–215.
Wolman, M.G. and Gerson, R. (1978) 'Relative scales of time and effectiveness in watershed geomorphology', *Earth Surface Process*, 3, 189–208.
Worsley, P. (1979) 'Whither geomorphology?', *Area*, 11, 97–101.
Yaalan, D.H. and Kalmar, O. (1972) 'Vertical movement in an undisturbed soil: continuous measurement of swelling and shrinkage with a sensitive apparatus', *Geoderma*, 8, 231–40.
Young, A. (1968) 'Material resources surveys for land development in the tropics', *Geography*, 53, 229–48.
—— (1973a) 'Soil survey procedures in land development planning', *Geographical Journal*, 139 (1), 53–64.
—— (1973b) 'Rural land evaluation', in Dawson, J. and Doornkamp, J.C. (eds), *Evaluating the Human Environment*, London, Edward Arnold.
Zingg, A.W. (1949) 'A study of the movement of surface wind', *Agricultural Engineering*, 30, 11–13, 19.

13
UNDERSTANDING AND PREDICTING THE PHYSICAL WORLD

_____ Antony Orme

What is physical geography? Does it have a respectable academic lineage or is it the illegitimate offspring of some dalliance at the margins of organized knowledge? Is it a viable intellectual and professional pursuit with clearly defined goals? Is it necessary to a modern geographical education? Does it meet the needs of society? Questions such as these are often asked both within and beyond academia's ivory towers, by town as well as by gown, and by students as much as by professors. The answers vary from time to time, and from place to place, reflecting both individual attitudes and national perceptions. I personally believe in physical geography, in its rightful place within a geographical education, and in its relevance to contemporary issues facing world societies. There are many geographers, but by no means all, who think similarly. There are others who disavow a physical contribution to modern geography, who view society as independent of the physical environment and its resource base. Such perceptions are a myopia perpetuated by those who are blind to the interrelationships that both simple and complex societies weave with their environment. My mission in this essay is to answer the above questions and thereby substantiate the role that physical geography can and must play in the future of geography. The precise scenario for this role will vary from one society to another, and from one intellectual milieu to the next, as befits a world of many peoples and nations, but geography's universal message must lie as much with understanding and predicting the physical environment as with explaining the vagaries of human behavior. Geography without a physical base is sociology.

PHYSICAL GEOGRAPHY DEFINED

Physical geography is the geography of our physical world, essentially a correlative natural science. Its study embraces to some extent the sky above and the earth below but its primary focus is the earth's surface, the interface for so many complex organic and inorganic reactions, the stage for so much human activity. Physical geography *sensu stricto* may be defined in terms of the physical processes and inorganic materials of our environment, but the term is often widened to embrace the biotic processes and organic matter that are the focus of biogeography. Indeed geography is well served by according similar emphasis to both physical geography and biogeography, thereby recognizing the close links between the physical and biological realms, for example in studies of soils, microclimate, and plant productivity.

Whereas the subject matter of physical geography may be framed by the sky above and the earth below, it is neither meteorology nor geology. The geographer should normally accept some observations from these sciences as given, so that attention can be focused more properly on or near the earth's surface. Thus the climatologist may accept values for the solar constant and the nature of the upper atmosphere while seeking to evaluate the complex interaction between solar radiation, the lower atmosphere and the ground, an interaction so important to human life and livelihood. Similarly the geomorphologist may accept certain postulates concerning the earth's internal structure and composition while seeking to explain landforms in terms of surface and near-surface materials and processes. Nevertheless, such is the nature of modern science that feedback across disciplinary bounds becomes not only desirable but also inevitable. For example, with the development of plate-tectonic theory over the past three decades, structural geology has reemerged as significant to geomorphology, and geomorphologists have had to reexamine their thinking with respect to tectonism, landforms, and time. Conversely, geomorphology has acquired new life as an investigatory field with much to offer structural geology, geophysics, and seismology, for example through the study of baseline features such as Quaternary marine terraces that serve to indicate the nature and rate of surface deformation and thus, to some extent, of seismic hazard (Orme 1980b). Thus, whereas physical geography may have as its principal focus the earth's surface, its students should be aware of the porous boundaries and feedback potential that exist with neighboring bodies of knowledge.

ACADEMIC LINEAGE

Physical geography possesses a worthy academic lineage whose seeds were sown by scholars of classical Greece and Rome, the early Islamic world, and eastern Asia, but whose further flowering was delayed, at least in terms of western civilization, until the Renaissance of the fifteenth century. Thereafter, new or rediscovered knowledge was given firmer roots during the scientific revolution of the sixteenth and seventeenth centuries. Exploration and discovery generated rapid growth in astronomy and cartography, while the need to explain new phenomena led to major advances in physics, chemistry, and biology. Scientific advances were further interwoven with the industrial and agrarian revolutions of the eighteenth and nineteenth centuries. During the present century, notably its latter half, physical geography has been greatly stimulated by quantification, computer technology, remote sensing, and other advances in instrumentation.

The above outline is well known to historians of science and relevant details should form part of every physical geographer's training. Whereas these details need not concern us here, two linked themes merit further discussion, namely the nature of scientific growth and the tendency for periods of observation to alternate with periods of synthesis.

Scientific growth commonly focuses on a small number of decisive periods and does not occur at a steady rate. If we plot cumulative growth against time, an idea usually shows a slow initial rise involving relatively few scientists, followed by a more rapid rise to some critical point of widespread acceptance, concluding with diminishing growth as the idea is developed to its limits or perhaps superseded. Such is the shallow S-shape of a logistic growth curve. If instead we plot the simple growth rate against time, a bell-shaped curve results as an idea rises to a peak and then declines. The emergence of physical geography is in essence a sequence of bell-shaped curves of varying dimensions and overlap, illustrated for example by the espousal of monoglacial theory or evolution. A useful exercise for all geographers, albeit somewhat subjective, is to plot logistic growth curves for various ideas. The result is both provocative and chastening for it warns us to beware the dogmatic acceptance of any one theory. Let me now illustrate this point by reference to the second theme noted above.

As with so much science, the growth of physical geography may be viewed as rhythmic, as periods of observation alternate with periods of synthesis. As scientific advances occurred during the seventeenth and eighteenth centuries, physical geography emerged through the observation of natural features, what we prefer today to call data collection and analysis. Rock units were mapped, landforms described, streamflows measured, climatic data compiled, and plant and animal specimens collected. With these new data, attempts at classification were inevitable, for instance in Linnaeus's *Systema Naturae* (1735), but synthesis was often premature. While observations proliferated, new ideas found little acceptance in the prevailing intellectual climate and innovative explanations were commonly encumbered by theoretical constraints such as catastrophism. Thus, in what Davies (1966) has called the first adequate British discussion of geomorphic principles, *Geography Delineated Forth in Two Bookes* (1625), Nathanael Carpenter indicated that most of the world's landforms had been shaped during the Creation, the remainder fashioned by the Deluge. Catastrophism in one form or another was to bedevil scientific synthesis for a further two hundred years or more.

During the late eighteenth and early nineteenth centuries, however, the development of physical geography gained fresh impetus as old ideas began to decay from overabstraction and false generalities. Fresh ideas espoused by the new enlightened scientists stimulated further observations. Using the present as the key to the past, James Hutton's uniformitarian approach to the earth sciences, first presented in 1785, was given wider currency by John Playfair and Charles Lyell, and came eventually to influence the evolutionary theories of Charles Darwin and Alfred Wallace. Meanwhile, the early nineteenth century's enthusiasm for exploration was personified by Alexander von Humboldt who strode as a colossus through the emerging realm of physical geography. Humboldt's objective, as expressed in *Cosmos* (1849), was to recognize unity among the vast diversity of natural phenomena, and his method was 'the art of collecting and arranging a mass of isolated facts and rising thence by a process of induction to general ideas'.

During the later nineteenth and early twentieth centuries, the intellectual climate swung strongly in favor of classification and synthesis. It was time to take stock. Classification was exemplified by the focus on weather spells, climatic types, relief features, physiographic provinces, and soil groups, and by such persons as Buchan, Köppen, Peschel, Powell, and Dokuchaiev. Synthesis was exemplified by descriptive modeling aimed at providing ideal explanations of real-world phenomena, most notably through W.M. Davis's cycle of erosion and Frederick Clements' theory of plant succession. This period of synthesis and classification was entirely defensible when viewed against the intellectual environment of the time. What went wrong was that these syntheses came to be seen not as way-stations on a pilgrimage toward understanding but as an end in themselves into which all later observations had to be fitted. Whereas Davis's cycle of erosion and Clements' concept of plant succession were imaginative enough schemes for their time, they later came to exert a stultifying impact on scientific enquiry. This was not the fault of Davis or Clements but of those disciples who blindly trod in their masters' footsteps. Thus in the earlier half of the twentieth century, both instruction and research in physical geography were hamstrung by adherence to constraining models like Davis's cycle, or to rigid classifications such as Köppen's climatic types. In an unguarded moment, a typical investigator of the period might well have admitted that believing was seeing, which is of course the very negation of good scientific method.

Toward the mid-twentieth century, classification and synthesis again gave way to observation, the latter now armed with more sophisticated exploratory and analytical tools, many the outgrowths of the Second World War. The atmospheric and marine sciences exploded with new observations which were to alter significantly our understanding of their environments. Fluvial geomorphology leaped forward through the hydraulics research of the United States Geological Survey in the 1950s, of such men as Leopold and Wolman. The search for generalization, for unifying laws in nature, continued, for example in the morphometric work of Horton and Strahler, but these enquiries were now supported by statistical techniques that tested the validity of data and hypotheses.

But something more fundamental was occurring in the 1950s. For more than a century physical geography had been dominated by evolutionary thinking. Hutton's concept of cyclic change and Lyell's brand of uniformitarianism had led Darwin to write *On the Origin of Species by Means of Natural Selection* (1859) and to Wallace's *Contributions to the Theory of Natural Selection* (1870). Evolutionary thinking was espoused by many branches of science, and synthesis, as shown by Davis and Clements, sought to fit observations into orderly continuous sequences of change. Such approaches were essentially timebound, relying on linear views of causation in which stage followed stage within a closed system. In the 1950s, however, physical geographers began rediscovering an alternative framework, one based on Rudolf Clausius's pioneer work in thermodynamics (1850) and on later research by J. Willard Gibbs and Henry-Louis Le Chatelier among others. The thermodynamic approach had not been entirely lost to the natural sciences for G.K. Gilbert's geomorphic model was based on the energy conservation principle of the First Law of Thermodynamics and on Le Chatelier's negative feedback homeostatic concept of 1884 (Chorley and Beckinsale 1980). But such was the dominance of evolutionary thought that it was not until the 1950s that physical geographers and their allies once again began to treat nature in terms of balance between external forces and internal self-regulation, with change occurring through pulses or as oscillations around some equilibrium. It was in this revamped idiom that Leopold and Maddock (1953) examined the hydraulic geometry of stream channels, work that was to have significant implications for geomorphology and hydrology.

Thus with the explosion of knowledge after the Second World War and with evolutionary constraints removed, a further period of synthesis soon arrived, based on the plethora of field and laboratory data that accumulated during the 1950s and 1960s. Strahler (1952) produced a revised geomorphic system grounded in the basic principles of mechanics and fluid dynamics. Leopold and Langbein (1962) introduced the concept of entropy in open geomorphic systems while Chorley (1962) advocated a role for general systems theory in geomorphology. Krumbein (1963) introduced the process-response model to the earth sciences while Whittaker and others developed an open systems concept of plant succession. Physical geography received further impetus from the ecologist Odum (1971, 1972) who examined ecosystems as open energy systems and devised graphic circuitry for presenting energy flows and storages in both ecological and social systems. Chorley and Kennedy's (1971) advocacy of a systems approach throughout physical geography was adopted by Terjung (1976) for climatology. Much attention focused on the nature of equilibrium, for example in Schumm's *The Fluvial System* (1977). But not all these developments occurred worldwide or simultaneously, and more traditional approaches persisted for many years, for example in Büdel's *Klima-Geomorphologie* (1977). By 1980, however, when Strahler carefully reviewed the role of systems theory in physical geography, it was becoming evident that this phase of synthesis was becoming hackneyed and played out. Systems theory, like Davis's cycle an artificial device for organizing facts, provided useful stimulus at the time but eventually came to constrain and befuddle the continuing search for explanation in the physical world. Graphic circuitry illustrated interrelationships well enough but the precise nature and rate of these interactions remained obscure.

In recent years, observation and synthesis have become closely linked for, with observational data proliferating at exponential rates, the theoretical significance of such work clamors for explanation just as, conversely, deductive mathematical models demand testable data. Physical geography needs observations such as those by Humboldt and Leopold, and it needs synthesists such as Davis and Chorley, though not necessarily in the same idiom. But if we are to learn anything from this academic lineage it is to beware of synthesis without all the facts and to beware of dogmatic insistence on a particular approach. Such is the nature of modern science that the reemergence of a domineering figure like Davis appears unlikely. Nevertheless, whereas we may be spared the stultifying impact of the dogmatists, we must also beware of assuming that we know all the answers or that one approach will necessarily lead us to nirvana.

From the foregoing discussion we should also beware of defining physical geography in terms of its practitioners. Some individuals such as Strahler and Chorley may indeed be professional geographers, but others may be geologists, biologists, meteorologists, or ecologists, wont as we are to pigeon-hole scientists into outmoded categories that bear little relation to the interdependence of modern science. A correlative natural science is not to be confined by pigeon-holes.

THE GOALS OF SYNTHESIS, INTEGRATION, AND PREDICTION

Is physical geography a viable intellectual and professional pursuit with clearly defined goals? If we agree with Mackinder that knowledge is one, its division into subjects being a concession to human weakness, then we can ignore the above question and blithely pursue whatever interests us. Such a glib retort, however, is unlikely to impress education authorities, university administrators, or potential employers. Thus, while reassured by Mackinder's bravura, we are obliged to answer the question.

A viable discipline is usually definable in terms of its subject matter and methodology. If defined by subject matter alone, physical geography does not have distinct status because its raw materials, like its practitioners, are shared with other sciences. Nor can methodology alone be invoked to justify physical geography because its methods – exploratory, statistical, cartographic – are also used elsewhere. But the subtle combination of subject matter and methodology does seem to create a unique and definable role for physical geography, namely as a spatial and temporal explanation of natural phenomena at or near the earth's surface, with particular emphasis on the interrelationships among phenomena and between these and society. Human geographers frequently defend their turf in terms of the supremacy of spatial analysis but, in my opinion, physical geographers must invoke both spatial and temporal perspectives in order to explain the earth's surface in satisfactory fashion.

With its emphasis on synthesis, physical geography plays an important role in the scientific world. Humboldt's quest for unity among the vast diversity of nature was one expression of this search for synthesis. Attempts by Horton and Strahler to define morphometric laws in the mid-twentieth century were further examples, albeit ultimately elusive. Indeed, nature is so complex that the search for explanatory laws, if indeed they exist, remains a major challenge to many geographers. Short of that goal, however, synthesis of the physical world remains a major objective of the well-trained physical geographer. As I have emphasized elsewhere (Orme 1980a), a good training implies a thorough familiarity with the pertinent research literature, a grounding in the natural sciences that underpin physical geography, an understanding of mathematics and statistics, and familiarization with the various tools of the trade – field and laboratory instrumentation, cartography and graphicacy, remote sensing, and computer technology. With the demise of purely descriptive syntheses after mid-century, physical geographers left the realm of hand waving, of research by debate, and entered the arena of mathematics and logic. The quantitative revolution of the 1950s was less dramatic in physical than in human geography because numeracy had always been an integral part of the former, especially of climatology, even when quantification was little developed and Davis was eschewing slope measurements. Even so, the renewed emphasis on numeracy over the past forty years has stressed the need for careful scientific design, including selection of variables, sampling plans, and statistical analyses, so as to extract maximum information from measurements obtained. Physical geographers were among the first natural scientists to pursue rigorous scientific designs in the 1950s and remain well placed to conduct worthwhile syntheses today.

Synthesis involves assembling separate bits of data into an understandable whole. The physical geographer should competently assemble data concerning geomorphology, climatology, and biogeography into a comprehensive expression of the physical environment. It is not, however, sufficient to rest there. The physical geographer must also be practical and keen to apply findings to such areas as resource management, regional development, and urban planning. Truism it may be but so many people now live in towns and cities that one of physical geography's greatest challenges lies in ensuring that urban authorities make correct decisions in such areas as flood control, hillside development, and air pollution.

This leads to the goal of integration. Society is well served by the pursuit of an integrated approach to the physical world that well-trained geographers are ideally suited to provide. Some of the most challenging problems on earth lie at the interface between the natural system and the human scene, problems often created by earlier misunderstandings of nature and by subsequent defiant attempts to engineer solutions. In southern California following the Second World War, for example, rapid urban development saw hillsides cut and filled in complete defiance of underlying rock properties, old landslide terrain or long-term climatic and hydrologic potential. In a semi-arid climate, these were relatively dry years and planners and developers alike, ignorant of disasters in 1916 and 1938, were lulled by a lotus-land mentality into believing that there would be no day of reckoning for their abuse of nature. In 1969 and again in 1978, 1980 and 1983, they were rudely awakened as series of intense, persistent, high-magnitude rainstorms reactivated old landslides and generated fresh debris flows among hillside communities, while valley communities were inundated by dangerous debris torrents. High seas accompanying these storms damaged or destroyed many coastal properties that in more tranquil times had been built on or below old landslides or constructed on the shore seawards of what was to become high-water mark. Warnings issued by physical geographers and other scientists in prior years had gone unheeded during the defiant land grab of the 1950s and 1960s. Fortunately, several physical geographers in the region have since made valuable integrative contributions to more rational watershed management and coastal planning, as they have elsewhere in the realms of timber harvesting, agricultural erosion, surface mining, and disease control. Geography as a whole is always strengthened when physical geographers can effectively demonstrate their commitment to careful synthesis and integrative problem solving.

The ultimate goal of scientific investigation is prediction, namely the acquisition and synthesis of sufficient data in order to predict future events. For example, research into streamflows or beach erosion should shed light on how a stream or beach will behave in future so that appropriate decisions can be made with respect to floodplain management and coastal planning. Studies of net solar radiation and other climatic elements at the earth–atmosphere interface should direct us toward optimum environments for crop growth. To stress prediction above all other goals may be extreme for there is much research that is conducted for its own intrinsic interest and, in universities at least, the postulates of academic freedom dictate that this work continue. Nevertheless, the cause of physical geography could be much advanced if more attention was given to the predictions needed by society. For example, the flurry of research into network analysis during the years 1945–65 came to be seen by many of its practitioners as an end in itself, whereas the origins of this work in the 1930s and the needs of society lay in a better understanding of the drainage network for the purposes of flood prediction and routing. Much research on glacial geomorphology focuses on regional descriptions of Quaternary depositional sequences, interesting enough work but far short of the challenge that lies in predicting the behavior of materials under stress, predicting climatic change from the glacial record, and predicting land-use capabilities. Let us not lose sight of the predictive goals of physical geography, even while we enjoy ourselves doing something that is personally interesting.

Some two hundred years ago, James Hutton postulated that the present is the key to the past (a belief in which he was not entirely correct). But the past can also be used as a key to the future. Hutton's retrodiction may serve the geologist concerned with interpreting some stratigraphic sequence from a knowledge of modern sedimentary processes, but the physical geographer is well versed to apply a sense of time and place to predict probable future events. This is particularly germane to the prediction of natural hazards, ranging in frequency from fairly common floods through less frequent forest fires to rare events such as high-magnitude earthquakes. Several human geographers in North America have already drawn attention to the problems and perceptions of natural hazards but they have gone just about as far as they can within the existing state of our knowledge. What is needed now is more research into the physical and statistical properties of past events, notably their magnitude, frequency, and spatial variation, so that communities may be forewarned of impending disaster or advised about land-use alternatives. Earthquakes, fire, flood, drought, tornadoes, hurricanes, blizzards, hailstorms, landslides, blight, disease, and like hazards are very much within the predictive capacity of the geographer.

Prediction is of course more than a matter of timing. It also concerns location and magnitude. Recent advances in earthquake prediction have achieved some success with respect to the probable location and magnitude of future events and, although the timing of major events remains somewhat elusive at present, prediction of the former goes some considerable way towards providing input for building codes and emergency services. In relative terms, hurricane prediction is farther advanced. The probable magnitude, landfall, and arrival time can be predicted within certain limits, such that the major challenge now lies in convincing local communities to take appropriate action. In this context, the social scientist has an important role in translating credible physical science into community response.

Physical geographers should also reevaluate the relevance of various conceptual models to the predictive role. In terms of magnitude and frequency concepts, for example, how relevant is the belief that average conditions achieve most of the work of erosion and deposition in a river valley? It has been argued that average conditions, modest rains or annual floods, are indeed most important in the eastern USA and other humid temperate lands. But such conditions have very little meaning in semi-arid and arid environments which cover more than one-third of the earth's land surface and where increasing numbers of people have come to live. In the western USA, for example, the landscape slumbers through long periods of quiescence, only to be rudely awakened by earthquake, fire, flood, or landslide. Averages have little meaning under such conditions and concepts formulated elsewhere may have little relevance. For example, most of the 9.0×10^6 tonnes of sediment delivered by southern California's Santa Clara River to the Pacific Ocean between 1933 and 1938 arrived in six days of flood in 1938. Furthermore, sediment discharged from the same river in the 1969 floods amounted to 47.6×10^6 tonnes, compared with the average annual yield of less than 2.0×10^6 tonnes over the previous thirty years.

The presently fashionable threshold concept is both stimulating and worrisome to the predictive goal – stimulating because the recognition of thresholds from studies of past events may indicate levels of tolerance to which future development may proceed, worrisome because the existence of thresholds implies sudden disruption in a system whose record, if short, may not permit prediction. Nevertheless, the threshold concept lends a useful added dimension to explanation in physical geography.

Synthesis of observed data, integration of these and other data into the environmental equation, and prediction of future events using this information – these are the primary goals of physical geography. These goals are an integral part of a geographical education.

THE INTEGRITY OF A GEOGRAPHICAL EDUCATION

Physical geography is a correlative natural science, the geography of the physical world. As such its future growth and that of geography as a whole are best served by integrating geomorphology, climatology, and biogeography with the various aspects of human geography. There has been too much fractionation in recent years, too much attention to research in narrowly defined areas that has stopped short of a broader correlative perspective, too much emphasis on the esoteric at the expense of the usefully predictive.

Ironically, as the number of professional geographers has increased, so communication between specialisms has diminished, and communication across the discipline of geography has almost ceased. National conferences are held but significant numbers of one specialism or another don't attend, claiming that general geography meetings have little to offer specialists. Geography is not of course unique. In most disciplines, the scientific explosion of the past forty years has stimulated increased specialization and the breakdown of traditional bounds. Healthy crossbreeding has promoted vigorous new avenues of enquiry, for example in biochemistry and geophysics. But for geography specialization is a mixed blessing because, while new studies can be pursued along narrower deeper lines, the traditional fabric of geography begins to crumble. In many countries, geography has reached a crossroads and to survive as a discipline it needs leadership from broadly based specialists who are committed to the kinds of synthesis and integration that the subject can offer so well. In the USA in particular, where geography has gained only grudging acceptance as an academic discipline and is frequently met with bemused tolerance or even ignorance beyond academia, physical and human geographers must pull together in the common cause.

Some geographers see no common cause, only diverging paths. For example, whenever physical geography is broken into discrete parts, some geographers and other scientists assume that geomorphology is really part of geology and thus best left to geologists, and that climatology is part of the atmospheric sciences best left to meteorologists. Nothing could be more wrong. Many roads lead to Rome and several different approaches are necessary for the growth of knowledge. Furthermore, we cannot assume that scientists in other fields will do a better job or indeed do the job at all. Like geographers, geologists and meteorologists have their own perceptions of their fields, perceptions that change from time to time. For example, many geology departments in US universities have never taught geomorphology while others have long since dropped the subject from their curriculum. Others saw fit to reintroduce the subject at about the time that society generally and government agencies in particular were becoming aware, belatedly, of environmental problems in the late 1960s. In the realm of geomorphology, the physical geographer has as much right as the environmental geologist to practice a trade and bring to the work a carefully nurtured understanding of the multivariate character of nature. In Britain, where geologists for so long focused on hard rocks, geomorphology has long been the province of physical geographers, often dominating geography as a whole in the 1950s and 1960s. Over the past twenty-five years, however, geographers of the British Geomorphological Research Group have incorporated sedimentologists, hydrologists, oceanographers, and others, a most natural and wise move for effective geomorphic communication but a move that has shifted many geomorphologists away from their parent geography. Human geographers, for some obscure reason the self-appointed guardians of the faith, have in turn questioned whether physical geography should be maintained within geography.

For many years, concern about the place of physical geography in a geographical education was largely an American problem (Ahnert 1962). While physical geographers remained stifled by Davis's dogma and by climatic classification, human geography prospered under Barrows's human ecology, Sauer's landscapes, and Ullman's spatial analysis. Kirk Bryan (1944, 1950) saw physical geography not only as separate from human geography but itself disintegrating into 'a group of specific sciences, each pursued for its own end'. When Leighly (1955) asked 'What has happened to physical geography?', Ackerman (1958) replied that physical geographers had 'concentrated more and more on the genetic steps, finally merging their work with that of neighboring sciences'. Ahnert (1962) painted a dismal picture of American physical geography in the 1950s and early 1960s when there were rarely enough papers offered at national meetings to schedule special sessions, when less than 10 per cent of doctoral dissertations addressed physical topics, when curricula revealed only one or two physical courses, and when there was scarcely a modern geomorphology text in English written by a geographer. Fortunately the situation was soon to change.

Today physical geography in the USA is more healthy and has more to offer than at any previous time during the twentieth century. For example, physical geography accounted for nearly 30 per cent of all papers presented in the 233 paper and poster sessions held during the annual meeting of the Association of American Geographers in Washington, DC, in April 1984. It was not without coincidence that the meeting's theme was 'Geography and Public Policy', a role in which physical geographers have much to say. Furthermore, despite the loss of a few university geography departments, physical geography has a stronger curricular presence in tertiary education than at any time in its history, while several successful attempts have recently been made to reintroduce geography to secondary-education programs in various states and school districts.

If we are to seek arguments for maintaining physical geography within the framework of geography, the reasons are both academic and practical. I have already addressed the academic reasons, namely the synthesis, integration, and prediction at which physical geographers can be so adept, and will return to these in my closing statement on society's needs. The practical reasons are three in number and lie at the roots of the recovery in physical geography within the USA noted above.

First, the environmental movement that began in its modern phase in the 1960s created a need for the kinds of training that physical geography provides. Before 1960, federal, state, and local governments within the USA were primarily concerned with encouraging and facilitating development at the expense of the natural environment. Environmental programs were mostly concerned with hazard mitigation, beginning with the creation of a national lighthouse system in 1789 and the Coast Survey in 1807 and later of a weather service to provide storm warnings and of work in flood control, harbor protection, erosion control, mosquito abatement, and the like. By the early 1960s, development in many parts of the country was largely out of control and it was public frustration and pressure, rather than government leadership, that led to the flurry of restrictive environmental legislation in the late 1960s and early 1970s. Above all, the National Environmental Policy Act of 1970 and similar state legislation created a demand for environmental impact reports, which in turn generated a demand for trained scientists among whom physical geographers were well placed. Colleges and universities responded by structuring environmental programs and geographers could once again claim pride in their profession (although not in all environmental impact reports).

Second, the 1950s and 1960s were years of remarkable growth in terms of the faculty, programs, and material resources of American university campuses as state and private agencies sought to accommodate the postwar baby boom. The bubble was pricked rather suddenly around 1970 and universities entered a period of retrenchment at about the same time that primary and secondary education was also experiencing major problems. Thus traditional outlets for geographers within the teaching profession were curtailed in the 1970s, forcing educators to redirect their students and graduates along other employment avenues.

Third, and clearly related to the first and second reasons, new employment opportunities developed in such areas as environmental management, regional development, urban planning, and similar fields which were able to absorb geography graduates. The latter have in turn often risen to positions of responsibility within public and private agencies, encouraging the appointment of more physical geographers.

I have chosen to argue for physical geography mainly from perspectives gained in the USA, but similar opportunities are to be found in other developed countries where geography has been able to demonstrate its relevance to contemporary needs. In many developing countries, the prospects for physical geography are both urgent and challenging. On recent visits to Somalia, I was impressed by the emphasis placed on a geographical education by government officials and educators who stressed the need for trained individuals in such problem areas as range management and dune stabilization. Similar demands for the expertise of physical geographers exist in such diverse countries as Tanzania, Saudi Arabia, Israel, and Morocco where the environment poses special problems for life and livelihood. In the Soviet Union of course geographers have long been involved in resource management and regional development, while in Japan and China many physical geographers are investigating the critical interface between neotectonic geomorphology and earthquake prediction. In short, across the world physical geographers can and do make significant contributions to environmental problems pertinent to national and local needs.

Of immediate import, the challenge for physical geographers lies not with convincing outsiders of their utility but with demonstrating their intellectual compatibility with other geographers within the profession. Reflecting on the growth of such specialisms as geomorphology, many human geographers have come to believe that physical geography no longer has a place within geography. Damn cheek! Having long believed in the unity of geography and, pragmatically, in the tenet that united we stand divided we fall, I would urge such harbingers of self-destruction to recognize that all geographers are concerned with environment and people. If physical geography's emphasis be the environment, it is important that its work contribute to the people. Conversely, without implied determinism, it would be a sorry day for geography if human geographers chose to turn their backs on the environment. If they did, pleas for the sanctity of spatial analysis notwithstanding, human geographers would become indistinguishable from other social scientists. To avert such schism, all geographers need to communicate their commitment to the common good, ultimately to the commonwealth of society.

MEETING SOCIETY'S NEEDS

Physical geography can best justify its existence if its product meets society's needs. In conclusion, therefore, I will outline some examples of such work that benefits from the correlative and predictive approach of physical geography.

The first example concerns a declining ecosystem along the Zululand coast of South Africa. At the close of the main Flandrian transgression some 5000 years ago, Lake St Lucia became a coastal lagoon 112 km long, over 40 m deep, and 1165 km^2 in area (Orme 1975). In this warm humid environment, a rich and varied ecosystem developed in which plants and animals from tropical Africa blended with species from temperate southern Africa. Today Lake St Lucia is 40 km long, less than 2 m deep, and covers only 312 km^2, and its biota are under great stress. To understand why this happened, it was necessary to investigate the interaction between the physical and biological systems and to consider human influences. Research revealed that, whereas most coastal lagoons are destined to be infilled over time, the processes of change at Lake St Lucia have been accentuated in recent decades, primarily by human activity. Present sedimentation rates are between two and three times the mean rate for the past 5000 years, a response in part to accelerated erosion farther inland. In addition, periodic low water levels and high salinities reflect diminished discharge from watersheds affected by erratic rainfall, irrigation agriculture, afforestation, and drainage diversions. Frequent closure of the outlet to the Indian Ocean by strong littoral drift had further inhibited the exchange of water and biota between the lagoon and the sea. While accelerated sedimentation curtailed mangrove growth and the breeding of many aquatic animals, high salinities caused reed swamps to die back thereby depriving hippopotami, crocodiles, and smaller life-forms of food and shelter. Now that the nature of the lagoon's decline is understood, a major counter program involving watershed management has been implemented in the hope of postponing the demise of this significant natural system.

If management of a distant natural ecosystem is a desirable conservation goal of ultimate benefit to society, the control of development close to major population centers is critical. One such area is the Lake Tahoe basin that straddles the California–Nevada stateline 300 km northeast of the metropolitan San Francisco Bay area. Lake Tahoe itself is a priceless natural resource and recreation area 2000 m above sea level, over 500 m deep and about 100 km around. During the freewheeeling 1950s and 1960s, California had permitted extensive residential and recreational development while to the east Nevada had promoted the construction of large casinos, golf courses, marinas, and other recreational facilities. As population increased and property owners manipulated their lake fronts, Lake Tahoe was in danger of becoming a polluted concrete-lined bath tub. Enough was enough and the bistate Tahoe Regional Planning Agency, created by federal decree because of the lake's complex administrative arrangements, commissioned a series of studies that have since led to the more effective management of the basin. Several physical geographers were instrumental in this work, coordinating various projects and providing for forest management and shore-zone planning. The latter linked an understanding of the natural system with land-use capability in producing a ranked series of shore-zone tolerances that were the basis for a definitive shore-zone plan (Orme 1972).

In another context, W.H. Terjung and his collaborators (1984) have produced a crop-yield model that uses climatic and environmental data to calculate yield and water consumption for a variety of major food crops in various areas of the world. Simulating actual and potential yield for rain-fed and irrigated maize in China, and making certain dietary assumptions, they concluded that the calculated maize production could potentially support between 400 and 700 million people, depending on the irrigation strategies adopted. If on the other hand the maize was used as feedstuff for beef production, only between 60 and 100 million persons could be supported. Using a similar model for the North

American Great Plains, Terjung and his colleagues were able to demonstrate the response of maize evaporation and irrigation water needs to a series of possible climatic changes involving air temperature, precipitation, and solar radiation. In a world in which future food increases will come at increasingly greater economic and environmental costs, mainly because the limits of potential food production appear to have been reached with present agricultural technology (Wittwer 1980), such studies by physical geographers assume growing significance.

Physical geography is alive and well. It will continue to flourish provided its practitioners are well trained, imaginative, inquisitive, and proud of the role that they can play. Physical geographers should implement their distinctive integrative role with vigor and enthusiasm and they should focus on prediction so that their work can be of major service to society. In the final analysis it is society, not another geographer, that is the arbiter of our success.

REFERENCES

Ackerman, E. (1958) *Geography as a Fundamental Research Discipline*, Research Paper No. 53, Department of Geography, University of Chicago.
Ahnert, F. (1962) 'Some reflections on the place and nature of physical geography in America', *The Professional Geographer*, 14 (1), 1–7.
Bryan, K. (1944) 'Physical geography in the training of the geographer', *Annals, Association of American Geographers*, 34, 189.
—— (1950) 'The place of geomorphology in the geographic sciences', *Annals, Association of American Geographers*, 40, 204.
Büdel, J. (1977) *Klima-Geomorphologie*, Berlin and Stuttgart, Borntraeger.
Chorley, R.J. (1962) *Geomorphology and General Systems Theory*, United States Geological Survey, Professional Paper 500-B, Washington, DC, US Government Printing Office.
—— and Beckinsale, R.P. (1980) 'G.K. Gilbert's geomorphology', in Yochelson, E.L. (ed.), *The Scientific Ideas of G.K. Gilbert*, Geological Society of America, Special Paper 183, Boulder, Colorado, 129–42.
—— and Kennedy, B.A. (1971) *Physical Geography, a Systems Approach*, London, Prentice-Hall International.
Davies, G.L. (1966) 'Early British geomorphology 1578–1705', *Geographical Journal*, 132 (2), 252–62.
Krumbein, W.C. (1963) 'A geological process-response model for analysis of beach phenomena', *Bulletin, Beach Erosion Board*, 17, 1–15.
Leighly, J. (1955) 'What has happened to physical geography?', *Annals, Association of American Geographers*, 45, 309–18.
Leopold, L.B. and Langbein, W.B. (1962) *The Concept of Entropy in Landscape Evolution*, United States Geological Survey, Professional Paper 500–A, Washington, DC, US Government Printing Office.

—— and Maddock, T. (1953) *The Hydraulic Geometry of Stream Channels and Some Physiographic Implications*, United States Geological Survey, Professional Paper 252, Washington, DC, US Government Printing Office.
Odum, H.T. (1971) *Environment, Power and Society*, New York, Wiley-Interscience.
—— (1972) 'An energy circuit language for ecological and social systems: its physical basis', *Systems Analysis and Simulation in Ecology*, 2, New York, Academic Press, 139–211.
Orme, A.R. (1972) *A shorezone plan for Lake Tahoe*, Tahoe Regional Planning Agency, California and Nevada.
—— (1975) 'Ecological stress in a subtropical coastal lagoon: Lake St Lucia, Zululand', *Geoscience and Man*, 12, 9–22.
—— (1980a) 'The need for physical geography', *Professional Geographer*, 32 (2), 141–8.
—— (1980b) 'Marine terraces and Quaternary tectonism, northwest Baja California, Mexico', *Physical Geography*, 1 (2), 138–61.
Schumm, S.A. (1977) *The Fluvial System*, New York, John Wiley.
Strahler, A.N. (1952) 'Dynamic basis of geomorphology', *Bulletin, Geological Society of America*, 63, 923–38.
—— (1980) 'Systems theory in physical geography', *Physical Geography*, 1 (1), 1–27.
Terjung, W.H. (1976) 'Climatology for geographers', *Annals, Association of American Geographers*, 66, 199–222.
——, Ji, H.-Y., Hayes, J.T., O'Rourke, P.A. and Todhunter, P.E. (1984) 'Actual and potential yield for rainfed and irrigated maize in China', *International Journal of Biometeorology*, 28 (2), 115–35.
——, Liverman, D.M., Hayes, J.T. and collaborators (1984) 'Climatic change and water requirements for grain corn in the North American Great Plains', *Climatic Change*, 6, 193–220.
Wittwer, S.H. (1980) 'The shape of things to come', in Carlson, P.S. (ed.), *The Biology of Crop Productivity*, New York, Academic Press, 413–59.

14
WILL GEOGRAPHIC SELF-REFLECTION MAKE YOU BLIND?
Peter Gould

IN WHICH YOU GET TO KNOW ME, SINCE I CAN'T GET TO KNOW YOU

Hi.[1]

That's an American and Canadian form of greeting not normally employed in serious academic writing, being considered much too casual. But I am using it in good faith, as a way of breaking the ice as you start this essay that you have been assigned, or perhaps have picked up casually. The editor of this book is a most enlightened fellow, and has encouraged me to write directly to *you*. And having been called 'enlightened', he now feels a bit awkward about censoring my style. So we've got him, haven't we?

I say 'we' deliberately, because I would like to think, as we read along together, that we are really doing the same sort of thing – talking to ourselves, and talking to each other. Now, talking to each other is considered perfectly respectable, while talking to ourselves is often considered a sign of on-coming madness. But you and I know that this really can't be right, and it's the others who have got it all wrong. Talking to yourself is actually a sign of *thinking*, and I challenge you to think without engaging in a sort of conversation with yourself. The problem is that too few people talk to themselves really seriously.

Before we start off together, I think I ought to tell you that I tend to be a polemicist, someone who has quite definite convictions (at least, for the moment), who is going to try to persuade you that his views are correct (at least, for the moment). This stance used to worry me greatly,

WILL GEOGRAPHIC SELF-REFLECTION MAKE YOU BLIND? 277

especially when I looked up the word *polemics* in the dictionary, and learnt that it came from the Greek word *polemikos*, meaning 'war'. I don't like wars, and I don't like the idea of being warlike. In any case, the word still has a rather disreputable tone to it, meaning that nice, well-brought-up academic people don't employ these forms of communication. Then one day I read in a book by the philosopher Ludwig Wittgenstein that polemics was really OK after all. It turned out that it was an ancient (so it *must* be respectable!) way of gaining someone's attention, so that at least they were aroused enough to think about what you were saying. 'Good old Ludwig,' I said to myself, and I have been an admirer of his ever since. Strange, isn't it, how we admire people who think like us and confirm our intuitive feelings?

Which brings me to Dr John Cole, Professor of Geography at Nottingham University, who once said to me, 'Peter, you really must not *preach* all the time!' But the problem is that I am much better at preaching to others than following my own advice, and, after all, we are all called to what we are called. In any event, you have had fair warning. Some of the things we are going to talk about may well be considered outrageous nonsense. But your professor, who may have omitted this essay from your reading list, can easily set you straight, and don't forget to give her the right answers when she asks you questions. These are the answers she gave you during the lectures. That's the whole point of the training you are receiving at your university, so remember always to give back the same truth that was given to you. In this way you will retire in forty year's time with a gold watch and a modest, but comfortable, pension. Of course, in the quiet of the night, you and I can think about other truths, preferably mine, which is, naturally, *the* truth.

With any luck, these two conflicting views of *the* truth may start you thinking for yourself, and you will reject both. Well, too bad ... no gold watch for you! Also, a sentence we read together, just now in the last paragraph, may raise another question – and these are always dangerous. Are you being trained, or are you being educated? Because if we are going to talk together and to ourselves about a geographical *education*, then I must ask you to let me take a step backwards, and place such an education in a larger, and for me essential, context.

LIBERAL EDUCATION AS A LIBERATING EDUCATION

Education, true education for me, liberates, and if you can't grant me that I am not sure where you and I go from this point. In other words, and even in a book like this, I am not going to extol geographic education as the be all and end all of life, but simply point to it as a part of a much greater, and actually lifelong, task. This is the task that the famous Swiss psychoanalyst Carl Jung called the process of *individuation*, the constant opening up and becoming of yourself. And I become more and more convinced that most universities don't do a very good job of educating, although the best may be able to expose people to examples of excellence so that they have a chance of learning how to educate themselves. Because – isn't it obvious once you think about it? – no institution educates *you*. When it really comes down to it, only you can educate yourself, only you can liberate yourself, and, I warn you, it's tough, hard, slogging work all the way. Perhaps that's why there are so few educated people around.

Training is a different matter, and we see trained seals all around us. So sit up straight, and clap your flippers in the prescribed manner, and society will throw a fish at you too. Not that efficient and effective training doesn't have its place. I would like to think that the mechanics working on my car are educated, and I wish them a liberating and continuing education as fellow human beings striving to reach the full richness of their human potential. But I also want them to be trained in the mechanical tasks they are doing on *my* car. Reciting from Dante's *Inferno* in unison, or humming the parts of the late Beethoven quartets as they fix my transmission is fine, but I hope they also know about thrust bearings, clutch plates and seals – transmission seals – in the right order. But notice that the picture I have deliberately painted for you reeks of *mechanism*: training always has a repetitive, mechanistic feel to it, a sense of repetition until the task becomes automatic. Which means, precisely, that you don't have to engage in *thinking* about it any more.

And that italicizing of *thinking* surely points to the difference between training and education, although be big enough to acknowledge that to be educated, to take up that continual opening up and development of yourself, means that there may be times when you have to train in certain tasks. But these tasks are always a means to a higher end, not an end in themselves. Musicians, artists, mathematicians, writers, even geographers, all undertake periods of self-training in certain, essentially technical, tasks. Otherwise you cannot undertake higher, more creative things. The superb technique of a Horowitz or Menuhin is necessary for their extraordinary artistic performances in which the sheer mechanics of finger movement plays such an essential part, even as it is thinkingly left behind. Similarly, a mathematician practises hard (good maths books are full of problems *with answers*), until the definitions and operations take on a subsidiary role to the creative possibilities opened up in whatever area of mathematics he or she is working in. Learning a new language requires constant, repetitive drilling, for only when the words and phrases of a new language become an unthinking part of us are the creative possibilities revealed.

But with that *caveat* and genuflection to training, and that overview that allows us to keep it in proper perspective, let's get back to education, the sort of education that only you can do for yourself.

If you are a university student in the English tradition (the Scots, as usual, are a bit better off), the highly constrained and specialized course in geography has little to do with education, except as certain bits and pieces provide you with opportunities to think, and perhaps follow up your thinking along lines not laid down by the curriculum. But most of the time things will be so structured that you will have little time to think for yourself, and so stray off course. And did you know that *curriculum* comes straight from the Latin, meaning a course laid out for a chariot race? No wonder some people come out of the university feeling like a well-trained and well-flogged horse! If you are a university student in the American tradition,[2] then the smorgåsbord of basic degree requirements and course distributions may come a bit closer to a liberal education (one can usually trace the ideal inspiring such a mish-mash), but most students quickly realize that the ideal has been distorted by departments being forced to fight for their lives by producing student credit hours, rather than educated people. In either system it is difficult to get an education, to learn how to educate yourself.

One of the problems is that the university is too often divided against itself, splitting itself in a schizophrenic fashion along lines labelled something like 'scientific' and 'humanistic'. These labels permeate our modern societies, and provide continuing evidence that we have increasingly lost the ability to reflect upon, to think about, ourselves. Inasmuch as you are caught up in your own university and society, you might start thinking about the degree to which you are an unthinking and unreflective product of them. And to help you, let me recommend a small book by Liam Hudson called *Contrary Imaginations*. Hudson worked as a teacher in an English boys' school (although what he discovered was not limited to this particular setting), and carefully investigated the way the 'system' channelled boys into scientific and artistic streams. At a pathetically early age, small tendencies in young boys were picked out, and ... as the twig is bent, so the branch grows. He discovered all sorts of interesting, and from my point of view quite horrifying, things which you can read for yourself, but let me pick out one. Two groups of boys, one 'arts', the other 'sciences', were asked to list all the things they could possibly do with an ordinary house brick. The lists from the boys in the arts were much longer than those in the sciences, demonstrating something about the different 'imaginations' of the two groups. But then the science boys were asked to play a game, to pretend they were artists in floppy velvet hats, poets with casual neckcloths, or long-haired musicians (notice the stereotypes!). Playing these pretend roles, they were asked again to list all the things they could do with a brick ... and the lists were much longer! Placed in a situation with a quite different set of expectations, their imaginations burst open to new possibilities. So perhaps we ought to reflect upon the sets of conditioning expectations in which we are embedded, not forgetting the 'expectations' of our geographical location,[3] and ask ourselves how we can get out of the trap. But the first thing, of course, is to realize we are in one.

WHERE WE COME FROM AS GEOGRAPHERS

We are thrown into a world not of our own making, and are shaped by that world. It cannot be otherwise, and even the idea, the possibility, of reflecting upon that world is itself a condition of the possibility of that world. Shall we run that sentence by once again? Right, and slowly this time: we live in a world where the possibility of thinking about that world is itself part of that world. In other words, the tradition (weak though it may be) of thinking about, of reflecting upon, our world is something we can meet within that world in which we have appeared. You may think this is naively obvious, but it is actually quite unusual. There are lots of other worlds, both today and in the past, in which such a tradition of self-reflection has not arisen, worlds of which *unthinking* acceptance of the seemingly unchangeable tradition is taken for granted. Even in our own world, in which the tradition of self-reflective thinking is available, how many actually release themselves to such a possibility?

But reflection, that is to say real thinking, is a tradition of our Western world (I do not have the temerity to speak for others), and has been a possibility (although not always realized) for the last 2500 years. I refer, of course, to the tradition of thinking that started in pre-Socratic Greece (say roughly 450 BC), one that runs like a slender and quite marvellous golden thread throughout the history that is our joint heritage, and at whose forward and changing edge into the future we both stand. And should you think by any chance that 'all that Greek philosophy stuff' has no real relevance, then ask yourself why the dialogues of Socrates cannot be taught in Prague today, why those teachers of philosophy who try to do so are dismissed from their university posts, why secret police hound the seminars in private homes, and why some teachers are forced to go into exile. How can the thinking of more than two millennia ago be so dangerous, and therefore so relevant, today?

We have to start with philosophy, because that is where, in a rather deep sense, we ourselves start from – as thinking men and women in the western tradition, and as geographers forming a subset of this larger ensemble. You see, 400 years ago we really couldn't talk of physics and chemistry, let alone biology, zoology and geology, and certainly not psychology, human geography and economics. Apart from medicine and jurisprudence (law), there was theology, which had most of the answers to people's questions, and there was philosophy which could ask questions provided the answers did not conflict with those of the theologians – as Galileo quickly found out. 'Science' had a rather different meaning, and what we call 'physics' today was then termed 'natural philosophy'. At Edinburgh University, one of the professorships in science is still called the Chair of Natural Philosophy, and one day other universities may glimpse the light again.

So what we see, very roughly, is the physical sciences hiving off from philosophy in the sixteenth–seventeenth centuries (although examinations in physics at Vienna were still based on Aristotle well into the nineteenth century!); the biological or life sciences hiving off in the eighteenth century; and what we might call the human sciences splitting away in the nineteenth century. I say 'very roughly' because these are obviously not exact dates, but broad tendencies. But they will serve well enough, the main point being that in hiving off from that old tradition of reflective thinking many of the sciences failed to carry it with them. Indeed, failing to carry that tradition of thinking about thinking with them, many felt that they could dispense with philosophy altogether, and they arrogantly elevated science to the primary (and, in some cases, the only) tradition of seeking the Truth. I would maintain that we have all paid a severe price for cutting that connection, and human geographers are paying that same price in their own way, shared, to a very large extent, by the other human sciences that developed as independent university disciplines about the same time. The problem is that we are all children of a scientific world, and as children of that world we fail to think very much about it. In marked contrast, there are certainly plenty of *unthinking* reactions to it, but these are as useless, and as dangerous, as the unthinking acceptances. What sort of world are we in?

A WORLD OF MECHANISM

We are thrown into a world of mechanism, which is to say that the *concept* of mechanism is the most outstanding characteristic of the world in which we live and have our being. Not just because we see machines everywhere, for these – everything from shavers and eggbeaters to moonrockets and computers – are just outward signs of something much deeper. Our thinking, that which we bring as human beings to things and ourselves, is shaped by mechanical images, so that we tend to look for descriptions, and seek for solutions, in mechanistic terms. Our natural language, which always shapes our thinking, is permeated with mechanical imagery to a degree literally unthinkable 300 years ago. The approximate date is significant: Newton's *Principia*, that great treatise on celestial *mechanics* suddenly appears; to be followed by Boltzmann's statistical *mechanics*; at roughly the same time that Darwin's theory of evolution requires ... a *mechanism* (what else!); that precedes Freud's psychoanalytic theories whose origins, we can see today, are deeply rooted in the all-pervading analogies of nineteenth-century *mechanics*. Freud, too, was a child of his time. We all, and always, are, and to point to this is not in the least to disparage or denigrate. Only a fool would disparage Newton, Boltzmann, Darwin and Freud. Rather, it should make us more sensitive to our own way of seeing, and make us realize that it too is historically contingent. A philosopher might refer at this point to our sense of *historicity*, and although the word is a translation from the German that seems a bit awkward at first in English, I think the word is useful. It means that our – what can I call it? – our intellectual *stance* is deeply conditioned by what has come before, and refers to how that 'coming before' opens or closes, brings out into the clearing or shuts away and conceals, possibilities for our own thinking. We are never wholly original, but always build upon and out from what is given to us by the world into which we have been thrown, even when we react against it.

What have been the consequences for the human sciences in general, and human geography in particular? When we investigate some aspect of the human condition in a scientific manner (whatever we might mean by that for the moment, but presumably excluding such things as listening to a beautiful piece of music, being moved by a lovely poem, taking part in a Japanese tea ceremony, or making love ...), we try to create knowledge that can be shared. And a major reason it can be shared is that evidence can be marshalled for a particular way of looking at, and describing, a part of our world so that others are convinced of its truthfulness. So the question is, how do we marshal evidence for our descriptions that is convincing, and this raises the further question of the 'languages' we chose to employ. I put *languages* in quotation marks, because I haven't made up my own mind about the propriety of using this word for two other important ways of communicating and marshalling evidence—namely, graphics and algebras. Yes, I know the word *propriety* has a rather stuffy, looking-down-your-nose-through-your-lorgnettes feel to it, but it means 'fitness' and 'rightness', with rather definite moral overtones. One day we can talk more about the morality of using certain words in certain ways, but if you know your George Orwell you also know my concern is not a trivial one.

So what languages do we use for scientific inquiry? Obviously, at least I think it is obvious, we use our natural language (in this particular case English), graphics, and algebras. Now as geographers, we are familiar, even intimately acquainted, with maps, and I take it we agree that these are useful. They certainly trap our thinking in certain ways (*every* 'language' does), but they were around a long time before mechanics and mechanism appeared on the scene. But we also use graphs, and by this I mean the usual two-dimensional pictures of Cartesian co-ordinates, with the axes labelled perhaps longevity (Y) and daily alcohol consumption (X). What we imply, of course, is that one thing, say longevity (Y), is a function of the other, say daily alcohol consumption (X). Change X, and Y changes too, or $Y = f(X)$.[4] OK?

Snap! The trap has just closed, and we didn't even feel it. For the form of this mathematical expression comes straight from the world of mechanism, whether it is the $F = ma$ of Newton, or the $E = mc^2$ of Einstein. In all cases it says, 'Turn the cogwheel (X) on the right hand side of the equation, and the cogwheel (Y) on the left hand side must turn too.' What could be more mechanistic? Of course, it's a bit more complicated than that; we may have lots of cogwheels on the right, and these are hitched together by things like $+$ and $\times$, things mathematicians call binary operations (because they work[5] on two things at a time, $x + y$, $a \times b$ ... and so on), but the underlying form of the function is purely mechanistic. This is hardly surprising when we recall that an enormous amount of functional analysis has been inspired by problems of describing the physical world. But — and this is where reflection upon what we are doing in the human sciences becomes important — but what happens when we unthinkingly borrow the mathematical forms of mechanism, forms used to describe purely unconscious and non-sentient *things*, to describe aspects of our human world? The mathematical language of mechanism is going to make our human world ... mechanical! Why? Because in the language we have chosen to describe it, that's all it can possibly look like. The language doesn't allow us to *think* anything else. And smearing the problem with a bit of statistical probability doesn't solve it — it just sweeps it under the rug.

If we are going to use algebras as descriptive languages for certain aspects of our human world (and if we are going to call ourselves 'human sciences' I think we must), then we have got to do what the best physical scientists have always done. And this is to sit down and *think* about the piece of the world we want to describe, and let our thinking about that part of the world suggest to us the form and structure of the algebraic language we really require to capture the truth we seek. And by form and structure, I really mean the things in the algebra, and the operations we are allowed to perform. The things may be various sorts of numbers, or lines and points, or even wiggly paths, and the operations may be such things as logical operators, wedge products, and path compositions. Most mathematics is geometric and qualitative today, not quantitative. Mathematicians, who are probably looking a bit pale around the gills if they are reading this cavalier description of their precious algebras, would refer to elements of sets, unary and binary operations, and vector spaces over fields, but these ideas we are talking about together are quite complicated enough in their general essence, let alone their specifics.

Is this happening? Are geographers beginning to think about other possibilities? Yes, here and there you can see glimmers of understanding that mechanistic descriptions, cast in the form of functions, are often so constrained that they actually form an unthinking parody of science. Are there any alternatives? Yes, bit by bit we are opening up the highly constrained functional forms of description, using such broader concepts as things called mappings and relations, and their geometric representations called simplicial complexes and lattices. We really cannot go into the details here (we have come right up to some research frontiers at this point), but I want you to be aware of what is happening – often against considerable opposition. As a professional geographer (and I assume I'm talking to a dedicated pro, not a casual amateur), these new ways of looking at things human and geographic will be part of your professional world in the decades to come, and you are going to have to do most of the hard learning yourself. I don't want to do a dramatic take-this-torch-from-my-faltering-hands act on you, but it is essentially *your* responsibility for where geography goes in the future. And remember it's all downhill once you're past 25. So first you feel like a well-trained and well-flogged horse, and now you suddenly have a great load of responsibility dumped on you. Who would become a human geographer? Only intellectual masochists enter the human sciences today. Well, you're quite right. Give it up. Go into something useful like winemaking.

Unless, of course, you're driven. Driven by those conversations with yourself called thinking, conversations that won't let go. All right, come and be mad then, but never say I didn't warn you. Welcome aboard, because it's an incredibly exciting time to be in the human sciences, and it's going to get better. You can see little chinks opening up through which ideas are evolving at last out of the old; the light at the end of the tunnel is still hellishly dim, and seems quite numbingly far away, but it's not all pitch-black any more. Those ties of reflective thinking are beginning to form once again – still slender, still fragile, still easily snapped, but they are there. I'm convinced of it, and to illustrate what I mean, let's look at three new words from that old tradition of thinking called philosophy, and ponder briefly what they mean for us who inquire into the human world. Throw away your preconceptions about Science for a moment, and let's start afresh.

CREATING AND INTERPRETING TEXTS

What do we really do when we inquire, when we undertake a piece of research? And I don't necessarily mean for a post-doctoral research programme, but also for a weekly essay we may have been assigned, a term paper, or a bachelor's thesis. The first thing we have to do is choose to observe, or to bring into our immediate realm of concern, some things and not others. That choice is a crucial one, and everything we can say thereafter hinges upon it. It is really a concern for definitions, and we have the obligation to think much harder about these and the sets of things we have chosen to define. Then we have to think about what we are going to do with the things we have defined. Map them? Count them? Relate them? Add, subtract, multiply and divide them? Once again, *how* we map them, *what* mathematical things we do to them, are crucial choices. Why? Because all these things belong to what the philosopher Jürgen Habermas has called the *technical* perspective, a concern for technical things like definitions, and choices, and algebras, and graphic representations. But why are we doing this? Why are we concerned with these technical tasks? Surely, it is because we are concerned with the manner in which we create 'texts', whether those texts are written words, maps, geometrical diagrams, or mathematical descriptions whose form is intimately bound up with the structures of the algebras we have chosen to employ. The technical choices determine the form and structure of our texts that are our descriptions.

But then what do we do with these descriptions we have created out of our technical perspective? I mean, there they are, sitting there – the maps, the diagrams, the equations, and so on. What do we do with them? Hang them on the wall and admire them? Obviously not: the 'texts' we have created in all their verbal, graphic and algebraic variety have got to be interpreted, so we land up right in the middle of the *hermeneutic* perspective. We have to place *meaning* on what we have observed and created, and we have to try to share that interpretation, the meaning we have given to the text, with others. After all, a fairly well-accepted definition of science is 'intersubjectively shared and verifiable knowledge', and although it is a bit limited, it will do for the moment. But if we thought certain aspects of the technical perspective were difficult, they pale beside the difficulties of the hermeneutic. Because to interpret a text convincingly, so convincingly it will be accepted (at least for the moment) by others, we have to bring every scrap of 'external' knowledge, every bit of experience, every bit of intuition and imagination we have to bear. Why? So that we can tell a convincing story about the text for the historically contingent moment through *rhetoric*, through persuasive argument so backed by evidence that others must say (at least for the moment) 'Oh, yes, I *see* . . . '. Which reduces science to story-telling, perhaps one of the oldest of human activities, but like all good story-telling it has to *illuminate* a part of our human condition.

But why do we do it? Why tell stories about our physical, biological, and human worlds? Presumably, because we are in a world ourselves which provides such expectations and possibilities. Many other worlds do not, and limit their story-telling to unchanging views that are preserved from one generation to another, often guarded zealously by a few who are responsible for passing on *the* Truth. In contrast, we seem to reach back to that old Greek tradition that drove people to question, to question even the unquestionable. And as each generation questions the last, we realize we may have found *a* truth, a truth for the moment. And today that questioning may well be based on something else besides pure curiosity, even though curiosity may still be the major driving force. Today we often want to know, to understand, so we can change the world, and at this point we have reached the third word, the *emancipatory* perspective. Sometimes it seems to have the connotation of a total standing apart, of an on-the-outside-looking-in stance that implies that somehow we can get outside of history and look in. Taken to this extreme, I think it is nonsense, and it produces all those mythologies of Marxism – that there are *laws* of history, and a privileged few can understand them. Again, this is not to denigrate Marx, only a fool would do that, but to recognize that Marx too was a child of his time. Mechanism was already well-established in the human imagination when he created his texts and interpreted them. If the physical world of things had laws of celestial mechanics governing the sun and the planets, so must the human world have laws – mechanical, of course, because that was all that *could* be thought. These are perhaps things we must talk about another time.

But that emancipatory perspective brings something else to the fore that has been missing – grievously missing. It is the concern and care that no longer lets us treat the human world as a mere thing for our curious attention, but recognizes that we are all a part of it. 'It' is us, and our questions lead us to ask how we can change it for the better. And as soon as we say 'better', our entire discourse is permeated with moral and ethical considerations. Marx is outraged,[6] and we should understand his rage, even as we might reject the limitations of his nineteenth-century interpretations. We are not very good at thinking from the emancipatory perspective, and our best intentions backfire with a distressing regularity, even as certain aspects of our increased knowledge raise soul-shredding questions of what is the right thing to do.[7] Many of the things that backfire do so because we did not do our job properly at the technical and hermeneutic levels: our texts of complexity were inadequate, perhaps too oversimplified, and our interpretations too quick and not properly informed. And so we see how all three – the technical, hermeneutic, and emancipatory – are not separate and distinct, but welded together in any person undertaking concerned inquiry.

TO BE CONTINUED, ALWAYS CONTINUED

One thing is certain: geography tomorrow is not going to be the same as geography today. How do I know this? Let's say I feel it in my bones, and, anyway, if tomorrow is far enough away I can't be wrong. Which makes my statement a tautology – which is what all profound truths are. All fields have periods of fluid thinking that then become congealed, like lava flows heating and cooling, flowing and hardening. Young Turks become Old Establishment too quickly, and you will see many who took part in the open methodological thinking of the 1960s and 1970s congealed to pretty solid positions in the 1980s. But here we really only face two choices: either the methodological millennium has already arrived, and we will be doing the same thing 100 years from now, or we will be doing something different. The lava is beginning to heat and crack again. What do *you* think?

I meant that seriously: what do you *think*? For the most important thing is to think, and think in that self-reflective tradition that knows it can never find an Archimedean point outside 'to move the world', yet constantly tries to gain some distance from the everyday task of inquiry. Philosophy is not an easy field to read in, for often the standards are extremely high, and the language may be difficult precisely because it is set down by men and women grappling, again and again, with some of the most difficult questions that have engaged our Western world for over two millennia. As someone engaged in human inquiry, I beg you not to turn aside: they are your questions too. But let me suggest two essays to get you started, and these may be the cracks through which the lava of your own thinking can flow to others. The essays are by Martin Heidegger; the first a memorial address for a composer, the second a lecture arranged by the Bavarian Academy of Fine Arts. They are difficult (at least I found them difficult), and you may have to read them slowly many times – something we're not used to. We tend to get cross if we don't understand something immediately, and it is only a short step to blaming the author for being obscure and failing to communicate. But remember, it is what *we* bring to a text that counts too.

These essays, like all of Heidegger's work, are invitations to thinking, not attempts to find closure to a system, not the final knots that tie the package up. Don't try to tie knots – in time they'll always come undone. Remember, even geographers are only right for the historically contingent moment. Aristarchos of Samos, 2500 years ago, said the earth moves around the sun. Another geographer, Ptolemy, said, 'Ridiculous! Any damned fool can see the sun goes around the earth!', and he was 'right', for about 2000 of those years. Do you get an inkling of how truths come in and out of concealment? And why Heidegger constantly and respectfully fondles the Greek word for truth – *aletheia*? *Lethe*, the river of forgetfulness in Hades, the nether-world, the concealed; our *lethargy*, and the weary lines in Keats's 'Ode to a Nightingale'.[8] But in the Greek the *a* of *aletheia* negates the forgetfulness and the concealment, and the truth now becomes that which is unconcealed. Stand open to new possibilities and a truth they may contain, even as you see and honour a truth in what has come before.

A PARABLE FOR OUR TIMES

Once upon a time, but not too long ago, I gave a polemical lecture at the geography department of an ancient English university. A distinguished geographer, on one of those awful commonroom point-scoring sprees, said at the end: 'But if you don't believe now the things you believed twenty years ago, then the things you have just been telling us you may not believe twenty years hence!'

It took me a few seconds to realize he was dead . . .

. . . serious.

And all I could do was answer 'Of course!'

Moral: Go to the university when you're 18, learn the truth, stay there the rest of your life, and don't change your mind. Above all, and in the name of the gods who have fled, don't think.

Hej!

NOTES

1 The same as *hej* in Scandinavia, but often used by the natives there as an exclamation upon leaving rather than arriving. Why they feel the need to greet you enthusiastically when they are about to leave is something the rest

of us will never know. Presumably, it must be the climate, particularly the long winters – as the geographer Hettner pointed out long ago.
2   If you are a Canadian student, do you get the best, or worst, of these two academic worlds?
3   For we are also prisoners of our geographic location, because that determines to a large extent the information we receive to shape into our opinions and prejudices. For some very simple, but I hope thought-provoking evidence, have a look at my 'Getting involved in information and ignorance', *Journal of Geography*, 82, 1983, 158–62, and 'Mental maps and information surfaces in Québec and Ontario', *Cahiers de Géographie du Québec*, 23, 1979, 371–98.
4   Fortunately a quadratic of the form $Y = a + bX - cX^2, a > 0$, which means (1) teetotallers croak (from the French *croquant*) first, and serve them right, (2) a little bit of what you fancy does you good, and (3) too much and your liver gives out. Quite a lot of interpretation carried by a little piece of mathematical structure, and quite a lot of philosophical implications – as we shall see in a moment.
5   Words again: remember that the Latin for 'work' is the *opera* of operations. So binary operations mean doing some work on two things at the same time.
6   I find it very puzzling that those who hold to any one of the positions in the broad spectrum of Marxism today never seem to discuss, examine, or acknowledge the moral foundations of their concern. Because of the innate arrogance of their position, a sort of standing-apart-and-watching-history-go-by, it almost seems as though they are embarrassed to acknowledge that they are human too. They remind me of certain schools of psychiatry whose practitioners can always fit you into their rigid, preconceived frameworks – frameworks that are increasingly coming apart today, as, indeed, any human construct must.
7   A concrete example may convince sceptics. A new test allows us to determine whether a person is carrying the gene of Huntington's chorea, a disease of the nervous system that kills in middle age. Do you tell a person? Do you tell the children, who have a 50 : 50 chance of carrying the gene, and the grandchildren who have a 25 per cent chance? Do you tell insurance companies? Employers? What are the ethics of such predictive tests? See S. Connor, 'Gene probes find Huntington's disease', *New Scientist*, 100, 1983, 399.
8   Do you remember them?

> My heart aches, and a drowsy numbness pains
> My sense, as though of hemlock I had drunk,
> Or emptied some dull opiate to the drains
> One minute past, and Lethe-wards had sunk:

ANNOTATED REFERENCES

Connor, S., 'Gene probes find Huntington's disease', *New Scientist*, 100 (1983), 399. A short article that raises explicitly the ethical dilemma of one aspect of

research in the human realm, that of medicine and forecasting certain death.
Couclelis, H., 'Philosophy in the construction of geographic reality', in P. Gould and G. Olsson (eds), *A Search for Common Ground*, London, Pion, 1982, 105–38. A very careful and thoughtful appraisal of what happens as we move from very detailed descriptions of urban systems, people and spaces to more and more general ones, with comments of deep philosophical concern along the way. Difficult, but worth reading many times and thinking about.
Gould, P., 'Getting involved in information and ignorance', *Journal of Geography*, vol. 82, 1983, 158–62. An example of a very simple exercise that raises some questions about the way we are all 'prisoners' of our location.
Habermas, J., *Knowledge and Human Interests*, Boston, Beacon Press, 1971. Go to his Inaugural Address, given when he accepted a chair at Frankfurt University, published here as an appendix. A slow and thoughtful reading is repaid many times over.
Heidegger, M., 'Memorial address', in *Discourse on Thinking*, New York, Harper & Row, 1966, 43–57. A memorial address given to a general audience about modern life and thinking, but touching a deep concern for a sense of place and home.
—— 'Science and Reflection', *A Question Concerning Technology*, New York, Harper & Row, 1977, 155–82. A very difficult essay that may have to be read and reread many times. *Chew* the words and sentences slowly.
Hudson, L., *Contrary Imaginations*, Harmondsworth, Penguin Books, 1967. A concrete description of the consequences of 'streaming' young boys into 'arts' and 'sciences'. Do you find your own educational history partially reflected here?
Olsson, G., *Birds in Egg/Eggs in Bird*, London, Pion, 1980. A strange, but deeply thoughtful, 'double' book of essays by a geographer whose concern for language, and what can and cannot be said, has been a paramount and continuing concern.

15
GEOGRAPHY AND SCHOOLING
John Huckle

I went out last night for inspiration, to the pub and friends of my age who went to similar or nearby schools. I asked them, 'Do you feel you were promised something in school you didn't get when you left?' One said, 'A job.' Another talked about them making our expectations too high. Another said we were promised nothing definite except maybe an ability to apply their ragbag of knowledge to the outside world. The last said that we had been given a fair education in an unfair society and that comprehensives ought to be as revolutionary as the public schools are elitist.

I just nodded at all the suggestions and comments, said I still didn't know what the promise was and that I would have another pint. (Roger Mills 1978)

Despite thousands of hours spent in school, hundreds of them in geography lessons, Roger Mills and his friends remain uncertain as to the purposes of it all. They and many others leave school with a profound sense of disillusionment which is rapidly reinforced as what they have learnt in the classroom proves of limited value in helping them cope with the world. Why is so much schooling profoundly anti-educational? Why do so few geography lessons develop a sensitivity to place and landscape and a critical awareness of one's location within economic and social structures which inevitably limit freedom? What are the alternatives and which are likely be be realized in the future?

The chapter attempts to answer such questions by relating school geography, mainly in Britain, to economic and social change. It argues that the future of a subject which owes its very establishment within higher education to the needs of schools, will be partly determined within school classrooms. Lessons taught here shape public images of the subject and the fortunes of academic geography are to a considerable extent linked to its popularity amongst older pupils. While the future of school geography will have a central bearing on the future of the subject, it will only be partly determined by geography teachers. That there are more powerful influences at work can be seen by considering the source of the disillusionment referred to above.

SOCIETY, STATE AND SCHOOLING

Each historical form of society has developed ways of raising and socializing children. In an advanced capitalist society such as ours, class-specific forms of education are largely controlled by those who control economic and social life. As part of its overall task of managing the economy and society on behalf of this group, the state establishes schooling as a mechanism of social reproduction and control. Education serves two main functions. It provides the general capabilities and vocational training necessary for different social classes to carry out different economic roles, and it transmits ideology which legitimates the existing form of class domination. State schools are therefore a public investment in labour and ideas which serves to lower capital's costs and legitimate its activity. They provide a publicly accepted mechanism for sorting future workers, and their work environments provide prior conditioning for life in the office or factory. Useful workers and citizens are created without raising their awareness of the inadequacies of the existing order. This is done by imposing a sharp divide between school and production and by using schools to create a hierarchy within the working class. The abstract knowledge which dominates so many lessons is used both to stratify pupils and to exclude genuinely relevant and critical ideas. The uselessness of school learning makes it a suitable preparation for alienated work while the hierarchies schools reproduce undermine class consciousness and are justified by reference to individual ability.

From such a view of schooling it follows that geography teachers fulfil both a general and a more specific role in social reproduction. Along with other teachers they sustain a hidden curriculum, or practical ideology (Sharp 1980), consisting of particular forms of social and technical relations, mediated by differing forms of language use. Social relations in classrooms vary between the coercive authority of the factory floor and the moral rationality of technocracy, while corresponding technical relations range from standardized routines designed to instil dependency to more independent learning and problem-solving designed to develop a degree of self-management. It is largely through taken-for-granted assumptions and procedures that schools help to reproduce society.

The specific role geography teachers fulfil is more related to the overt curriculum and theoretical ideology. The reality, rather than the rhetoric, of school geography, suggests that the majority of lessons cultivate a voluntary submission to existing social, spatial and environmental relations. The subject is generally presented as a body of unproblematic facts; many of them dull, boring, or redundant. Pupils are given a dehumanized and depoliticized view of the world and their success or failure depends largely on their ability to reproduce ideas, skills, and attitudes which sustain the status quo. There is little reference to economic and social processes which could explain the phenomena being studied, and what is generally offered as explanation is mere description. Ideas and material critical of capital are largely excluded, 'theories' are not placed in an historical and social context, and pupils are encouraged to see institutions, processes and knowledge as pre-given, neutral and static. Problem-solving and decision-making are usually cast within a consensus view of society, conflict is regarded as dysfunctional, and little attention is given to radical social alternatives. By failing to draw upon humanistic and structuralist philosophies school geography fails to develop its potential for cultivating environmental sensitivity and social awareness. As far as pedagogy is concerned, didactic teaching still predominates. Although this is made tolerable by a range of audio-visual media and other resources, there is little pupil-initiated enquiry or extension of classroom work into the community.

Geography teachers do then contribute to the process of hegemony where behaviour and common sense are shaped to conform to the necessities of capitalist production. Most of them accept this role due to their class position, the hold of hegemony both inside and outside the school, and their acceptance of educational ideology which offers notions of professionalism and political neutrality (Harris 1982). Further refining our view of society, state and schooling, suggests that geography lessons are but one element in a larger process, and that other teachers resist such a role by offering content and activities which are counter-hegemonic.

Marxist theories of education and the state (Carnoy 1982) vary in the emphasis they give to the economic or cultural role of schooling and in the degree of autonomy they allow education within the overall social dynamic. First, schools act with other agents of hegemony such as the family, the media and the workplace. Carnoy suggests that the imperatives of technology and bureaucracy in the workplace, threats of unemployment, and real increases in wages, living standards and welfare, have been more significant than schooling in persuading people to accept existing social relations and their role within them. Post-war welfare capitalism and economic growth produced a prolonged period of social consensus during which schooling's role in sustaining society was relatively minor compared with that of state intervention in the economy. Second, hegemony is both contested and dynamic. Schools reflect the resistances, antagonisms and struggles which characterize the overall process, and alternative practices, which undermine the existing order, are therefore found within them. Schooling must adapt to capital's changing economic and cultural needs and its continuing attempts to mediate or deal with evolving problems or contradictions. The state of the education system and school geography at any one time therefore represents a dynamic settlement between different interests and it is their power and actions which determine the course of educational change.

THE NATURE OF EDUCATIONAL CHANGE

Contrary to the beliefs of many geography teachers, changes in the nature of schooling, curriculum content, and methodology are not then simply a response to the growth of knowledge or the changing preoccupations of geographers and educationalists. A dialectic view of education and society suggests that debate on the nature of schooling will be particularly intense in periods of rapid economic and social change when the dominant class seeks to adapt education to new needs and subordinate classes can use the breakdown of the existing form of education to put forward radical alternatives. While the education system has some autonomy to interpret and shape the resulting demands in its own interests, it should be remembered that capital has core problems which form permanent items on the state's agenda and place recurring limits on educational change. Dale (1982) has identified three such problems and it is the last of these which hints at the primary function of school geography. The state must maintain support for the capitalist accumulation process, guarantee the context for its continued expansion, and ensure the legitimation of the capitalist mode of production, including the state's own part in it.

Salter and Tapper (1981) provide a model of educational change which focuses on the varied contexts within which economic and social demands are translated into educational practice. They recognize that such institutions as the examination boards, the Department of Education and Science, the subject associations, and the former Schools

Council, are bureaucracies with their own preferences and ambitions. Their need for routine procedures, desire to appear rational, and lust for power, means that they can acquire a developmental logic of their own which may significantly affect the nature of change. Attempts to alter the geography curriculum must be negotiated within contexts which have their own needs and priorities and it is therefore simplistic to seek an overdetermined correspondence between the needs of capital and school geography. Nor should it be thought that capital's needs form the only set of demands upon the educational system. Patriarchy, racialism and tradition also give rise to strong social demands as do such contradictions as the need to reproduce inequality while at the same time satisfying popular demands for greater equality of provision and outcome within secondary education. Change is facilitated and contradictions conccaled by the propagation of ideology which legitimates education's power and the social inequalities it helps to perpetuate. This ideology embraces new or modified images of the educated person and the role of education in a desired social order. While the early work of correspondence theorists in education appeared pessimistic, subsequent modifications allow significant scope for teachers and others to formulate and promote ideas and practices which oppose capital's continuing attempts to restructure education according to its needs. For geography teachers, one of the most significant contexts for political debate has been the subject association. Its power in determining the past and future form of school geography can only be realized by considering geography as a school subject.

GEOGRAPHY AS A SCHOOL SUBJECT

Geography's history as a subject is one of aspiration. As we shall see, the subject started with low status in the schools and eventually became established as a university discipline. Goodson (1983) presents a social history of geography, together with other school subjects, in order to show that by promoting their subjects as academic disciplines, some subject groups have had a considerable influence on the course of curriculum change.

Such groups are not monolithic entities but shifting amalgamations of subgroups and traditions which give the subject changing boundaries and priorities. In the case of a field of knowledge as broad and philosophically diverse as geography, there is a constant identity problem and threat of fragmentation which means that periodic attempts to redefine the subject and unite its practitioners are necessary. Goodson shows how in order to become established as an academic subject, geography sought to associate itself with the academic tradition in education and dissociate itself from alternative utilitarian and pedagogic traditions. Its pursuit of status and resources was best served by gaining acceptance amongst the other areas of high status, abstract knowledge taught in the universities and associated with a classical, liberal education. It was largely to promote geography as an academic discipline that the Geographical Association was established in 1893 and its success can be measured by the rapid establishment of geography within both the universities and school examinations over subsequent decades. Recognition as an academic subject brought geography teachers access to the brightest pupils, a share of the more generous resources allocated to their education, and also improved career prospects. Their desire to promote academic, rather than utilitarian or pedagogic versions of school geography, has to be seen as a response both to existing demands on the education system and their own material interests.

School geography teachers needed university geography in order to legitimate their claims to academic status, and its associated rewards, within the school curriculum. Once established, however, university geography developed its own needs and the school subject was increasingly shaped from above. The Geographical Association's role now became that of mediator between geography as researched and taught in the universities and geography as taught in schools. The status of the subject remained low in the universities long after it had been accepted at the highest levels in schools, and school geography was to continue to reflect the material aspirations, and associated shifts in philosophy and methodology, which characterized the university discipline. At both levels it is possible to recognize subgroups promoting different interpretations of the subject and associating these with different educational traditions or ideologies. Those likely to be most successful are in harmony with the material interests of the subject's scholars and teachers and the major vested interest groups within education and the broader society. They largely determine the subject's history.

THE EARLY YEARS

Geography gained a strong hold on the school curriculum in the nineteenth century due to growing recognition of its utilitarian value by the middle and upper classes. Suiting the vocational needs of future merchants, clerks, statesmen and strategists, school geography also served to facilitate increased exploitation of the physical environment and legitimate nationalism and imperialism. The resulting demand for teachers explains the survival of academic human geography which was earlier in a state of crisis, and its establishment in the universities by the turn of the century (Capel 1981). The state did not provide education for all children until 1870, by which time it was necessary in order to qualify workers for new production processes, free parents for factory labour, and provide a means of ideological control which would counter the self-education of the working class. In addition to basic skills and a strong moral code, the new elementary schools taught identification with nation and Empire, and a new vision of the world of work. Geography entered such schools rapidly after 1875 serving to counter children's 'magical' conceptions of the world and instil a view of economy and society supportive to capital. This can be clearly seen in the school readers of the time.

The rise of industrial society produced a range of ideological and utopian thought concerning alternative forms of education and society. Romantics, such as Rousseau, proposed a child-centred form of education which was to shape all subsequent advocacy of progressive education. At his experimental school in New Lanark in the early nineteenth century, Robert Owen combined progressive ideals with the anarchism of Godwin. Like later anarchists Godwin was alert to the social control function of state education and wished children to be able to resist the ideology transmitted by the school. Geography teaching at New Lanark was based on the real environment and curiosity of the children and was designed to help them understand the natural and social world. Geography and history were used to provide an insight into the economic and political relationships of society, to develop character, and to counter narrow prejudice.

While Owen's experiment was much valued by later socialists, including Marx and Engels, it failed to reflect two principles which are central to socialist theories of education. Owen failed to involve the workers in decisions relating to their children's education and failed to involve the pupils themselves in productive work. The working class would need to gain control of education if it was to become the political instrument of ordinary people, and pupils would need to be involved in work in order to learn that labour is essential to self-realization. Using schools to break the division between mental and manual work would also erode the basis of class domination.

From the ideas of Marx and Engels developed the concept of polytechnic education (Castles and Wustenberg, 1979) which has been applied, in various forms, in socialist states throughout the world. Such education aims to produce fully developed human beings capable of productive work and of understanding and controlling the present and future nature of society. Within the general component of such education, pupils are taught the scientific foundations of the production process and sufficient economics and social science to enable them to understand the mechanisms which shape society. Geography can make an important contribution to polytechnic education but it will need to be integrated with other areas of the natural and social sciences, with productive work, and with growing participation in political decision-making.

While Marx regarded the centrally controlled socialist state, with polytechnic education, as a transitional stage to communism, other socialists have advocated the reform of society and education from below. Two leading theoreticans of late nineteenth century anarchism, Peter Kropotkin (Breitbart 1981) and Elisée Reclus (Fleming 1979; Dunbar 1981) were both geographers who wrote a great deal about education. Reclus was encouraged to give increased attention to education as an instrument of revolutionary strategy, by state repression and his dismissal of political activity within the state. He regarded education as a form of consciousness-raising which could lift the masses from their state of prejudice and ignorance. In a circular of 1876 he outlined a scientific socialist education, outside religious, national and political influence, and suggested that this could best be encouraged by the provision of alternative textbooks. Aware that existing education amounted to 'bourgeois indoctrination', he proposed not a counter-indoctrination but an education free from indoctrination. Amongst the projects which Reclus encouraged the Vevey section of anarchists to promote was a geography project designed 'to expose the laws regulating the planet, to study the species inhabiting it, the races which quarrel over it and whose common property it is'. The project was to represent a scientific argument in support of universal brotherhood but unfortunately it never materialized, partly due to Kropotkin's belief that money would be better spent on political tracts. While as aware as Reclus of school geography's role in spreading imperialist ideology, generating disrespect for other cultures, and stifling independent thought, Kropotkin did not share a faith in 'scientific education' within existing society. By 1882, Kropotkin had persuaded Reclus that attempts to establish libertarian or integral education within a capitalist society would only be diversionary; that the creation of a libertarian society must come first. Nevertheless Kropotkin's statement 'What Geography ought to be' (Kropotkin 1885) remains one of the clearest and most influential statements on radical geographical education. Its advocacy of an anti-militarist, anti-imperialist, anti-capitalist education which examines issues from the point of view of the working class, fosters social harmony and mutual aid, and involves pupils in the life of the community, was to become part of a libertarian movement in education which subsequently challenged orthodox geography on numerous occasions (Smith 1983).

GROWTH AND DEVELOPMENT

Strong advocacy of socialist education took place during the educational reforms of the 1920s. By then the ending of imperialism, the onset of prolonged economic crisis in the 1890s and the economic and social legacy of the First World War, meant that elementary schooling was altogether too harsh and rudimentary a form of socialization. Its reform was necessary both to provide the increased number of skilled workers required by capital's second technological revolution and to temper growing class conflict. Capital, state and labour were to find in progressivism and the expansion of opportunity, the language and policies of reform which dominated educational debate for the next fifty years (Jones 1983). Imported from European and American philosophers, progressivism offered the twin tenets of child-centredness and social relevance. The ideas of Froebel, Dewey and others were used to justify a more humane and acceptable pedagogy and a more pragmatic, utilitarian and socially relevant curriculum. By providing the image of self-government in an organic unified society based on collaboration, Sharp (1980) suggests that progressivism helped to adjust people to new economic and social forms without threatening the underlying order. It did introduce some criticism and dissent into education, but its utopian foundations have generally ensured that this is readily absorbed and tamed by prevailing interests. Progressivism was coupled with policies to first expand, and later equalize, educational opportunity. In the debates of the 1920s those who wished to see state education become socialist education linked to the interests of the working class were defeated by those who merely wished to impove access to education as a means of expanding occupational choice. The TUC and Labour Party were converted to a meritocratic, or social democratic, view of education which overlooked issues of content and control and therefore left teachers and others with much autonomy concerning curriculum decisions. Geographers were amongst those who were relatively successful in exploiting this situation.

By the 1920s a great deal had been done to overcome school geography's lack of intellectual credibility and specialist teachers. In 1903, Mackinder had proposed a fourfold strategy of reform; university schools of geography to train teachers, school geography to be taught by specialists, the formulation of an efficient and progressive pedagogy, and the promotion of examinations set by geography teachers. Together, he and Herbertson were largely instrumental in ensuring its success and the space created by the social democratic consensus continued to be exploited by geographers until the 1970s. Since a truly technical or vocational alternative to the specialized academic curriculum failed to develop, growing numbers of pupils were taught and examined in class-specific forms of school geography. While this suited the interests of the school and university geographers, examination of successive text-books and examinations would show that it also suited dominant interests. School geography continued to act as ideology by drawing pupils into a unified national experience.

The state of 'enlightened traditionalism' which school geography had reached by the 1960s has been well described elsewhere (Beddis 1983; Walford 1981). This decade saw a revival of social democracy in education and a series of reforms designed to further expand and modernize provision as one means of sustaining post-war economic growth and political consensus. By now academic geography had gained status by applying itself to 'the technics and mechanics of urban, regional and environmental management' (Harvey 1974), and geographers were to use opportunities for curriculum reform to tighten the correspondence between school geography, university geography and the labour needs of the corporate state. Far from being a revolution or crisis (Graves 1975), the infusion of positivism into school geography, which became known as the 'new' geography, was profoundly adaptive and conservative. It was an élitist exercise from the start; an attempt to render the schooling of a minority of pupils more technocratic and vocationally relevant. Using such agencies as the examination boards, School Council projects, the Geographical Association, Her Majesty's Inspectorate, and text-book publishers, the advocates of reform were able to create a new orthodoxy in school geography within a decade. Their success owes much to the climate of the times and to a renewed appeal to geography teachers' self-interest. Curriculum reform was sold as 'new professionalism' (Tolley and Reynolds 1978) and many of its strongest advocates are now in positions of considerable influence over school geography. Educationalists in university departments and colleges of education played their part by combining positivist geography with rational curriculum theory to provide the professional perspectives on which new kinds of in-service education were based. The 'new' geography had most effect on 'able' pupils but others did not escape the climate of curriculum reform it brought in its wake. Attempts to identify key ideas (HMI 1978), common criteria for assessment at sixteen-plus (Joint Council 1982) and a new syllabus framework for 16–19-year-olds (Geography 16–19 1980), all contained elements of ideology associated with positivism. By the late 1970s school geography had come of age but was little influenced by the philosophical debates which had so enlivened its academic counterpart.

School geography's growth and development was not smooth. It had to be periodically disciplined and redefined from above in response to its expansive and fragmentary tendencies, and it also had to fight off claims to its territory from such competitiors as Social and Environmental Studies. It is difficult to trace a history of continuing socialist contributions but libertarian ideas re-emerged in the late 1960s to contribute to new forms of environmental, development and urban education. The mounting crises of capitalism were by then requiring interdisciplinary solutions and the resulting debate on integrated curricula allowed radicals to challenge liberal proposals. While the majority of school geographers were preoccupied with the 'new' geography, others were employing humanistic and structuralist philosophies to design lessons on such topics as environmental issues, global inequalities and urban redevelopment. It was the crisis in education which eventually encouraged a wider recognition of these developments amongst the majority.

THE CRISIS IN EDUCATION

Economic recession, and capital's attempts to restore profitability, caused the break-up of social democracy in education along with the wider decline of political consensus. While the academic curriculum had contributed much to this consensus, its poor record with regard to the economy was the focus for the 'Great Debate' on education launched in 1976. 'Thatcherism' built on ground first neglected and then created by Labour (Sarup 1982; Wolpe and Donald 1983). Under the guise of offering higher standards, greater accountability, and more choice, educational expenditure was cut back and that which remained made more functional for capital. As far as the curriculum was concerned new ways had to be found of purveying the ideologies, attitudes and behaviours necessary for loyal and disciplined workers during a time of economic crisis and high youth unemployment. Those jobs produced by capital restructuring require less skill of the majority of school leavers and new courses reflected this by inculcating 'social and life skills' which ensured submission to alienated work and the authority of the state. The Manpower Services Commission and other agencies first introduced the new vocational education into further education and then into the schools (Hart 1982). The Certificate of Pre-vocational Education and the new Technical and Vocational Education Initiative are in their early days but their concern to transmit 'economic and social awareness' to growing numbers of pupils represents a direct attack on school geography's long-established role. A traditional academic curriculum is being reasserted alongside the new vocationalism but Sir Keith Joseph's attacks on the social sciences, and DES ambivalence concerning geography's place in the core curriculum (Walford 1982), do not augur well for the subject's continuing status and level of support in terms of pupil numbers. The restructuring of secondary education is also designed to ensure a tighter control over the curriculum. Such measures as the abolition of the Schools Council, the establishment of common criteria for assessment at sixteen-plus, a stronger role for Her Majesty's Inspectorate and more restricted forms of entry to teaching, seek a stronger correspondence between schooling and the economy and seriously erode the autonomy which teachers have enjoyed for much of the century. Their work is also affected by spending cuts, increased scepticism on the part of the pupils, greater demands from parents and administrators and a general decline in morale induced by falling rolls, school closures and redeployment. The crisis in education is serving to radicalize a growing number who no longer regard themselves as professionals above politics, but as workers who share interests with large sections of the communities within which they teach.

In this new harsh climate geography teachers gradually became more aware of the opportunities presented by developments in the universities. Following some early initiatives (Lee 1977), parallel developments in educational theory provided the basis for new forms of humanistic and radical geographical education by the early 1980s (Huckle 1983). Awareness of these alternatives was heightened by research which revealed the most acute symptoms of school geography's role in social reproduction (Hicks 1981; Gill 1983). The early debate was preoccupied with issues of ethnocentric and racist bias but it was soon realized that there was a more general disease. By 1983 a new subject association had been formed (Association for Curriculum Development in Geography 1983), largely on Dawn Gill's initiative, and debate at the secondary level began to resemble that which had taken place in higher education several years before.

TOWARDS A SOCIALIST SCHOOL GEOGRAPHY

As the study of people's active construction and transformation of their physical and social environment, geography has a central role to play in a critical and emancipatory education. Geography lessons should help pupils to understand how societies are made and remade, and how landscapes and human–environment relations change in the process. Roger Lee (1983) explains why the dialectic between social structure and human agency is central to such understanding and how a theoretical and practical grasp of this dynamic would better enable pupils to create their own histories and geographies. Curriculum content should be based on the realities to be transformed; such material conditions as youth unemployment, technological change, environmental deterioration and lack of social justice which confront young people daily (Donnelly 1980). Through a process of dialogue, teacher and pupils would seek a critical awareness of their own identity and situation, would analyse causes and consequences, and would then examine ways of acting logically and reflectively to transform that reality (Friere 1972). Geography teaching would then become a co-operative exercise in reclaiming stolen humanity and reconstructing society. Much of what is presently taught can be adapted to this purpose, but it will need to be shaken free of its purely empiricist or positivist presentation, and integrated with other social sciences within an overall materialist framework. Study of our own society should be complemented by that of others where people are transforming nature and themselves in different ways. The integration of people and nation states within wider economic, strategic and political frameworks should also be explored. Such aims are unlikely to be realized unless school geographers make much more use of humanistic and radical geography with its potential for humanization and liberation.

A socialist school geography should not substitute one form of indoctrination for another. While teachers should reject stances of neutrality which inevitably leave existing patterns of power and inequality undisturbed, their commitment to justice should require the pursuit of truth as a duty (Wren 1977). In helping pupils to recognize structural oppression and exploitation, critically assess alternative political and social arrangements, and develop the ability co-operatively to pursue political aims, teachers should strive for scientific rigour, integrity and honesty. Robin Richardson (1982) offers suitable guidelines, designed to protect the pupil from the teacher's powers of persuasion, allow space for doubts and differing viewpoints, and prevent teachers from starting other people's revolutions. Such safeguards are an integral part of an alternative pedagogy (Norton and Ollman 1978) based on the concept of dialogue mentioned above, and offering an active and experiential alternative to the didactic methods which currently dominate classrooms. Reformed content and method can seriously challenge the existing practical and theoretical ideology of geography teaching, but is likely to face much oposition.

At a time when the state finds it increasingly necessary to link learning with productive work and raise economic and social awareness, there are significant opportunities for socialist teachers to exploit. The rhetoric of relevance, critical thinking, vocationalism and citizenship, which is being used to legitimate the restructuring of education, allows us to argue for genuine polytechnic education. At the same time the mounting contradictions of schooling, particularly the credibility gap between its promises and outcomes, create a climate in which liberal and radical alternatives are more acceptable (Husen 1979). While there is a continuing threat that emerging radicals will be tamed and co-opted by the system, growing numbers of geography teachers are looking for the type of curriculum which this final section has begun to outline. The history of our subject suggests that they now need more allies in higher education for only through joint action can the power of the examination boards, text-book publishers, Geographical Association, Her Majesty's Inspectorate and the DES be challenged. A growing coalition within the subject community should also look for allies outside education, in Labour-controlled local authorities, in trade unions, amongst parents, and in certain groups concerned with political education in the widest sense. The struggle to construct and implement a socialist school geography will face many setbacks as it has in the past, but it remains part of the overall struggle for a counter-hegemony and an alternative future.

REALIZING THE PROMISE OF SCHOOLING

Remember Roger Mills and his friends. Perhaps the one who suggested that comprehensives ought to be as revolutionary as the public schools are élitist, could have explained the others' disillusionment with school more fully. We will never know. What we do know is that his remark hinted at the potential of schooling for social reconstruction which has still to be realized. The success of school geography teachers in meeting that challenge will play a small but not insignificant part in helping to create the future of geography and the geography of the future.

REFERENCES

Apple, M.W. (ed.) (1982) *Cultural and Economic Reproduction in Education: Essays on Class, Ideology and the State*, London, Routledge & Kegan Paul.
Association for Curriculum Development in Geography (1983) 'An introduction to contemporary issues in geography and education', *Contemporary Issues in Geography and Education*, 1 (1), 1–4.
Beddis, R. (1983) 'Geographical education since 1960: a personal view', in Huckle, J. (ed.), *Geographical Education: Reflection and Action*, London, Oxford University Press, 10–19.
Breitbart, M.M. (1981) 'Peter Kropotkin, the anarchist geographer', in Stoddart, D.R. (ed.), *Geography, Ideology and Social Concern*, Oxford, Basil Blackwell, 134–53.
Capel, H. (1981) 'Institutionalization of geography and strategies of change', in Stoddart, D.R. (ed.), *Geography, Ideology and Social Concern*, Oxford, Basil

Blackwell, 70–80.
Carnoy, M. (1982) 'Education, economy and the state', in Apple, M.W. (ed.), *Cultural and Economic Reproduction in Education: Essays on Class, Ideology and the State*, London, Routledge & Kegan Paul, 79–126.
Castles, S. and Wustenburg, W. (1979) *The Education of the Future: An Introduction to the Theory and Practice of Socialist Education*, London, Pluto.
Dale, R. (1982) 'Education and the capitalist state: contributions and contradictions', in Apple, M.W. (ed.), *Cultural and Economic Reproduction in Education: Essays on Class, Ideology and the State*, London, Routledge & Kegan Paul, 127–61.
Donnelly, P. (1980) 'Ways of learning geography as social education for youth years 9–12', in *Focus on the Teaching of Geography*, National Conference Proceedings AGTA, Adelaide, 25–58.
Dunbar, G.S. (1981) 'Elisée Reclus, an anarchist in geography', in Stoddart, D.R. (ed.) *Geography, Ideology and Social Concern*, Oxford, Basil Blackwell, 154–64.
Fleming, M. (1979) *The Anarchist Way to Socialism*, London, Croom Helm.
Friere, P. (1972) *Pedagogy of the Oppressed*, Harmondsworth, Penguin.
Geography 16–19 (Schools Council Curriculum Development Project) (1980) *The Advanced Level Syllabus*, London, Schools Council.
Gill, D. (1983) *Assessment in a Multicultural Society Project; Subject Report: Geography*, London, CRE.
Goodson, I.F. (1983) *School Subjects and Curriculum Change*, London, Croom Helm.
Graves, N.J. (1975) *Geography in Education*, London, Heinemann.
Harris, K. (1982) *Teachers and Classes: A Marxist Analysis*, London, Routledge & Kegan Paul.
Hart, C. (ed.) (1982) *The Geographical Component of 17+ Pre-employment Courses*, London, Schools Council.
Harvey, D. (1974) 'What kind of geography for what kind of public policy?', *Transactions, Institute of British Geographers*, 63, 18–24.
Her Majesty's Inspectorate (1978) *The Teaching of Ideas in Geography*, London, HMSO.
Hicks, D. (1981) 'The contribution of geography to multicultural misunderstanding', *Teaching Geography*, 7 (2).
Huckle, J. (ed.) (1983) *Geographical Education: Reflection and Action*, London, Oxford University Press.
Husen, T. (1979) *The School in Question: A Comparative Study of the School and Its Future in Western Society*, London, Oxford University Press.
Joint Council (for 16+ National Criteria) (1982) *Draft National Criteria for Geography*, Manchester, Joint Matriculation Board.
Jones, K. (1983) *Beyond Progressive Education*, London, Macmillan.
Kropotkin, P. (1885) 'What Geography ought to be', *Nineteenth Century*, 18, 940–56.
Lee, R. (ed.) (1977) *Change and Tradition*, London, Queen Mary College.
—— (1983) 'Teaching geography: the dialectic of structure and agency', *Journal of Geography*, 82 (3), 102–9.
Mills, R. (1978) *A Comprehensive Education*, London, Centerprise Trust.

Norton, T.M. and Ollman, B. (eds) (1978) *Studies in Socialist Pedagogy*, New York, Monthly Review Press.
Richardson, R. (1982) 'Now listen children ...', *New Internationalist*, 115 (September), 18–19.
Salter, B. and Tapper, T. (1981) *Education, Politics and the State: The Theory and Practice of Educational Change*, London, Grant McIntyre.
Sarup, M. (1982) *Education, State and Crisis*, London, Routledge & Kegan Paul.
Sharp, R. (1980) *Knowledge, Ideology and the Politics of Schooling: Towards a Marxist Analysis of Education*, London, Routledge & Kegan Paul.
Smith, M.P. (1983) *The Libertarians and Education*, London, Allen & Unwin.
Stoddart, D.R. (ed.) (1981) *Geography, Ideology and Social Concern*, Oxford, Basil Blackwell.
Tolley, H. and Reynolds, J.B. (1978) *Geography 14–18: A Handbook for School Based Curriculum Development*, London, Schools Council and Macmillan.
Walford, R. (ed.) (1981) *Signposts for Geography Teaching*, Harlow, Longman.
—— (1982) 'British school geography in the 1980s: an easy test?', *Journal of Geography in Higher Education*, 6 (2), 151–9.
Wolpe, A.M. and Donald, J. (eds) (1983) *Is There Anyone Here from Education? Education after Thatcher*, London, Pluto.
Wren, B. (1977) *Education for Justice*, London, SCM Press.

16
GEOGRAPHY, CULTURE AND LIBERAL EDUCATION
J.M. Powell

The mediaeval liberal arts are not adapted to the task of liberating men today; and the humanising arts are not the 'humanities', or the 'social sciences', or the 'cultural sciences', or any interdisciplinary amalgam in which humanists will learn the second law of thermodynamics and responsible governmental posts will be created for scientists. Some thought must be given, first, to discovering what arts liberate and humanise in the present world situation. More thought and invention must be employed, second, to develop those arts for use in the revolution in education which will transform our schools and colleges and graduate schools during the next few decades. (McKeon 1964, 172–3)

The last quarter of this century will be judged a primary conjuncture in the cultural development of mankind; it is therefore a period of pivotal significance in the history of western education, as McKeon and others proclaimed twenty years ago. And McKeon was right again, though a trifle conservative, when he cautioned that the kind of education emerging after the revolution would be 'put together capriciously in response to pressures and counterpressures within communities' unless it was wisely designed from the outset. The confusion he predicted is already with us, because we failed to ask enough questions of ourselves and of our academic subjects. *Never* too late to ask searching questions of scholars and teachers: a truism? If so, then it contains another – discomforts accumulating from procrastination lead to despair.

In our own day, despair and redundancy incessantly conspire to torment our tired consciences. So we urgently begin: what sort of society for what sort of geography; or, if you prefer a more familiar version, what sort of geography for what sort of society? These crucial questions cannot be separated, and they cannot be debated often enough. Put either way choices are now being made very rapidly throughout the world for every academic subject, to the bewilderment of the majority of our educationists. Of course the speed of change is itself confusing, but it may not be the chief cause for alarm. Perhaps we are more fundamentally disturbed by what appears to be an exhausting, anarchistic display – so many kites, under so little control: pedagogical and research structures which may be bold, seductive, flimsy, dangerous, self-destructing. And they are seldom anchored, so it seems, in recognizable territory.

Is that the key? We accept that education promotes, reflects and addresses cultural change. But that rich word 'culture' is variously seen to connote ways of training and improving, or civilization's intellectual aspect, or indeed both of these; and it is also employed in our descriptions of distinctive modes of living – with 'ways of life' and the forces underpinning them. Thus, for the sin of surviving a 'liberal' training, educators were traditionally assigned the ill-defined task of transmitting and defending the 'culture'. It has resulted in a way of life which has become increasingly complicated over the past few decades, and is now somewhat hesitatingly pursued in the absence of any guiding consensus on the definition of an 'educated person' (Bouwsma 1975; Goodlad 1975; Kaplan 1980; Sarup 1982).

The strident demands for 'relevance' in education are already being fulsomely met in a bewildering array of sincere and committed responses. These responses reflect our various personal and professional attitudes towards the continuing but uneven influence of powerful traditional models of the educated person, differences in our interpretations of the needs and preferences of the mixed societies from which those models are by no means purged, and the impact within and beyond the classroom and lecture theatre of certain consuming images of the future. By definition, therefore, all of our recent responses, including the present consciousness-raising discord itself, are 'relevant' for students, teachers and the wider community. The old equilibrium has gone. Therefore, when the only certainty is uncertainty, it is prudent indeed to retain the 'liberal' requirements of self-knowledge and a trained ability to develop critical analyses of the contexts for action.

'All life is adjustment and transition' – so we were reminded by Spate (1960, 377) – 'and only people and institutions already dead have no further crises to face'. For our own insecure generation, it surely behoves professionals *as citizens* to seek ways of initiating or extending an honourable, mutually productive dialogue between each of their chosen academic fields and the wider community. Without some such involvement entire areas of study will be crudely but deservedly refashioned, even jettisoned, according to the mismanagement, foibles, prejudices and honest misperceptions which characterize the directionless society. Applying these observations strictly to geography, my main contention is that the recent revival of what I choose to call 'humanistic' perspectives – following what I believe is, of necessity, a more generous description of the term than others may allow – has contributed towards keeping us solvent, both intellectually and socially. In my experience these concerns are far more usefully assessed as *expressions of a shared orientation* than as still more shrapnel from the cannons of indulgent individualism. No more 'sub-disciplines', please: there just isn't the market for them.

THE PRESENCE AND PROSPECTS OF 'HUMANISTIC' GEOGRAPHY

But Knowledge to their eyes her ample page
Rich with the spoils of time did ne'er unroll;
Chill Penury repressed their noble rage,
And froze the genial current of the soul. (Gray 1751)

An elegy on the modern graduation ceremony, musing once more on the fate of each 'mute inglorious Milton' paraded before us like an accusation, might call back an old question – 'Where did we go wrong?' If geographers feel any of that less keenly than their colleagues, they should think again. In some large measure, academic geography in its more or less recognizable modern guise owes its very origins to educational reform, and it is too late to expect to be insulated from today's deep-seated transformations. Thankfully, one healthy characteristic of the subject down the years has been the ability of many of its practitioners to adapt to broader developments in mission and method which have sent shock waves through western education systems.

That quality was especially prominent, naturally, in the subject's formative years; but it was also a feature of those strong reorientations in content and approach during the 1960s and 1970s which are too often inaccurately described as the reflections of autonomous growth and tension within geography itself. In fact the other extreme position might be marginally more acceptable – that is, that the subject as we know it is the somewhat artificial creation of modern education movements. In either case, few professional geographers could choose to stand aloof from the social and political forces which rocked education to its foundations; none were unmoved by the arguments of Fromm (1978),

Holt (1977), Illich (1970) and others, although those arguments were received second- or third-hand all too often. Commenting on the past decade or so in *Signposts for Geography Teaching*, Walford (1981) distinguishes four competing but not mutually exclusive groups on the basis of current ideological concerns. For convenience these may now be related to broader reform movements and to Bouwsma's (1975) descriptions of the 'educated person' down the ages.

Professionals supporting the *liberal-humanitarian* (loosely, the comparison is Bouwsma's literate 'scribe') tradition are portrayed as socially conservative. On the whole they have been concerned with continuities, with passing on a respectable cultural heritage of information and ideas, implying that they uphold the principle of a set curriculum and favour the retention of strong subject areas or disciplines. The push by scientism towards quantification and the like in the 1960s and 1970s is one example; it was opposed by supporters of similar scribe-type ideals who argued that the move was unsuited to the mass of students, and that the required companion subjects were in any event traditionally blocked against geography. Second, the *child-centred* or *student-centred* tradition, with support from psychology and romantic naturalism, respects self-autonomy and self-development. Geographers holding strongly to this view uphold the old 'bridging' ideals of the subject, which they contend will provide integrating 'experiences' for the student. Fieldwork and individual and team research projects are deemed essential, and stress is laid on defining and cultivating personal views – on the development of 'effective personalities', not unlike the ancient 'aristocratic' ideal. The *utilitarian* tradition, a mutation from the scribe tradition, concentrates instead on equipping students with employment-orientated skills. Despite the classical arguments in support of vocational education (*inter alia* Keller 1948) which liberals choose to ignore to their cost, it is true that when this perspective takes over all non-vocational subjects are queried and that a clear susceptibility to state control, and to some extent an enslavement to a kind of 'superindustrial' imagery, may be discerned. Lastly, there is the *reconstructionist* view, a highly politicized relative of the scribe and 'civic' or citizenship ideals which is principally motivated by an acceptance of the value of education as an agent for social change: environmental issues and social injustices are vigorously sought out and students are urged to become 'involved'.

All persuasions claim strong moral, even humanist bases, and although the 'humanistic' label might be awarded most fittingly to the second category, it needs to be said that the multi-faceted contribution of truly humanistic perspectives extends to an insistence on an empathetic appreciation of *all* viewpoints, as well as to an elevation of the student and the placing of a high premium on supportive teaching and learning environments. In practice the most humanistic professionals, simply because of their far less exclusive occupation of personal or group domains, may be unusually well placed to nurture inter-subject and intra-subject dialogue.

What this is intended to suggest is that humanistic perspectives in geography, as in any other subject, must be primarily concerned with searching questions of *professional orientation* and not with the construction of a new fleet of diverting and unreliable bandwagons. In its full expression in the larger realm of the humanities the entire interest is in man (mankind). All the preferred 'models' are undeniably human and other conceptualizations are held to be secondary. The 'free agency' of man is assumed to offer the prime research and teaching goals, but each of his works is seen to be profoundly affected by the material conditions of its time. *Contexts* are therefore central, but the main propositions also subsume the principle of 'reflexivity' – so the teacher/researcher, also a free agent, is enjoined not only to reflect on his own professional life in its relevant temporal location, but also to be fully prepared to accommodate radical changes of direction from time to time. In the mature professional the resulting tendency towards increased tolerance and sensitivity encourages improved self-consciousness and suitably critical/analytical approaches to professional tasks. Within geography the interrelated contributions of this humanistic perspective can be conveniently listed in eight simplified divisions.

EPISTEMOLOGY

Resurgent humanistic geography insists that the indisputable professional need to examine the accumulation of distinctive kinds of knowledge is omnipresent, not recurring or episodic. The concept of the individuality of scholarship is particularly honoured and oversimplistic classification into 'schools' of thought is resisted, but there has also been a heightened interest in what were vaguely described for many years as 'philosophical' matters (Johnston 1983a, 1983b). These include a definite methodological (as opposed to a mere 'techniques') orientation, the rebuilding of the old relationships between physical and human geography, a renewed investigation of linkages with other subject areas, the historiography and sociology of geography, and so forth. The same general influences, in geography as in most other academic fields, have recently underlined the issues of professional ethics and responsibilities, areas in which the application of modern 'practical' philosophy has dramatically sensitized all sections of the academic community (Mitchell and Draper 1982; but see also Haan *et al.* 1983).

SPACE

'Spatiality' or the 'space-fetishness' of the 1960s and 1970s, particularly in human geography, is variously ignored and openly challenged. Many humanistic professionals appear embarrassed by the old routines and terminologies and prefer not to make repeated reference to 'spatial expression', 'spatial organization' and the like, although interest is shown in the uses made of such terms by social scientists. The point seems to be that 'contrived' frameworks have far less value than those reflecting observed human experience: indeed there is a commonly declared preference for 'experiential' viewpoints, in which a stress on 'realism' or 'given meanings' is notable. In this context a further link with epistemological concerns may be seen in the importance attached to existentialism and related philosophical modes. And in sum this helps to explain a little of the characteristic neglect of quantification in humanistic scholarship.

PLACE

It is fair to say that there has been a widespread rediscovery of place as the hub of geographical enquiry. In humanistic terms 'place' is preferred to 'space' because it is seen as a centre of action and intention, a focus where we locate the 'relevant' events of our existence (Relph 1976). Considered in this light places are necessarily embodied in the intentional structures of all human consciousness and experience. The essence of a place can never be said to arise from mere location or by reference to the 'functions' ascribed to it according to one abstract formulation or another. If geography has anything to do with the earth as the home of man then it has everything to do with this business of *intentionality*, which describes relationships of 'being' between man and the world. Perceptions, preoccupations, fears and aspirations, all fundamental to the human condition, invest the world with meaning and in the process define places. So the need to discover how people other than social scientists define and relate to places is not merely intuitive: consider the abundant cultural/historical work on 'images' and landscape perception. Therein lies the best explanation and justification of the rising demand for imaginative new regional courses – it addresses present and future requirements, and is in no way connected with conservative reaction or impotent nostalgia.

TIME

The singular importance of context, including the reflective analysis of research and teaching situations, has already been outlined. This is inevitably associated with an even greater degree of historicity since key *events* and *periods* are allowed unusually intensive geographical treatment, and political and economic *decisions* may be subject to close investigation. The resulting literature proclaims its geographical essence in both its means and ends, yet it is true that non-specialists presently identify a few fields more easily than others as our certain domain – the diffusion of selected innovations; differential regional expressions of hardship, health and leisure-seeking over time or at particular times; conflicts in resource appraisal and environmental management; the standard temporal profiles of natural and quasi-natural hazards distinguishing pre-, during- and post-*event* phases. And I suppose it need not be underlined that the pervasive interest in Marxist interpretations in all fields is heavily dependent on careful historical analysis. (Selected examples for this section include Heathcote 1983; Meir 1982; Powell 1977; Sheail 1976.)

PEOPLE

It is no longer sufficient to deal with people in the aggregate. It is all too easy for that to become another mystification ritual distancing the professional from the paying public – or it can be described in that way, which may be just as bad. The reflexive component of the humanistic perspective requires a full-bored 'humanizing' of the research enterprise, and so it is now far more common to find actual individuals and groups as the subject and the destination of geographical enquiry (Ley and Samuels 1978). For so many years the only people clearly nominated in most social science publications were those named in the bibliographies; they were of course the 'authorities' cited. In contrast today's historical geographers (for example) may be concerned with the roles of prominent individuals and families, formal and informal associations of pioneer settlers, and religious, scientific and cultural groups, political alliances, protest movements, and the like: in fact, closer communion with social historians and sociologists is more apparent every year (for a briefing on one changing context see Powell 1981). Without question, there are many common interests in the interpretation of changing ideas and motivations and in their contemporary modes of communication. Yet again, however, that can and does extend to reflexive historiographical critiques of one's academic subject. Contextual and biobibliographical studies of geographical thought have become quite well established over the past ten years, and we are beginning to evaluate the contributions of leading figures, distinctive schools of geography, and the influential concepts and innumerable blind alleys which have characterized the development of the subject (for instance Buttimer 1983; Stoddart 1981).

VALUES AND INTERFACE CO-OPERATION

If it is true that time is now out of joint, that socially prevalent values and higher humane aspirations are in contradiction, then the humanities' old claims to nobility take on new meaning. Only the humanities have consistently claimed that they can provide individuals with some understanding of themselves. They protest against the 'loss of the person' in the social sciences and natural sciences alike, emphasize the destructive and constructive potential of man, and repeatedly state that their own aim is to help to mould rounded persons who will be liberated cultural witnesses in their own age. The kinds of humanistic arguments I have tried to describe respect these grand themes, but suggest that some rapprochement is long overdue between the warring branches of learning, particularly between the humanities and social sciences. Despite their occasional excesses the main thrust of the latter interests is now more firmly directed towards issues of general public concern. Within geography this extension features most prominently in an admirable quest for 'values' in our teaching and research situations (Huckle 1983).

Value systems involve the most profound orientations and notions of priorities. In general, they are exceedingly slow to form and highly resistant to change. When communities fragment into disorderly confusion and family life, a proven foundation, is blindly or viciously undermined, educators are asked most unfairly to carry an immense burden. With an increasing participation from much older students we can expect to be able to challenge some of their values, adult to adult. And for the younger people in the schools, perhaps we can only hope that the provision of better kinds of information, and more opportunities for exploration in oral and written expression, will assist towards the formation of humane views and responsible citizenship. Secondary students' orientations towards one another and to the world about them are presumably far from fixed, and it cannot be pointless to encourage them to state, explain and share differences of opinion. On the other hand, so many huge questions demand the kinds of information and life experience which are normally beyond the comprehension of younger students, and we could easily close more doors than we open simply by making premature demands – or by a return to obfuscation, the psychic numbing of the failed scientism of yesterday (Powell 1980).

Established literature courses may function more effectively in this area, if only because tradition proclaims it part of their constituency. Uninhibited humanistic approaches will always favour inter-subject cooperation, however, and the massive promise of cross-curriculum or inter- and intra-faculty teaching enterprises – including joint English-history-fine arts-architecture-geography projects, for instance – seems to indicate a potential for several kinds of 'interdisciplinary' growth. Certainly that applies to underexploited liberal arts/liberal science environments in Australia, New Zealand and North America; even in Britain some of the stubborn academic barriers to innovation have been breached. In a few corners of the geography/humanities interface, recent studies in each of these countries indicate that a bolder campaign may soon be launched on a wider front (interesting examples varying in scope and quality are: Barrel 1982; Cosgrove 1982; Hudson 1982; Karan and Mather 1976; Pocock 1981; Spolten 1970; Zaring 1977).

Geography's capacity for producing good general literature was severely reduced in the 1960s and 1970s, and in this regard Meinig's (1983) trenchant criticisms of the profession at large were well deserved. Regrettably, some of our purportedly humanistic contributions – more 'rigorous' and self-consciously 'sub-disciplinary' – were also lost in the black depths of the social science pit, so unkindly described by Nisbet (1982, 229):

> First, be fruitless and reify. Snuff out the lives of particulars through suffocation by structures and systems. Agree with Blake: 'I must locate a System or be enslaved by another Man's.' Regard individuals and events as so many masterless dogs needing impoundment in a theory. The rage to reification, for pretending that life and meaning exist in the most boneless of abstractions, is a special mark of ages in history such as our own. *Grau ist alle theorie*, wrote Goethe, but you would never know it from the coloured ribbons with which we bedizen our pet principles, laws and axioms in the human sciences. With considerable reason has our epoch been called the Age of Reification.

DIALOGUE

Our previous sections prompt a simple additional observation: scholars and teachers must be fully conscious of their own values or opinions *and of the predilections imbibed from their chosen subjects.* Where dialogue is a prime educational goal, as it must be, one should beware the wasteful monologue of professional discourse. Above all, perhaps, one's research career path or favoured teaching topics may often clash with what students need to know.

At the very beginning of a steep inflation of confusion in the social sciences, one of Spate's (1960, 393) celebrated asides, 'it is never wrong to plug your own line' (cf. Spate 1965), was souvenired from context and flown before a mad band of lemmings hell-bent on publishing before they perished. Nothing very liberal or liberating in any of that – not for the majority of students, nor for the handful of winners, our frenetic Shamans. Yet the trance continues and now even the simplest calls to professionalism may be indistinguishable from the throbbing of a sacrificial drum. Quite as much could be said of the sad parade in the postgraduate world of the 1960s and 1970s, where the demand for 'sound technique' frequently led to the selection of minor (sorry, 'bounded', or 'do-able') topics. But rigorous precision may be best achieved within the limits of the smallest themes, and it is now obvious that many were well recompensed for their swift descents into triviality. See this from A.D. Hope, one of Spate's Canberra neighbours who must have shared his dismay:

> The scabs scratched off by genius, sought with care
> Stuck back again earned Doctor Budge a chair;
> And now, Professor Budge, his claim made good,
> He works like dry rot through the Sacred Wood;
> Or like dead mackerel, in a night of ink
> Emits a pale gleam and a mighty stink. (Hope 1970)

Hope's *Dunciad Minor* was mainly concerned with the sterility of literary criticism in the 1950s, but I am not the first to claim that it holds good as a description of western academia over the next two decades. Consider not only the careerist thesis-mongering but also the demented proliferation of esoteric articles, properly destined to remain forever in pristine condition on library shelves which were built to be busy: 'Let me shed light on things both dark and dense / Yet never move them into common sense.'

Long-suffering undergraduates who were eventually rewarded with school-teaching positions were saddled with the task of marketing the mass-produced goods of these academic researchers, until the 1980s gave them a chance to throw off the bonds. Look therefore at some cases on our own home ground. For example, the detailed contribution of central place theory to the explanation of the growth of urban 'systems' in the nineteenth century undoubtedly provides a fascinating intellectual challenge for some people, but more urgent faculty-wide tasks may make other demands – that someone should, perhaps, pursue different research and offer less narrow historical-geographical courses, to break down the artificial barriers between geography and history; or that the traditional 'environmental' core should be rebuilt because the image of geography held by staff in other subject areas may make for special expectations in the servicing of an environmental management/resource management area for which there are increasing demands. So the decision about *what* to teach or research is never less important than ideas about *how* to perform these functions. Again, if 'schooling' really describes a drawn-out, multi-levelled process addressing the 'national interest' (Dale *et al.* 1981), then the specifications of each situation insist on the social and intellectual value of a creative, integrated regional geography which picks up the pieces and delivers coherent syntheses appropriate to their cultural/educational context. First World, Third World, Fourth, or any other culturally conditioned enumeration we employ; new nation, old nation – whatever: their special characteristics and their locations in an interdependent world must be appraised and communicated. Since complexity can only increase through the advanced levels of education, the firm cultural appeal of regional synthesis will not diminish. Rather, today's questions will be projected into the future in aggravated form: who is sufficiently prepared or sufficiently talented to do it, and how should it be done?

Teachers in most of our schools have fewer opportunities for indulgence than their tertiary colleagues who, until recently, were all too prone to forms of ego massage. In today's universities it is seldom easy to find the line between indulgence and service, but it is imperative that we define new ways of rewarding innovative teaching, for this effort at dialogue is just as vital as the research effort in our beleaguered institutions. Unless more forceful attempts are made to redress the balances in these areas, students' freedoms are seriously curtailed and community confidence is further diminished.

EDUCATION REFORM

All the evidence gathered thus far indicates that no evaluation of the true *interests* of a fully humanistic (or humanized?) geography is possible without considering the passionate concern of its champions for the protection and promotion of formal and informal education throughout every community. But note again that *self*-interest is not intended, because that is ultimately connected with the 'having' and not the 'being' modes of existence. Examine instead the Latin root, *inter-esse* – to be in it, even *within* it. The aim is to be so actively involved in educational processes that one fully 'belongs' to – is possessed and transformed by – that engagement. In short, in its full expression it is the cornerstone of one's professional life in the community, the source and product of ethical frames of reference; it suggests nothing less than a mode of living. That too may be charged as another personal conviction, but allow me to press on towards its logical end: a greater openness to the full range of humanistic influences should alert us to the meaning of the current ferment in western education, which is the need to submit to as much change in ourselves as we seek to promote in wider society.

The various expressions of reform in modern education necessarily reflect differences in social traditions and perceived requirements around the world, and it is dangerous to rely too heavily on convenient individual samples. I am naturally more confident about commenting on changes in my own Australian community, however, and furthermore it seems likely that recent developments in Australia do contain useful lessons for colleagues elsewhere. To the casual observer, Australia's education systems have been firmly based on the most 'liberal' of traditions in the arts and sciences, and this makes the unusual vehemence of our current 'anti-academic' movement hard to grasp. Even in the universities, narrow specialization has not been encouraged. The complex dynamics of our changing national situation forbid deep analysis here, but it is fair and useful to say that a similar mixture of forces can be detected in Australia as in Britain or North America – widespread disenchantment or confusion about the aims, methods and products of education; the imposition or interference of any number of government directives and incentives; a rapid politicization of the teaching profession linked to a multiplicity of school-based assessment procedures, teacher autonomy and school-based curriculum reform. No single statement can be offered as a reliable summary of these developments, but we are very frequently asked to prepare for 'the negotiated curriculum'.

One of the messages is that those who claim to be 'professing' individual subjects must now expect to be asked what they are bringing to the negotiating table – if indeed they have the good fortune to receive an invitation. Another important notion anticipates a fundamental restructuring of tertiary education. The casual view of 'liberal' education in Australia is ridiculously superficial. In fact, the humanities and social sciences are variously regarded as foppish adornments, the poor white trash of the system, expensive luxuries for the élite, wasteful diversions from 'real' science and technology. Until recent years a fundamentally shaky community support was hidden by the sure demand for schoolteachers in a young and growing immigrant population, but the astonishing boom in teaching employment has receded and as a result those who blithely rejected the jibe of 'jumped-up teachers' colleges' now regret that they cannot boast a well-endorsed vocational prop like their competitors in the other faculties. The entrenched hierarchical structure in Australian universities puts medicine, specialized science and engineering at the apex. General educational subjects occupy the base, and humanities and social science staff are not permitted even the small compensation of a viable graduate school to take them up the ladder occasionally; in return, they protest that scientists have not been 'kosher' about their own research and teaching. Recent public debates highlight the possible legitimation of some such two-tiered environment, generalist-specialist, but the saving feature is that there are signs of an emerging consensus on the need to boost the 'liberal' sector by a vastly augmented intake incorporating the rapidly increasing influx of 'mature-age' and part-time students. That is where we may find an excellent platform for geography's old claims for well-balanced 'integration' providing an entrée to a host of narrow specialisms, vocational and otherwise. The argument will not be judged complete, however, until we take steps to improve our contribution in secondary education where the image of the subject is still ambiguous.

Geography in its twentieth-century British and European representation is a comparatively new arrival in Australian schools, and it was not until the late post-war era that it managed to gain a toehold in most of our universities. It does not have the wide base from which its assertion of a long-established affinity with the 'new' education aims of the 1980s can be promoted, yet the merest glance will disclose that some of these earnest proposals have an astonishingly familiar ring about them. Perhaps the best example is a statement from the Curriculum Development Centre in Canberra declaring that *'our traditional way of packaging knowledge into required subjects no longer satisfies society or students'* (Curriculum Development Centre 1980, 7). *All* established school subjects, even the 'hard' sciences, were thenceforth challenged to reorganize in order to address a curriculum based on a Core of Learning and a Core of Knowledge and Experience. The first cites ways of organizing knowledge, dispositions and values, interpersonal and group relationships, learning and thinking techniques, skills or abilities, forms of expression and practical performances. The second is concerned with nine areas – arts and crafts; environmental studies; mathematics; social, cultural and civic studies; health education; science and technology; communication; moral reasoning and values; and work, leisure and lifestyle. Who would seriously suggest that the content and approaches of modern geography do not *already* address virtually all of these 'core' areas? I suppose we need a better press, not some latter-day Griffith

Taylor, to seize this golden opportunity to stake a claim for another kind of 'correlative science'. If that is the way things are likely to go in the schools, then we are further down the track than any of our colleagues in other subject areas.

The rapid disappearance of formal academic subjects in the Australian school systems has been accompanied by the emergence of any number of loose federations, together with predictions of comparable developments in the non-vocational faculties of the universities and colleges in the near future. Conservatives in the humanities and social sciences are discussing ways to 'respond' to the new situation, while startled or indignant supporters of the physical sciences debate their 'reaction'. But the point is that many students have been entering tertiary education with minimal 'academic' preparation for some years, if we are to judge by the standards of earlier generations – to the extent that stern prerequisites in specified subjects have been largely abandoned except for the vocational areas. About three-quarters of the first-year geography intake at Monash University, for example, will not have studied the subject at the higher school level, and in our mixed general degree system only about half of that intake will continue with the subject in later years. Clearly, the universities do need to put their own houses in order. The geographers' practical contribution to that end draws on a relevant and distinctive inheritance incorporating the experience of a lively federation of interests in which the bonding elements are broad educational objectives rather than tightly defined team-research missions or a guarded public allegiance to some unexamined or unexaminable 'disciplinary' purity.

Something similar to that could be said for other national systems, but I have less information about their influential policy statements on curricula. For Britain, the educational aims of Labour's Green Paper, *Education in Schools* (1977) and the Tories' *A Framework for the School Curriculum* (1980) appeared admirably liberal and humanistic only in a rhetorical sense. The notion of a 'core' was narrowly confined in both to the thinnest of thin-lipped academic lines – English, mathematics, science, a foreign language – and I cannot comment on the implications for geography in those contexts. The trend towards teacher autonomy and the like is surely inexorable and universal, however, and several facets of our Australian situation will be immediately recognized in Britain.

RECOGNIZING UNCERTAINTY

The educated man is someone who has come to care about his own well-being in the extended sense which includes his living a

GEOGRAPHY, CULTURE AND LIBERAL EDUCATION   321

morally virtuous life, this latter containing a civic dimension among others. Whereas other recent accounts of him have made his possession of *knowledge* his chief characteristic, this one makes *virtues* more central. The educated man is a man disposed to act in certain ways rather than others. He possesses the general virtue of prudence, or care for his own good (as well as subordinate virtues like courage and temperance). This, being in an extended rather than a narrow sense, includes within it the more specifically moral virtues like benevolence, justice, truthfulness, tolerance and reliability. It includes the lucidity needed to sort out clearly the complex conflicts of value which face him, and the wisdom needed to reflect on these conflicts and try to resolve them within as broad a framework of relevant considerations as possible. The educated man, prizing autonomy, will be independent-minded himself and sympathetic to independent-mindedness in others. . . . (White 1982, 121)

Twenty years ago a distinguished scholar wrote, 'In a sense the social sciences are fighting for life, the humanities against death' (Plumb 1964, 10). The current doom and gloom represents a marked deterioration: in the conventional wisdom these poor relations are companions in misery, scrambling for every crumb that falls from the scientists' table. It is high time they worked more purposefully together to question the educational worth or credibility of the traditional assumptions on which the more arrogant of our scientists continue to found their many claims; and they must resolve to invest their best energies in a concerted effort to improve their own performances as educators. Critical educational aims can never be baldly listed, not without great distress and controversy anyway, but those who are in broad sympathy with the long-winded passage quoted above should be prepared to measure the contribution of every voracious and self-perpetuating subject against this description, or some close alternative. For too long, scientists and non-scientists alike have been permitted to fly blind before plunging themselves into esoteric specialization. Fundamental educational freedoms are thereby curtailed and the teacher-student-teacher circle is usurped.

Whatever we choose to do (and choose to call ourselves) in our specialized research, as educators we must also address the more accessible and communicable educational aims of a kind of geography which sets a high premium on the skilful blending of humanities, social science and natural science approaches, with the hope of contributing towards the formation of the educated person. One valiant effort to measure the achievements of current liberal arts programmes in higher education offers nine major goals or groups of effects – critical thinking;

learning how to learn; independence of thought; empathy; self-control; self-assurance; mature social-emotional judgement and personal integration; equalitarian or liberal values; and participation in and enjoyment of cultural experience (Winter et al. 1981). The list describes long-established aims, but it also resembles some of the most recent statements on curriculum reform, where the appeal of non-vocational education remains surprisingly strong. Few subject-specialists engaged in the general education of undergraduates would not recognize their own professional goals in this list, but there can be no doubt that well-balanced geography programmes offer dramatically understated bonuses for humanities, science and technology students by extending their choice of content and method. If the term describes the training of the free and the freeing of the trained, then geography already provides a peculiarly relevant 'liberal' education. Over the short term, our greatest gain may be from the humanities areas: but that may not apply to the more specialist departments in Britain, for example, as well as it does to most of our Australian contexts.

The past is nothing if not relevant: it warns us that 'industrialization' has far too much to answer for already, that we would be foolish to trust to it again. Our preference for a genuine *post*-industrialism scenario is based on its promised maintenance and extension of vital human qualities – the primacy of people loyalty, mutual interdependence, individual dignity, the need for creative expression, respect for environmental integrities. The frightening success of the promoters of the super-industrial world rests in part on the very familiarity of the assumptions on which their promised millennium rests. The current discord is related to this marketing phenomenon and it is common to a large number of established subjects. It must also be related to the partial survival of earlier models of education and to the chorus of dissent raised against the powerful new industrial imagery; in addition, within geography it reflects a laudable empathy with modern reform movements. Unfortunately, there are so many voices, following many different scores. We need to distil some order from this cacophonous display and the message of the humanistic approach asserts that, because of our natures, we can do no less than try.

It is not extravagant to expect more imaginative management to improve the short-term viability of school geography quite dramatically. Longer-term prospects are particularly bleak for those who would measure academic health only by the seeming continuity of autonomous subject-based research and higher-level studies generally, and it must be repeated that the same contextual arguments apply across the board to economics, history, sociology and other fields: they must all change or be changed. The analysis and possibly the correction of the erosion of our constituencies in the schools might be seen as a first priority, if only because a large investment there is both honest and sound. Somehow the sins of the universities have been visited on the schools, and collectively, geographers and their academic colleagues have yet to complete their penance. A firm purpose of amendment has been demanded and it is at last apparent in practical efforts at review and consolidation which will build for a better future of service. If we can read aright all these signs of change, our teaching and learning environments have already moved into a new phase.

The oldest and best justification of the humanistic perspective is the degree of self-awareness it injects into the study and practice of any subject. For geography especially this should also lead to a renewed belief in the primacy of our educational roles. And therefore it emphasizes that we need a sustained reinforcement of the philosophical and historiographical foundations of the profession, a revivification of regional syntheses at whatever cost to the purest of highbrow researchers, an improved co-operation between 'human' and 'physical' camps which is not restricted to the realms of instruction, the promotion of inter-subject linkages, and in general a heightened regard for the teaching function.

REFERENCES

Barrel, J. (1982) 'Geographies of Hardy's Wessex', *Journal of Historical Geography*, 8, 347–61.
Bouwsma, W.J. (1975) 'Models of the educated man', *American Scholar*, 44, 195–212.
Buttimer, A. (ed.) (1983) *The Practice of Geography*, London, Longman.
Cohen, A.A. (ed.) (1964) *Humanistic Education and Western Civilisation*, New York, Holt, Rinehart & Winston.
Cosgrove, D. (1982) 'The myth and the stones of Venice: an historical geography of a symbolic landscape', *Journal of Historical Geography*, 8, 145–69.
Curriculum Development Centre (1980) *Core Curriculum for Australian Schools*, Canberra, CDC.
Dale, R., Esland, G., Fergusson, R. and MacDonald, M. (eds) (1981) *Education and the State*, vol. 1: *Schooling and the National Interest*, Bascombe Lewes, Falconer Press/Open University.
Fromm, E. (1978) *To Have or To Be?*, London, Jonathan Cape, reprinted Abacus 1979, 1980, 1981.
Goodlad, J.S.R. (ed.) (1975) *Education and Social Action. Community Service and the Curriculum in Higher Education*, London, Allen & Unwin.
Gray, T. (1751) 'Elegy Written in a Country Churchyard'.

Haan, N., Bellah, R.N., Rabinow, P. and Sullivan, W.M. (eds) (1983) *Social Science as Moral Inquiry*, New York, Columbia University Press.
Heathcote, R.L. (1983) *The Arid Lands: Their Use and Abuse*, London, Longman.
Holt, J. (1977) *Instead of Education. Ways to Help People do Things Better*, Harmondsworth, Penguin.
Hope, A.D. (1970) *Dunciad Minor. An Heroick Poem, Profusely Annotated by A.A.P. and A.P.*, Carlton, Melbourne University Press.
Huckle, J. (ed.) (1983) *Geographical Education*, London, Oxford University Press.
Hudson, B.J. (1982) 'The geographical imagination of Arnold Bennett', *Transactions, Institute of British Geographers*, NS 7, 365–79.
Illich, I. (1970) *Deschooling Society*, Harmondsworth, Penguin.
Johnston, R.J. (1983a) *Geography and Geographers. Anglo-American Human Geography since 1945*, London, Edward Arnold (first published 1979).
—— (1983b) *Philosophy and Human Geography: An Introduction to Contemporary Approaches*, London, Edward Arnold.
Kaplan, M. (ed.) (1980) *What Is an Educated Person? The Decades Ahead*, New York, Praeger.
Karan, P.P. and Mather, C. (1976) 'Art and Geography: patterns in the Himalaya', *Annals, Association of American Geographers*, 66, 487–515.
Keller, F.J. (1948) *Principles of Vocational Education. The Primacy of the Person*, Boston, Heath.
Ley, D. and Samuels, M. (eds) (1978) *Humanistic Geography: Prospects and Problems*, London, Croom Helm.
McKeon, R.P. (1964) 'The liberating arts and the humanising arts in education', in Cohen, A.A. (ed.), *Humanistic Education and Western Civilisation*, New York, Holt, Rinehart & Winston, 159–81.
Meinig, D.W. (1983) 'Geography as an art', *Transactions, Institute of British Geographers* NS 8, 314–28.
Meir, A. (1982) 'A spatial-humanistic perspective of innovation diffusion processes', *Geoforum*, 13, 57–68.
Mitchell, B. and Draper, D. (1982) *Relevance and Ethics in Geography*, London, Longman.
Nisbet, R. (1982) 'What to do when you don't live in a golden age', *American Scholar*, 51, 229–41.
Plumb, J.H. (ed.) (1964) *Crisis in the Humanities*, Harmondsworth, Penguin.
Pocock, D.C.D. (ed.) (1981) *Humanistic Geography and Literature: Essays on the Experience of Place*, London, Croom Helm.
Powell, J.M. (1977) *Mirrors of the New World. Images and Image-Makers in the Settlement Process*, Folkestone, Dawson, and New Hampden, Archon; reprinted 1978, Canberra, Australian National University Press.
—— (1980) 'The haunting of Saloman's house: geography and the limits of science', *Australian Geographer*, 14, 327–41.
—— (1981) 'Wide angles and convergences: recent historical-geographical interaction in Australasia', *Journal of Historical Geography*, 7, 407–14.
—— (1984) 'Curriculum reform and the "constituency" challenge: recent trends in higher school geography', *Australian Geographical Studies*, 22, 275–95.
Relph, E. (1976) *Place and Placelessness*, London, Pion.

Sarup, M. (1982) *Education, State and Crisis. A Marxist Perspective*, London, Routledge & Kegan Paul.
Sheail, J. (1976) *Nature in Trust. A History of Nature Conservation in Britain*, Oxford, Clarendon Press.
Spate, O.H.K. (1960) 'Quantity and quality in geography', *Annals, Association of American Geographers*, 50, 377–94.
—— (1965) *Let Me Enjoy. Essays, Partly Geographical*, Canberra, Australian National University Press.
Spolten, L. (1970) 'The spirit of place: D.H. Lawrence and the East Midlands', *East Midlands Geographer*, 5, 88–96.
Stoddart, D.R. (ed.) (1981) *Geography, Ideology and Social Concern*, Oxford, Basil Blackwell.
Walford, R. (ed.) (1981) *Signposts for Geography Teaching*, Harlow, Longman.
White, J. (1982) *The Aims of Education Restated*, London, Routledge & Kegan Paul.
Winter, D.G., McClelland, D.C. and Stewart, A.J. (1981) *A New Case for the Liberal Arts*, San Francisco, Jossey-Bass.
Zaring, J. (1977) 'The romantic face of Wales', *Annals, Association of American Geographers*, 67, 397–418.

17
TO THE ENDS OF THE EARTH
R.J. Johnston

Geography is writing about the earth. As an academic discipline, its origins and early development reflected nineteenth-century interest in the acquisition of information about the great variety of environments, peoples and places on the earth's surface, in part to serve the needs of mercantile adventurers, but much of it merely to satisfy the curiosity of the educated, affluent classes. Today, that curiosity is strangely absent. It has been replaced, especially among human geographers, by a focus on assumed general laws of human responses to environment, space, place and people that ignores the diversity of contexts – physical and human – which is the overwhelming feature of the earth's surface. This shift of orientation among geographers is both myopic and dangerous. There is a desperate need for a rekindling of geographical curiosity about all parts of the earth's surface, not for any simplistic, voyeuristic reason but because without a major educational and academic discipline which removes the blinkers on the societies that it serves, world understanding, and ultimately world peace, will become even more threatened than it currently is.

DISENGAGEMENT

Underlying my thesis is the belief that geographers, especially but not only human geographers, have become parochial and myopic in recent decades. They no longer direct their eyes to the ends of the earth, but concentrate their attention solely on their home countries and regions.

This statement calls immediately for a qualification; it is directed only at geographers in the UK and North America. The discipline may be similarly characterized in other countries (though not in West Germany, it seems), in which case the arguments will be relevant there too; my attention, however, is directed only to those geographers who share my language, common cultural base, close academic links, and philosophies of social and natural science.

The disengagement of British and North American geography from close field contact with much of the world has proceeded continuously over several decades. It was pungently debated by Buchanan (1962) and Spate (1963) more than two decades ago, and ten years later Mikesell (1973, iv) characterized foreign area research as having undergone 'a recent transition from "boom" to "bust"'. The extent of this transition can be illustrated in a variety of ways. For example, the Institute of British Geographers – the learned society for academic geographers in the UK – has eighteen study groups, none of which is concerned with any defined area of the earth's surface: the implication is either that each group is global in its interests or that attention is concentrated on the home country. Similarly, the Association of American Geographers has thirty-five specialty groups, of which five are concerned with particular parts of the earth's surface (African; Asian; Canadian; Latin American; Soviet and East European); the total membership of the specialty groups is 5619, whereas the five named contain only 420 members. In a recent survey of 'areal proficiencies' of the Association's membership (published in its *Newsletter*, 1 April 1984), 5275 responses referred to areas outside the USA and Canada. There were 5474 members, and 9367 'proficiencies' claimed. This suggests that some 3000 of the Association's members had overseas interests.

The material presented in the above paragraph is somewhat contradictory. A probable explanation for this is that whereas many academic geographers have teaching interests in overseas countries – especially in the context of North American generalist undergraduate degrees – there is much less research interest. This is clear in analyses of publications. A recent analysis of the three leading US journals *(Annals of the Association of American Geographers, Economic Geography* and the *Geographical Review)* shows that in the years 1960–4, 49 per cent of all articles covered foreign areas (excluding Canada) compared to 24 per cent in 1978–82; at the AAG annual conference, about 13 per cent of papers over the period 1980–3 covered foreign areas (Swearingen 1984): American geographers, it is concluded, have a 'dangerously parochial focus' for their work, combined with a 'belief that geography should search for patterns independent of particular place' (p. 75). For British geographers, it seems, the percentages have always been low. Analysis of the *Transactions of the Institute of British Geographers* shows that during the years 1963–5, 26 per cent of the papers published were concerned with non-UK areas; a decade later (1973–5), the percentage was 18; and for the most recent period (1981–3) it was 17. Farmer (1983) attributes much of the decline in the British commitment to overseas work to financial stringencies and the absence of opportunities. But this must be only part of the reason. It is clear, as Farmer himself recognizes, that much of the decline reflects 'a premium on universalism in theory and practice that had no use for specialism by area' (p. 77).

Finally, and following on from the research interests, teaching programmes – especially at the tertiary level – also indicate a clear disengagement from the outside world. Three decades ago, most British university geography departments offered a wide range of regional courses covering many if not all parts of the world; in some, regional courses were compulsory, especially in the final year of the undergraduate degree, with the claim that regional studies synthesized the various systematic investigations and gave geography its particular role within academia (Johnston and Gregory 1984). Today, regional courses are few, and compulsory ones – especially if they deal with foreign areas – are rare.

Disengagement from the ends of the earth by geographers produces myopia. It may lead to the development of xenophobia. Geography, by virtue of its academic history, has long played a vanguard role in the education of people as world citizens, informed about the variety of environmental situations. But academic disengagement has been followed by educational disengagement, crucially at the secondary as well as the tertiary level. Geography is no longer fulfilling its traditional role. The universities, polytechnics, and colleges are not producing graduates skilled in the interpretation of foreign fields, and the nature of school education is changing accordingly. Further, because of the power of the academic discipline over the school curriculum, syllabuses are being rewritten which institutionalize geographical myopia. No longer does a school geographical education provide a foundation of knowledge for a life in world society; instead, it suggests that the world is a series of examples of a few simple geographical laws. Everyone is like us.

WHY HAVE WE DISENGAGED?

This retreat into our parochial homelands has occurred, paradoxically, during a period when the ability to travel widely has extended very substantially, when the integration of the world into major economic and political blocs has proceeded apace, and when the threat to world peace and survival has never been greater. Why, then, have we disengaged, and what have we replaced our traditional concerns with?

A variety of reasons has been suggested for the decline of regional geography in recent decades. Some refer to the poverty of what was practised under that heading – a criticism made mildly by Freeman (1961) and pungently by Gould (1979). Others refer to the attractions of systematic studies – both intellectually and in terms of their relevance to the local needs of the sponsoring society. Financial and logistic support for overseas research has declined in the countries under consideration – although there is now much greater transatlantic interaction than ever before – and a major shift of geographical emphasis has taken place (Johnston 1983a).

Modern (i.e. post-1960) human geography is characterized by systematic studies; most human geographers use adjectives such as economic, urban, transport and social to describe their research interests and their undergraduate courses, and most physical geographers are geomorphologists, biogeographers, or climatologists; few profess detailed knowledge of the various elements that make up the geography of a particular region or country. Most of their work continues the empiricist tradition of traditional (i.e. 1920–60) regional geography; the stress is on observation. But whereas that regional geography was *exceptionalist* – stressing the singularity of each place – modern geography is *positivist*. This latter philosophy stresses generalization; within human geography it is assumed that behaviour is subject to the operation of general laws of cause and effect, and the nature of these laws can be identified by the process of hypothesis-testing against empirical evidence (Johnston 1983b). Within this context, human geographers have focused their attention on laws of spatial behaviour and spatial arrangements.

This positivist focus on laws underpins the great majority of geographical work at the present time. Even where it is not explicit, descriptions of particular situations are frequently interpreted as case studies of processes which can be extrapolated and used as the basis for generalizing about other places – if not the whole world. Such a positivist approach is implicit throughout a society which yearns, it seems, for models that will encapsulate human behaviour within a few equations, descriptive devices that can also be used as prescriptive tools for social engineering.

The positivist model was introduced to human geography from the natural sciences. It contains a strong universalist element: behaviour observed in one place is assumed typical of all behaviour. Whereas this may be entirely justifiable as a foundation for studying the physical and biological (non-human) sciences, its relevance to the study of human societies, characterized by their cultural and historical depth and the decision-making of intelligent beings, is dubious. Nevertheless, it pervades the disciplinary literature, particularly the text-books on which school and undergraduate education is based. Texts in the systematic human geographies are built around general themes relating to spatial behaviour and organization, and 'relevant' examples are selected to illustrate these from virtually any part of the earth. Where the example is taken from is largely irrelevant. British and North American situations have been considered entirely equivalent by many authors – so much so that the literature of urban geography assumed for a number of years that if it was true for Chicago, Cedar Rapids and Toronto, it must be true for everywhere else; texts on American urban geography even included British examples! And it is always neat if an African or Asian example can be found to illustrate the general themes too.

The universalist element of the natural sciences is clearly indicated by their text-books. These, as Kuhn (1970, 10) describes them, 'expound the body of accepted theory, illustrate many or all of its successful applications, and compare these applications with exemplary observations and experiments'. In other words, they define the field; they summarize what is known (what puzzles have been solved), what puzzles this leaves to be solved, and how puzzle-solving is to be pursued. In human geography, as described above, such summaries usually imply that our knowledge and methods are universal. They do this, in large part, because of the acceptance of the positivist philosophy, and they are encouraged by the commercial text-book publishers who want as large a market as possible and so encourage authors to write books about everywhere – and so to some extent they write about nowhere too.

The universalist assumption has come under increasing criticism in recent years, from geographers of a variety of philosophical persuasions (Johnston 1983b) who are convinced that such an approach is invalid. They wish to promote instead a discipline which recognizes and emphasizes the variety of human responses to environment, space, place and people and which presents the world as a complex mosaic of different places, not as a series of examples of some general models of behaviour. But they are in a minority; the positivist approach dominates the research literature and student texts.

DISENGAGEMENT, IMMATURITY AND THREAT

Why should the positivist model be downgraded in importance, if not overthrown? Is it just to give a new generation of geographers a fresh set of causes on which to launch their career aspirations? Or is it more than mere academic debate? My case is that without re-engagement with the complexity and variety of the world, geographers will contribute to the ultimate demise of that world.

This argument is based on a belief that the universalist element in positivist human geography has promoted not knowledge, nor even information, but ignorance, especially ignorance of the world beyond one's national boundaries. The models of spatial behaviour and organization central to this approach assume that everybody is like us, or should be: if they are not like us, they are deviants and should be 'educated' so that their 'errors' are removed. In this way, we either eliminate from our consciousness — as communities and national societies as well as individuals — the variety of cultural and other mosaics that comprises world geography, or we manipulate it, producing a 'purified' image of 'them and us' which polarizes situations and encourages conflict.

This 'purification' process is central to Richard Sennett's analysis of the causes of urban conflict in his *The Uses of Disorder* (1970). Individuals and societies perceive the need for identity, for an image of themselves. In creating this self-image, 'other-images' are also created, stressing the differences between 'us and them'. This becomes a process of stereotyping and labelling emphasizing within-group commonalities and between-group differences; a person is judged by his or her membership of a particular stereotyped group, not by his or her personal characteristics.

This process of stereotyping is associated by Sennett with adolescence and the failure of the individual to mature. Adolescence, he claims, is generally characterized by 'wandering and exploration' (p. 22) but at the same time it comprises 'those tactics of evasion and avoidance of unknown, painful experiences that give rise to the desire for purity and coherence'. We seek to impose a fixed order on our lives, limiting our individual freedoms (and, by the sanctions of our communities and societies, the freedoms of our fellows) to avoid the 'unknown, and therefore potentially threatening, experiences in life' (p. 24). And yet, by distancing ourselves from others in order to create our own purified, coherent identities we create the conditions for conflict. These are based on ignorance and stereotypes.

> By imagining the meaning of a class of experiences in advance or apart from living them ... [we are] freed from having to go through the experience itself to understand its meaning ... [we make] up the meaning in isolation ... it is coherent, it is consistent ... [because we have] learned how to exclude disorder and painful disruption from conscious consideration. (pp. 27–8)

The result is a fixed set of images of the world. Exploration is unnecessary because we know what it will produce. In the context of the adolescent about whom Sennett wrote: 'the forces behind purification, forces of fear, lead the young person to enter adult life in a state of bondage to security, in a self-imposed illusion of knowledge about the outcome of experiences he has never had' (p. 30).

Sennett used this model of purified identity in adolescence to analyse the problems of conflict within urban societies, with particular reference to the American race rioting of summer 1968. He argued that groups within society avoid potentially unpleasant experiences – contact with other groups – by the processes of distancing, especially in the creation of urban residential patterns, thereby evading contact with the unknown, and replacing it by images which are frequently both negative and false. In this way, people act communally to create a joint purified identity, as a defence against the potential disorder of experiencing other identities;

> people draw a picture of who they are that binds them all together as one being, with a definite set of desires, dislikes and goals ... collapsing the experiential frame, a condensing of all the messy experiences of social life, in order to create a vision of unified community identity. (p. 38)

The consequences of this ideology of sameness include: an absence of participation outside the purified community, with no confrontation of differences; the repression of deviants, either excluding them from any contact or insisting on their conformity; and the promotion of conflict that can escalate into violence.

> Essentially, communities whose people feel related to each other by virtue of their sameness are polarized. When issues within or without the community arise that cannot be settled by routine processes ... they feel that the very survival of the community is at stake.... Individuals in the community have achieved a coherent sense of themselves precisely by avoiding painful experiences, disordered confrontations and experiments .... Having so little tolerance for disorder in their own lives, and having shut themselves off so that they have little experience of disorder as well, the eruption of social tension becomes a situation in which the ultimate methods of aggression, violent force and reprisal, seem to become not only justified, but life-preserving. (pp. 44–5)

Sennett's thesis relates to the polarization of communities within cities, the creation of purified identities within such separate territorial structures, the stereotyping – and its frequent emphasis on negative aspects of the 'others' – of members of other communities, and the strength of the reactions when communities come into contact and conflict. The race riots in American cities during the 1960s were thus related to the distancing that promoted residential segregation, the mutual fear of blacks and whites which was exacerbated by this separation, and the tensions that erupted when the two groups came into conflict over the unequal distribution of wealth within American urban society. The solution, according to Sennett, is to remove the purified identities and so to promote contact and mature accommodation, by the creation of cities characterized not by purified communities but by anarchic disorder: 'the experience of living with diverse groups has its power. The enemies lose their clear image, because every day one sees so many people who are alien but who are not all alien in the same way' (p. 156).

Sennett's analysis can be applied at larger spatial scales, although his solution is not readily translated too. The creation of nation states has been an exercise in the production of purified identities, promoting feelings of 'us-ness' and emphasizing our differences from 'them'. 'They' are seen as threats, as a collective danger against whom we must defend ourselves. Nowhere has this been clearer in recent decades than in the 'cold war' of east–west relations, in which each of the major parties on either side of the Iron Curtain portrays the other as an enemy, as a threat to national identity and security. But it occurs at many other scales too. The human geography of the world is a mosaic of mutually mistrusting, if not antagonistic, national and regional communities. It has always been so, but never before has the violence that can erupt from exacerbation of these antagonisms carried with it the threat of total destruction.

Sennett's anarchic solution to such problems at the city scale is not easily adapted to a broader canvas. Other mechanisms are needed to break down the purified identities, to remove mistrust and fear and replace them by accommodation and mutual respect. For these, we need information and understanding; we must appreciate the diversity of humanity, accept it and live with it, not seek to erase all differences as a means of removing potential threats.

But it is not only in community conflict and the ghetto mentality that the creation of purified identities and the negative labelling of others takes place. It occurs in most aspects of life, at all scales – between two individuals within a marriage, for example, between 'management' and the 'workforce' within a factory and an industry, and so on. Occasionally, some form of crisis and potential breakdown of the social order will occur, and attempts are made – by marriage guidance counsellors and by industrial conciliators in the above examples – to solve the problems and re-create harmonious working and living arrangements. Such attempts frequently involve role-playing, whereby the conciliator, working separately with the various parties, invites the people concerned to present the 'case for the opposition'. This promotes understanding of the other point of view, breaks down the ignorance barriers of the purified identities and negative stereotypes, and paves the way to accommodations via mutual understanding rather than mistrust.

At the international scale, similar processes can be involved in seeking to reconcile antagonists – sometimes via public 'shuttle diplomacy'. But this seeks to solve the problem once it has arisen, not to prevent it happening. The latter should be possible (or more possible than is the case now) if the current ignorance barriers were broken down and societies (national, regional and local) were made much more aware of each other. For this fundamental task within society, human geography can play a vital role. It must promote knowledge, understanding and trust, for to survive in the world we must present it to ourselves in its full diversity. Geography is a major educational path to international awareness and peace.

SO HOW DO WE DO IT?

Geography's *raison d'être* should be to develop appreciation of the great variety of cultures that comprise the contemporary world, and to show how in each society these have evolved – and are evolving – as specific responses to environment, to space, to place and to people. Students, at all educational levels, should not be deprived of experiencing, albeit usually at second hand, the richness of the world mosaic, and they should not be left with the impression that 'we' are right, some others are like us and so are right too, but others again act differently – wrongly – and should have their errors corrected. And researchers, to promote this vital awareness, should be encouraged to explore the diversity of the earth and to appreciate it, not to reduce it to yet further examples of Christaller and von Thunen.

But there are dangers. This is not a case for geography as travelogue, let alone for a return to the much-criticized 'capes and bays geography' and its successor 'traditional regional geography'. Arguments for such a return have been made recently (Steel 1982; Hart 1982) in papers which, although soundly motivated, offer no coherent view of a revived regional geography, and promise little more than a return to the exceptionalist tradition. Exceptionalism is at one pole of a continuum of geographical philosophies and positivism is at the other: the latter assumes that all is general, whereas the former assumes that nothing is general. As is so often the case, a middle ground is needed.

This middle ground must combine the general and the particular in a way which will advance understanding without falling into one of two traps. The first of these is the *singularity trap*. In this, every region – with its assemblage of environmental, cultural, economic, political and social characteristics – is presented as an individual phenomenon, which can only be understood as such. Analogies with other places are irrelevant, and there are no general laws which are in any way applicable. Thus every region has its own specialists, and no generalists are required. Set against this is the *generalization trap*, which presents the characteristics of every region as no more than the consequence of a particular combination of general laws. No other place may display that same combination of laws, but understanding of each place can be achieved by, first, identifying the general laws and, second, identifying how the particular combination of laws came about. (The latter itself may be the result of 'higher-level laws'.)

Avoidance of these two traps involves treating every place and region as *unique*, as the product of individual responses to general processes. The responses are made by freely-acting agents, operating either individually or collectively. The processes they are responding to are general, but their interpretation is individual. The respondent does not come to the processes unprepared, however. The whole process of socialization provides a context within which the processes can be interpreted, and sensible responses identified; in turn, the responses become part of the context for future socialization, additions to the infrastructure of the local society within which its members act.

In the contemporary world, the general processes in most regions relate to the operations of the capitalist mode of production. This provides a materialist base to the daily life of individuals and societies; it is driven by the necessities of accumulation of wealth – obtained through profits on the sale of goods and services – and is based on conflict between the (minority) accumulators of wealth on the one hand and the (majority) sellers of labour on the other. The entire functioning of capitalist societies is built around this imperative and its basic division of the population into classes.

The capitalist mode of production is not a deterministic mechanism, however, although some interpretations of it fall into the generalization trap. Its imperative has been interpreted in a great variety of ways by individual agents, producing immense variety in the nature of capitalist societies – in the individual social formations that characterize countries, regions and places. These interpretations are etched in the collective memories of the social formations, as parts of their cultures, and provide the context within which future decision-making is socialized. They are also etched into the landscape, producing one of the contexts that both constrains the evolution of the social formation and enables decisions to be made.

Our goal as geographers must be to foster appreciation of these social formations, in all their diversity. But we must not fall into the singularity trap, focusing on the particulars of our chosen societies and failing to identify their relationships to the general processes. Our work must contain two major elements, therefore: (1) a *theoretical appreciation* of the general processes of the capitalist mode of production (or other modes where relevant); and (2) an *empirical appreciation* of the particular social formations. Only by combining the two can we advance world understanding: without the former, we fall into the singularity trap and without the latter we are sucked into the generalization trap. The appreciation of contexts is crucial. Social formations have histories and cultures; these must be fully appreciated if the contemporary society is to be understood.

Not all societies are capitalist, and most of those that are were built on pre-capitalist foundations. Those modes of production must be understood too. In many cases the initial context for the transition to capitalism was provided by the pre-capitalist social formation, for example, and that context has been carried forward into the contemporary social formation and its landscape. Today, there are societies which have attempted to break away from the capitalist mode of production and to establish a socialist alternative. This mode of production also varies in its characteristics from place to place, reflecting different local interpretations of its imperatives, interpretations conditioned by the local, pre-socialist context and the links between that society and others.

To promote the sort of societal understanding argued for here, geographers must be catholic in their approach, drawing on a variety of disciplinary sources for their stimuli. Their theoretical appreciation requires close attention to political economy, especially of historical materialism. Their empirical appreciation demands careful study of the history, culture and institutions of the societies being analysed. Their interests may lead them to focus on the activities of individuals – but these must always be set in context and the singularity trap avoided. Alternatively, they may seek to generalize about the aggregate behaviour of many individuals; statistical procedures may be the ideal descriptive tools here, but those descriptions must not be interpreted as laws, leading into the generalization trap. For uniqueness to be appreciated, a variety of methodologies could be required. As methodologies, they are to be welcomed. To be avoided, however, are the philosophies commonly, though erroneously, associated with those methodologies – positivism with quantification, for example, and idealism with the study of individuals.

IN SUMMARY

Ignorance in some contexts may be bliss. In the context of present threats to world peace ignorance is dangerous. It leads to bad decisions, based on false stereotypes. Too often, situations are polarized, because the protagonists oversimplify the positions of others. This is a sign of immaturity, of an unwillingness to face the complexity of the world and accommodate to it. The alternative to accommodation is frequently conflict. The consequence of conflict is often death. For the next conflict, it may be the death of all.

To promote accommodation we must promote understanding. We must appreciate what other people think and do. For this we need geography, not an arid, placeless geography in the positivist tradition, nor a voyeuristic, structureless geography in the exceptionalist tradition (including its recent idealist successors), nor a mechanistic geography which precludes freedom of individual action. We need a regional geography which is contextually based, which locates decision-makers in their historically produced cultural environments (which include attitudes to the physical environment) and in the imperatives of their mode of production. And this must be the basis of our teaching, for without the understanding that it can bring, the future of our societies is in peril.

Most generations of geographers have been attracted to the ends of the earth. My generation disengaged itself from such an attraction. If geography is to be a relevant subject, that change must be reversed. We must educate the world about the world, remove myopia, counter xenophobia, and promote peace through mutual understanding and respect. The world is our oyster; unless we promote its proper study, it may be nobody's.

REFERENCES

Buchanan, K.M. (1962) 'West wind, east wind', *Geography*, 47, 333–46.
Farmer, B.H. (1983) 'British geographers overseas – 1933–1983', *Transactions, Institute of British Geographers*, NS 8, 70–9.
Freeman, T.W. (1961) *One Hundred Years of Geography*, London, Duckworth,
Gould, P.R. (1979) 'Geography 1957–1977: the Augean period', *Annals, Association of American Geographers*, 69, 139–51.

Hart, J.F. (1982) 'The highest form of the geographer's art', *Annals, Association of American Geographers*, 72, 1–29.
Johnston, R.J. (1983a) *Geography and Geographers*, London, Edward Arnold,
—— (1983b) *Philosophy and Human Geography*, London, Edward Arnold.
—— and Gregory, S. (1984) 'The United Kingdom', in Johnston, R.J. and Claval, P. (eds), *Geography since the Second World War: An International Survey*, London, Croom Helm.
Kuhn, T.S. (1970) *The Structure of Scientific Revolutions*, Chicago, University of Chicago Press.
Mikesell, M.W. (1973) Preface, in Mikesell, M.W. (ed.), *Geographers Abroad*, Department of Geography Research Paper 152, Chicago, University of Chicago, iii–vii.
Sennett, R. (1970) *The Uses of Disorder*, Harmondsworth, Penguin Books.
Spate, O.H.K. (1963) Letter to the Editor, *Geography*, 48, 206.
Steel, R.W. (1982) 'Regional geography in practice,' *Geography*, 67, 2–8.
Swearingen, W.D. (1984) 'Foreign languages and the terrae incognitae', *The Professional Geographer*, 36, 73–5.

INDEX

abstraction, 170
accessibility, 200f
accommodation, 333
agency, human, 16f, 174, 302
aggregation problem, 198
applied geography, 20f, 100f, 212, 219, 221, 229, 245
applied geomorphology, 236, 242
applied problems, 237
applied research, 211, 217f, 222, 235, 244, 246, 248
areas/areal, 191, 194, 197, 202, 223
autocorrelation, 193, 197, 217

Bayesian methods, 215

chi-square, 197
classification, 14, 261, 269
class struggle, 63
cognitive mapping, 202f
conjecture(s), 6, 121, 123f, 127
crisis (in education), 301
critical rationalism, 120
critical sciences, 19
critical theorist, 213
curricula, secondary-school, 6

de-definition, 60f, 219
deduction, 115, 117f, 126f, 160, 214
determinism, 178, 272

determinist mechanism, 335
development, 60, 82, 108, 160, 166, 169, 226, 271, 301
dialogue, 315f
diffusion, 205f
disengagement, 326f, 329, 330, 337
distance, 191, 201f
distancing, 331f

earth sciences, 28f, 40
earth scientists, 34
ecology, 49, 51
ecosystem(s), 49f
education: 8, 10, 20, 23f, 28, 93, 99, 104, 277, 278, 279, 291f, 299f, 307f, 316f, 320f, 330; liberal, 22, 277, 296, 307, 318, 322
efficiency, 191
emancipation, 22, 24
empirical-analytic science, geography as an, 19f, 95f
empirical generalization(s), 115
empiricism, 13f, 133, 303
entropy, 53
environment, 13
environmental determinism, 22, 125
environmental impact reports, 271
epistemology, 75, 113, 117, 311
exceptionalism, 13f, 137, 329, 334, 337
existentialism, 146f, 312

explanation(s), 15f, 19, 44, 66, 117, 118f, 132, 153, 159, 160, 162, 164, 170, 175f, 268, 293
exploratory data analysis, 216
extensive research questions, 166f

fallibilism, 15, 114, 120f, 126
falsifiability, principle of, 120f
falsification, 120, 123, 126, 134f
fetishism, 213
field experiment, 138
flows, 191
fragmentation, 19, 78, 207, 218
functionalism, 176f, 220

Gaia, 52f
generalization trap, 25, 335f
Geographical Association, 29, 31, 296, 300, 304
geologists, 39, 267
geology, 27, 30f, 35, 37, 40f, 225, 229, 246, 269, 281
geomorphology, 11, 27f, 36, 40, 225f, 228f, 234f, 237, 246, 248, 259, 262f, 268f, 272

hazards, 225f, 242, 244f, 267
hegemony, 179, 293f, 304
hierarchical structure, 191
historical-hermeneutic sciences, 19f, 95f, 106, 285f
historical materialism, 72, 74, 76f, 80, 107, 336
historicity, 282
holism, 12f, 20, 45f, 48f, 50f, 52, 54, 57, 60, 77, 84, 96f, 97, 99f, 103, 105, 106, 108, 109
human agency, 16f, 174
human geography, 11f, 16f, 21f, 27, 30, 32, 37, 129, 136f, 152, 156, 162, 168, 213, 222, 245, 264, 267f, 272, 281f, 284, 296, 311f, 326, 329, 331, 334
humanistic geography, 45, 49, 103, 105f, 108, 119, 141, 146, 143f, 146f, 161, 165, 177, 279, 302f, 309f, 315, 317f
humanistic philosophy, 16, 18, 21, 61, 301
hypothesis, 15, 44, 47, 52, 66, 122, 131, 134, 137f, 141, 162, 165f, 206, 215, 221, 262
hypothesis-testing, 138, 141, 147, 214, 236, 329

idealism, 66, 70, 72, 76, 106, 119, 177–8, 337
297, 300f, 310, 332
ideographic approach, 115f, 159f, 163, 172, 223
ideology, 18, 24, 47f, 62f, 67f, 77, 79, 81, 139, 146, 150, 175, 177f, 292f, 295, 297, 300f, 310, 332
incommensurability, 136
individualism, 174f, 177f
individuation, 278
induction, 15, 113f, 122f, 127, 133f, 137f, 160, 212f, 218f
inference, 132, 167, 211, 214, 216f, 219f, 222
innovations, diffusion of, 205
intensive research questions, 166f, 171, 172
intentionality, 176f, 312
interpretation, 155

knowledge, 6, 9f, 12, 15f, 44f, 47, 60, 64, 84, 95f, 100f, 113, 118f, 122f, 133, 136f, 139, 143, 151f, 181f, 217, 222, 258, 260, 263f, 267, 287, 292, 295, 311, 319, 330f, 334
kriging, 193f

landscape, 145f, 154, 160, 181f, 226, 228, 268, 270, 291, 302, 336
language, 11, 24, 45, 60, 102, 130f, 146f, 278, 282f, 287, 292, 299, 327
laws (in geography), 14f, 44f, 56, 102, 115, 117f, 122f, 126f, 130f, 133, 136, 137, 140, 159, 161, 163, 212, 219, 262, 265, 286, 326, 328f, 335f
legitimacy, 180, 183f, 294f, 319
liberal pluralism, 71
lines, 191
location, optimal, 191f
logic, 130f, 133f, 140

Marxism, 17, 66, 70, 71, 74, 77, 107f, 176, 177, 179, 219, 286
Marxist: analysis, 69, 109; conceptions, 63; interpretations, 313; methods, 129; position, 106; practice, 61; research, 161; sociologists, 213; theory, 48, 293
materialism, 17, 60, 106, 303, 335; historical, 72, 74, 76, 77, 78, 80, 107
meaning, 16, 19, 145, 152f, 155, 165, 285

measurement, 132f, 136f, 169, 244
measuring devices, 141
mechanisms, 16f, 24f, 77, 121, 125, 163f, 166f, 172, 180, 187, 281f, 298, 333; deterministic, 335
methodology, 3, 13f, 17, 19, 56, 99, 103, 143, 152, 159f, 166, 168, 170, 172, 213, 218, 222, 242, 264, 336
models, 15, 50, 132, 160, 168, 177, 219, 237, 261, 267, 273, 331
modifiable areal unit problem, 216f
myopia, 326, 328, 337

narrative, 153f, 187
nature, 169, 263
NERC, 35f, 41
nomothetic approach, 115f, 159, 161f, 171f
normal science, 135
nuisance fields, 199

objectivity, 136f, 150
optimal location, 191f
order, 16, 70, 146, 160f, 168f, 182, 331

paradigm, 135f, 212
participant observation, 148
people's geography, 21f, 72
phenomenology, 120, 141, 146f, 216
philosophy: departments, 146; of geography, 3, 13, 46, 56, 71, 77, 97, 101, 123, 143, 159, 172, 281, 336; Greek, 49; holistic, 103; positivist, 330; of science, 133
physical geography, 11f, 14, 17f, 20, 27f, 36, 37f, 46, 99, 129, 137, 139, 213, 222, 245, 258f, 263f, 270f, 274, 329
place(s), 48f, 78, 108, 143, 145f, 148f, 151, 182, 213, 221, 223, 267, 291, 312, 326, 334f; singularity of, 13f; spatial relations between, 74; uniqueness of, 13f, 24f
pluralism, 12; liberal, 71
policy: public, 19; social, 47
populism, 72
positivism, 14f, 18, 24, 66, 70, 75, 81, 96, 97, 103, 105f, 106, 108, 119f, 123, 133, 143f, 150, 152f, 161, 212, 218f, 220, 222f, 233, 300, 303, 329, 331, 334, 337
prediction, 19, 119, 132, 141, 266f, 270f

processes, 15, 18, 146, 154, 161, 163, 178, 213, 222, 225, 259, 329, 335f (see also mechanisms)
'purification' process, 331f, 332f

Q-analysis, 47
quantitative revolution, 159, 211, 265

radicals, 19, 61, 103f, 154f, 160f, 169, 176, 211, 218f, 223, 298, 302f
realism, 16f, 66, 159, 161, 163, 165f, 168f, 220
reappraisal, of geography, 3
recession, 8, 10, 93, 100f, 103f, 301
reductionism, 12, 44, 46, 47, 48, 54, 55, 130f, 131, 136, 138, 140
refutation(s), 121f, 127
region, 49, 56, 75, 101, 125, 145, 161, 171f, 223, 335
regional geography, 13f, 30, 51, 102, 105, 108, 159, 172, 317, 328f, 334, 337
regularity, 133, 162f, 170, 175f
relevance, 211, 217, 233, 258, 303, 308
relevant: curriculum, 299; geography, 19, 152, 245, 312, 330; research, 234f
remote sensing, 243
reporting, 147
resource(s), 12, 20, 24, 34; squeeze, 221
rhetoric, 286, 293

scale, 217
schools, 291f, 294; geography in, 292f, 296, 299f, 328
scientific methods, 113, 129, 133, 137, 139, 141
scientism, 132, 139
self-awareness, through geography, 24
singularity, 329; trap, 335f
space, 18, 73f, 76, 78, 103, 145, 171, 190f, 201f, 205, 212, 219, 222f, 312, 326, 334
spatial: analysis, 18f, 67, 103, 105, 144, 190f, 201, 205, 264, 270, 272; fetishism(s), 64, 73, 75; interaction, 162, 190; organization, 13; patterns, 124, 212f; process, 213f, 216; science, 47, 191, 206, 211f, 216f, 221, 223; structure, 216; view, 213
specialization, 12f, 95f, 100, 102f, 269, 318
statistical inference, 135

stereotypes, 331, 333, 334, 337
structuralism, 70, 174f, 178f, 301
structuration, 78, 174f, 178f, 183, 187f
structure, 16f, 22, 163, 168, 172, 174f, 182, 185, 187, 206, 284, 291, 302; hierarchical, 191; in Marxism, 187
structure-action, 175
students, 10f
subjectivism, 16, 150f, 217
surfaces, 191, 199, 201
symbols, 181, 186
synthesis, 171, 222, 226, 261f, 265, 270, 317
systematic geographies, 14
systems, 49f, 134, 137f, 153, 161, 163, 164, 166f, 171, 175, 178f, 186, 220f, 236, 272f; analysis, 45f

teleology, 49, 79, 176f
test, 131f, 134, 137, 215, 263
texts, 63, 67, 181, 183f, 187, 285f, 330
theory, 15, 44, 46, 64, 67f, 77, 100, 120f, 124f, 130f, 137, 139, 140, 154, 159f, 162, 168f, 172, 187, 214f, 219, 221, 223, 233f, 293, 330
thinking, 129, 276, 278, 280f, 283f, 287f, 321; intuitive, 55; self-liberation through, 24
threats, 330, 333, 337
threshold concept, 268
training, 277f, 292
truth, 277, 281, 284, 286, 288

understanding, 15f, 18, 25, 147, 154, 165, 237, 297, 314, 333f, 337
uniqueness, 45, 75, 115, 130, 137f, 159, 162, 212, 268, 335f
universalism, 329f

value-free research methods, 223
values, 150f, 178, 211, 218, 314
verification, 120, 134f, 151
visual arts, use of, 149

xenophobia, 328, 337